Introduction to Structural Mechanics and Analysis

Donald A. DaDeppo

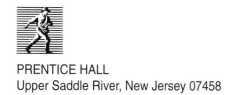

PRENTICE HALL
Upper Saddle River, New Jersey 07458

Library of Congress Cataloging-in-Publication Data

DaDeppo, Donald A. (Donald Adam)
   Introduction to structural mechanics and analysis. / Donald A.
DaDeppo. — 1st ed.
   p.   cm.
   Includes index.
   ISBN 0-13-859794-4
   1. Structural analysis (Engineering) I. Title.
TA645.D24   1999   98-10926
624.1'71—dc21   CIP

Acquisition Editor: Bill Stenquist
Editorial/Production/Composition: Interactive Composition Corporation
Editor-in-Chief: Marcia Horton
Assistant Vice President of Production and Manufacturing: David W. Riccardi
Managing Editor: Bayani Mendoza de Leon
Manufacturing Manager: Trudy Pisciotti
Full Service/Manufacturing Coordinator: Donna Sullivan
Creative Director: Jayne Conte
Cover Designer: Bruce Kenselaar
Editorial Assistant: Meg Weist
Copy Editor: Lori Stephens

 © 1999 by Prentice-Hall, Inc.
Simon & Schuster / A Viacom Company
Upper Saddle River, New Jersey 07458

All rights reserved. No part of this book may be
reproduced, in any form or by any means,
without permission in writing from the publisher.

The author and publisher of this book have used their best efforts in preparing this book. These efforts include the
development, research, and testing of the theories and programs to determine their effectiveness. The author and publisher
make no warranty of any kind, expressed or implied, with regard to these programs or the documentation contained in this
book. The author and publisher shall not be liable in any event for incidental or consequential damages in connection with,
or arising out of, the furnishing, performance, or use of these programs.

Printed in the United States of America

10 9 8 7 6 5 4 3 2 1

ISBN 0-13-859794-4

Prentice-Hall International (UK) Limited, *London*
Prentice-Hall of Australia Pty. Limited, *Sydney*
Prentice-Hall Canada Inc., *Toronto*
Prentice-Hall Hispanoamericana, S.A., *Mexico*
Prentice-Hall of India Private Limited, *New Delhi*
Prentice-Hall of Japan, Inc., *Tokyo*
Simon & Schuster Asia Pte. Ltd., *Singapore*
Editora Prentice-Hall do Brasil, Ltda., *Rio de Janeiro*

To Dude,
Your dad loves you.
Papa

Contents

Preface   **viii**

Chapter 1   Analysis of Framed Structures   **1**

1.1	Introduction	1
1.2	Force Analysis	2
1.3	Principle of Superposition	2
1.4	Terminology	2
1.5	Stability and Determinacy	3
1.6	Practical Application When $n = m$	5
1.7	Statically Determinate Structures	6
1.8	Plane Framed Structures	7
	Problems	10

Chapter 2   Trusses   **15**

2.1	Stability and Determinacy	16
2.2	Forces of Interaction/Internal Force Resultants	17
2.3	Force Analysis of Statically Determinate Trusses	19
2.4	Method of Joints	19
2.5	Method of Sections	19
2.6	Conventions and Practice	19
	Problems	27

Chapter 3   Internal Force Resultants   **37**

3.1	Internal and External Force Functions and Diagrams	38
3.2	Shear and Bending Moment Functions and Diagrams for Beams	39
3.3	Relationships Between Internal and External Loads	43
3.4	Shear and Bending Moment Diagrams for Frames	47
	Problems	49

Chapter 4   Moveable Loads/Influence Functions and Lines   **57**

4.1	Influence Functions/Influence Lines	58
4.2	Application of Influence Line Concept	63

	4.3	Generalized Application	64
	4.4	Floor Girders	67
	4.5	Bridge Trusses	69
	4.6	Extreme Effects Due to Fixed Loads of Variable Placement	72
		Problems	80

Chapter 5 Energy Methods (Part 1) 93

	5.1	Introduction	93
	5.2	Basic Concepts	93
	5.3	Virtual Displacement/Virtual Work	95
	5.4	Contacting Surfaces and Forces	96
	5.5	Forces	97
	5.6	Rigid Displacements	97
	5.7	Small Displacements	98
	5.8	Rectangular Components	99
	5.9	Instantaneous Center	99
	5.10	Principle of Virtual Work	100
	5.11	Virtual Work of Forces on a Rigid Virtual Displacement	100
	5.12	Vector Representations and Work Formulations	102
	5.13	Example Problems	103
	5.14	Influence Lines	108
		Problems	115

Chapter 6 Displacement-Deformation Analysis of Structures 126

	6.1	Trusses	128
	6.2	Beams	129
	6.3	Moment-Area Theorems	130
	6.4	Notes for Application of the Moment-Area Theorems	134
	6.5	Procedures for Application of the M-A Theorems to Deflection Analysis	135
	6.6	Frames	143
	6.7	Procedures	144
	6.8	Combination Approach	156
		Problems	158

Chapter 7 Energy Methods (Part 2) 173

	7.1	Introduction	173
	7.2	Displacements by Virtual Work	173
	7.3	Procedure	175
	7.4	Evaluation of δW_σ	175
	7.5	Basic Applications	178
	7.6	Beam	179
	7.7	Truss	180
	7.8	Displacements Due to Applied Loads	181

7.9	Beams and Frames	182
7.10	Trusses	187
7.11	Combined Loads	191
7.12	Strain Energy and Internal Force-Deformation Relations	194
7.13	Conservative Applied Forces	195
7.14	Principle of Stationary Potential Energy	196
7.15	Principle of Minimum Potential Energy	197
7.16	Complementary Strain Energy and Force-Deformation Relations	198
7.17	The Special Case of Hooke's Law	199
7.18	Betti's Theorem	200
7.19	Maxwell's Reciprocal Theorem	201
7.20	Castigliano's Theorems	203
7.21	Theorem on Deflections	204
7.22	Theorem on Deflections Specialized for Hooke's Law	206
7.23	The Theorem of Least Work	207
7.24	Application of Castigliano's Theorems	209
	Problems	211

Chapter 8 Statically Indeterminate Structures   231

8.1	Introduction	231
8.2	General Procedure of Analysis	233
8.3	Member Rigidity	233
8.4	Solution Strategies	233
8.5	Force Method	234
8.6	Displacement Method and Kinematic Indeterminacy	234
8.7	Flexibility and Stiffness: Flexibility Coefficients	235
8.8	Stiffness Coefficients	237

Chapter 9 Force Methods   239

9.1	Method of Consistent Deformation	239
9.2	Use of Superposition	241
9.3	Example Problems	241
9.4	Prescribed Displacements	257
9.5	The Three-Moment Equation	261
9.6	Application Procedure	267
9.7	Combined Methods	271
9.8	Method of Least Work	273
	Problems	280

Chapter 10 Displacement Methods of Analysis   308

10.1	Slope-Deflection Equations	310
10.2	Applications	314
10.3	Note on the Equations of Equilibrium	315
10.4	Restatement of the Slope-Deflection Equations	315

10.5	End-Rotations	316
10.6	Examples of Beams and Frames with Prescribed Joint Displacements	316
10.7	Frames with Sidesway	326
10.8	Modified Slope-Deflection Equations	330
10.9	Applications	330
10.10	Sidesway Problems and Superposition	333
10.11	Moment Distribution	339
10.12	One Independent Joint Rotation	343
10.13	Two Independent Joint Rotations	344
10.14	Several Independent Joint Rotations	346
10.15	Non-Zero Prescribed Displacements and Moment Distribution	354
10.16	Sidesway Analysis and Moment Distribution	356
10.17	Equations of Equilibrium by Virtual Work: Matrix Methods	359
10.18	Hooke's Law	366
10.19	Reactions and Virtual Work	369
10.20	Strain Energy of Bending	373
10.21	Potential Energy of Prescribed Loads	376
10.22	Principle of Stationary Potential Energy: Equations of Equilibrium	378
10.23	Matrix Formulations—Summary	379
10.24	Application Procedure	380
	Problems	387

Chapter 11 Influence Lines — 415

11.1	Introduction	415
11.2	Müller-Breslau Principle	415
11.3	Influence Lines for Deflection	417
11.4	Practical Application	417
	Problems	421

Appendix A — 424

A.1	Supplemental Notes on the Principle of Virtual Work	424

Appendix B Sidesway Analysis — 428

B.1	General Formulation and Superposition	428
B.2	General Formulation	428
B.3	Superposition	430

Answers — 432

References — 451

Index — 453

Preface

This book is intended for use in a modern two-semester course in structural mechanics and analysis. The essential preparation consists of traditional courses in statics, elementary mechanics of deformable bodies, mathematics up through vector calculus, and a course that includes an introduction to the elementary operations of matrix algebra.

The book is focused on the fundamental principles of mechanics and the basic assumptions that are the heart of the linear theory of structures. It was written with the conviction that a good understanding of principles and how they apply to the mechanics and analysis of structures is essential to a good understanding of structural behavior. The writing was undertaken with two primary goals in mind: (1) to show how basic ideas together with the prerequisite mathematics can be used to derive the working principles/theorems that are commonly employed in the practical analysis of structures, and (2) to provide a foundation for additional study in advanced structural mechanics, including matrix analysis of structures and finite element methods. Toward these ends derivations are carried out in detail sufficient for the student to follow the flow of thought and develop a feel for how derived principles evolve from rigorous but relatively uncomplicated reasoning. Sprinkled throughout the text are reminders about the principles involved and the fundamental assumptions of the theory. Practical application to concrete problems is essential to understanding. Accordingly, numerous examples are included to illustrate key points and to provide the student with a guide to efficient practical application of the theory. Several example problems are solved by proceeding along different lines of reasoning, but always based on the same principles. In this way, the student is given the opportunity to evaluate the efficacy of specific solution techniques. No attempt has been made to be comprehensive. By design there is no discussion of arches, approximate methods of analysis of statically indeterminate structures, conjugate beam method of deflection calculations, non-prismatic members, effect of shear deformation, or special methods of analysis.

Class notes used in teaching the foundation courses in statics and strength of materials as well as courses in the analysis and design of structures for more than twenty-five years form the basis for *An Introduction to Structural Mechanics and Analysis*.

Neither this book nor a career in education that spanned thirty-seven years would have been possible without the guidance of two special individuals. Dr. James M. Paulson, my mentor while I attended Wayne State University, encouraged me to go on to graduate study. Dr. Bruce G. Johnston (deceased), my Ph.D. advisor at the

University of Michigan, provided the opportunity and encouraged me to take up teaching as a profession.

Diane M. Basile was instrumental in calling the draft of this book to the attention of Bill Stenquist whose kind words and support played an important role in bringing this work to completion. My thanks go out to Diane and Bill and to all of the unnamed individuals at Prentice-Hall who contributed to the production of this book.

I thank all members of my family, especially my dear wife Shiela on whom the greatest burdens fell, for their understanding and endurance.

CHAPTER 1

Analysis of Framed Structures

1.1 INTRODUCTION

A *framed structure* is understood to be one that is composed of long, thin, beam-like members—which may be curved or straight—joined together and to a foundation to provide mutual support. In the great majority of practical situations, the members are straight. Because they are long and thin, the members of a framed structure are usually represented by their respective reference lines. This representation is consistent with the assumed geometry and with the theories used to describe the mechanical behavior of such members. Full-body representations are used occasionally but, in general, such representations are not practical in analysis.

As treated in structural analysis, a *joint* or a *node* is a point at which the reference lines of two or more members are connected in some way. The method of connection constrains all members coming into a joint to have common values for certain components of displacement. The common values of the displacements are called the *joint displacements*. Analytical statements of the equality of the displacements at the end of a member to the displacements of the joint are called *compatibility equations*.

As a consequence of the connection, forces of interaction are developed at the ends of each member coming into a joint. Knowledge of the forces at the ends of a member and of the loads between the ends is sufficient to determine the internal forces (stresses) over the length of the member. Key elements in the development of a successful design are calculation of the internal forces and calculation of displacements as are required by design codes. Accordingly, the general problem of structural analysis is stated as follows:

> Given a framed structure and the prescribed (applied) loads, determine the reactions at all supports, the forces of interaction at all joints, and displacements at prescribed points.

The practical analysis of structures is carried out with the aid of certain fundamental assumptions that are justified by practice and experience. These assumptions, which are first introduced in statics and in the mechanics of deformable bodies, are reviewed here as a preliminary to solving problems.

Assumption 1: The deformations and rotations are (infinitesimally) small.

Consequences:

a. The equations of equilibrium may be written with respect to the undeformed geometry of the structure. Therefore, the equations are linear in the applied forces and the unknown forces of reaction and interaction.
b. The equations relating deformations to displacements are linear.

Assumption 2: Hooke's law is valid.

Consequence:

The force-deformation relations, which are material properties, are formulated as linear equations.

1.2 FORCE ANALYSIS

The fundamental problem of the force analysis of a structure is understood to be that of calculating all forces of reaction at the supports and all forces of interaction at the joints. The general procedure of force analysis, which does not depend on the previously stated assumptions, and which applies to all structures, is to formulate and solve simultaneously:

1. the equations of equilibrium
2. the displacement-deformation equations
3. the force-deformation equations
4. the compatibility equations (these are always linear equations that relate displacements at the ends of the members to the displacements of the joints).

In practice, these equations are usually combined in a standard way to eliminate unknowns and thus facilitate generation of the solution.

1.3 PRINCIPLE OF SUPERPOSITION

Because all of the pertinent equations are linear, what is called the *principle of superposition* is valid. The principle of superposition is not fundamental; it must be proved. This principle asserts that if C_1 and C_2 are admissible choices of independently prescribed forces and displacements and $E_1(p)$ and $E_2(p)$ is the force or displacement (or strain, etc.), at a point p when C_1 and C_2 act alone, then due to the simultaneous action of $\alpha_1 C_1$ and $\alpha_2 C_2$, where α_1 and α_2 are constants, the net force (or displacement, etc.), at p, $E(p)$, is given by

$$E(p) = \alpha_1 E_1(p) + \alpha_2 E_2(p)$$

1.4 TERMINOLOGY

Frequently a structure will be referred to as being *fully restrained*. A common understanding of the term is essential. Consider first a single body. The statement that it is fully restrained is understood to mean that the body, considered as being rigid, remains at rest for arbitrary applied loads. This means that the acceleration of the body is zero, whatever the loads. Recall that a *structure* is understood to be an assembly of members connected to each other and to a foundation so as to provide mutual support. A structure is understood to be fully restrained if, and only if, every member of that structure is fully restrained.

As a consequence of this definition, it follows that for any fully restrained structure, there always exists at least one set of values for the unknown components of the forces of reaction and interaction such that the equations of equilibrium are satisfied for arbitrary applied loads. It also follows that if a structure is not fully restrained, there exists at least one set of applied loads such that no set of values can be found for the unknown force components such that the equations of equilibrium are satisfied. A structure that is not fully restrained is said to be partially restrained, or kinematically or geometrically unstable. In general, a structure is designed with the intent that it be fully restrained.

Fully restrained structures are classified as being either *statically determinate* or *statically indeterminate*. By definition, a statically determinate structure is one for which all of the unknown forces of reaction and interaction can be found by writing and solving only the equations of equilibrium.

1.5 STABILITY AND DETERMINACY

All questions concerning full or partial restraint, statical determinacy and statical indeterminacy are answered unequivocally from an analysis of the pertinent equations of equilibrium. Of primary interest here are the number of independent equations of equilibrium that may be written and the character of the equations. Counting schemes can be set out to determine the number of independent equations of equilibrium that relate the unknown forces for any given situation; however, such schemes are not taken up here. The number of independent equations and the number of unknowns are taken as given quantities. Let:

$N_e = n$ be the number of independent equations of equilibrium that can be written;

$N_u = m$ be the number of unknown components of forces of reaction and interaction;

$R_1, R_2, \ldots, R_m$ denote the unknown components of forces of reaction and interaction.

Then, the equations of equilibrium for a framed structure always take the form:

$$a_{11}R_1 + a_{12}R_2 + \cdots + a_{1m}R_m = b_1$$
$$a_{21}R_1 + a_{22}R_2 + \cdots + a_{2m}R_m = b_2$$
$$\cdots$$
$$a_{n1}R_1 + a_{n2}R_2 + \cdots + a_{nm}R_m = b_n$$

The coefficients $a_{11}, a_{12}, \ldots, a_{nm}$ are constants determined by the geometry of the structure and the types of connections. The coefficients $b_1, b_2, \ldots, b_n$ depend on the loads. All of the b's are zero if there are no applied loads. For arbitrary loading the b's can be made to take on any values. There are three cases to consider according to the relationship of n to m.

1.5.1 Case 1. $n > m$

In this case the equations of equilibrium have no solution for arbitrary load. The structure is incompletely restrained and is said to be geometrically unstable. The structure may be loaded so that it will be in equilibrium; however, there is at least one mode of loading such that equilibrium cannot exist. Useful structures are designed to avoid this situation.

1.5.2 Case 2. $n = m$

In this case let det (A) be the determinant of the coefficients of the unknown forces of reaction and interaction in the equations of equilibrium; i.e., let

$$\det(A) = \begin{vmatrix} a_{11}, a_{12}, \ldots, a_{1n} \\ a_{21}, a_{22}, \ldots, a_{2n} \\ \ldots \\ a_{n1}, a_{n2}, \ldots, a_{nm} \end{vmatrix}$$

With respect to det (A) there are two possibilities to consider:

a. det $(A) = 0$. Here, the equations of equilibrium have no solution for arbitrary loads. The structure is incompletely restrained and geometrically unstable. The structure may be in equilibrium for special loadings. Useful structures are designed to avoid this situation.

b. det $(A) \neq 0$. Here, the equations of equilibrium have a unique solution for arbitrary loading. In other words, for any specified load there is exactly one set of values for the unknown forces of reaction and interaction that satisfies all of the equations of equilibrium. Because equilibrium exists for arbitrary loads, the structure is fully restrained and because all of the unknown forces can be found by solving only the equations of equilibrium, the structure is said to be *statically determinate*. The force analysis of statically determinate structures is of primary interest in the first part of a complete course in structural analysis.

1.5.3 Case 3. $m > n$

For discussion, assume that the mechanical elements which generate the reactive/interactive force group consisting of $R_{n+1}, R_{n+2}, \ldots, R_m$ are removed and it is found that det $(A) \neq 0$. From the discussion of Case 2, it is concluded that $R_1, R_2, \ldots, R_n$ are sufficient for equilibrium whatever the applied loads. It follows that the structure is necessarily fully restrained even when the mechanical elements that develop $R_{n+1}, R_{n+2}, \ldots, R_m$ have been removed. Because $R_{n+1}, R_{n+2}, \ldots, R_m$ are not essential for equilibrium, these reaction components and the mechanical devices that develop them are *redundant* with respect to the condition of equilibrium. The statically determinate structure that remains after the mechanical devices that develop $R_{n+1}, R_{n+2}, \ldots, R_m$ have been removed is called the *primary* or *cut-back structure* corresponding to the selected redundant forces. In this case the best that can be done with the equations of equilibrium is to solve for $R_1, R_2, \ldots, R_n$ in terms of the applied loads and $R_{n+1}, R_{n+2}, \ldots, R_m$. Because the structure is fully restrained but specific values cannot be determined for all of the unknown forces in terms of the loads alone just from the equations of equilibrium, the structure is said to be *statically indeterminate*. The degree of indeterminacy is, by definition, $d = m - n$. Because of coupling of the equations, the force analysis of a statically indeterminate structure requires that the equations of equilibrium, the displacement-deformation equations, the force-deformation equations, and the compatability equations be solved simultaneously.

It may happen that det $(A) = 0$. Here, no general conclusion can be drawn without further work. It may turn out that by retaining some combination of n reactive forces other than all of $R_1, R_2, \ldots, R_n$, a set of equations is obtained for which the determinant corresponding to det (A) is non-zero. If this is so, the structure is fully restrained and still statically indeterminate. At this point it should be noted that any grouping of n reactive elements such that the determinant corresponding to det (A) is non-zero defines a particular primary or cut-back structure, along with an associated set of redundant mechanical restraints and their associated redundant reactions.

Now suppose that all possible combinations are examined and it is found that there is no way in which n reactive elements can be selected so that the determinant corresponding to det (A) is non-zero. Then, effectively, it has been established that there is no set of values for the unknown reaction/interaction forces such that equilibrium will prevail for arbitrary loading. The conclusion is that the structure is not fully restrained; it is geometrically unstable. Practical structures are designed to avoid this situation.

1.6 PRACTICAL APPLICATION WHEN $n = m$

With $n = m$ the structure is either statically determinate or geometrically unstable. If the structure is statically determinate the unknown forces are uniquely determined by the equations of equilibrium and it makes no difference how the unknown forces are determined. It is this feature of the statically determinate structure that makes the commonly employed piecemeal methods (e.g., methods of sections and joints in the analysis of trusses) of solution possible.

In the great majority of practical cases, it is clear by inspection whether a structure (with $n = m$) is statically determinate. In the borderline cases one can write out the complete set of equations and examine the determinant of the coefficients of the unknowns to establish determinacy or instability unequivocally. Ordinarily, in the case of manual calculation because of the total work effort involved, one tries to avoid this approach. An alternate, and perhaps somewhat preferable, approach that is almost equivalent (but leaves room for doubt) is to simply attempt to carry out a conventional piecemeal force analysis. A successful analysis for a not-too-special case of loading can

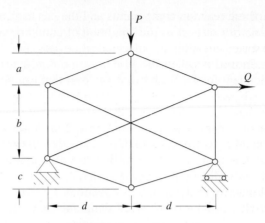

FIGURE 1.1 A stable truss?

be taken to imply that the structure is statically determinate. The reason is that even when using the piecemeal approach for a general case of loading, at some point in the analysis one carries out a division by det (A). If the loading is too specialized, it may be compatible for equilibrium even though det $(A) = 0$. In the absence of numerical errors, a forced division by zero is clear-cut evidence that det $(A) = 0$. It is also noted that if there is any loading for which the equations of equilibrium cannot be satisfied, the structure is necessarily geometrically unstable.

Even when the loading is arbitrary, care must be taken in drawing general conclusions about stability and determinacy of a structural type when piecemeal methods of solution are used to calculate unknown forces. The truss shown in Figure 1.1 illustrates the point (note: the members cross but are not connected at the center of the truss). The relative values of the dimensions a, b, c, and d determine the value of det (A). Successful analyses by piecemeal methods for several different values for the dimensions could lead one to the conclusion that such a truss is always statically determinate. In reality, however, the truss is unstable when $a = c$.

1.7 STATICALLY-DETERMINATE STRUCTURES

The analysis of a statically-determinate structure is intrinsically less involved, and demands less effort, than an analysis of a statically indeterminate structure. The relative ease with which a statically determinate structure can be analyzed stems from the fact that here the equations of equilibrium, the displacement-deformation equations, the force-deformation equations, and the compatibility equations are decoupled. Therefore, a complete analysis, including displacement calculations, can be accomplished as the solution of a sequence of independent problems. Ease of analysis is but one of many practical considerations in structural design. Often, after all design aspects have been considered, statically determinate construction is indicated. Skill in the analysis of statically determinate structures is essential because of the importance of statically determinate construction, and because the ability to analyze statically determinate structures is essential to the analysis of statically indeterminate structures. Statically determinate structures are the focus of the material in Chapters 1–7, while statically indeterminate structures are taken up in Chapters 8–10.

The starting point in an analysis consists of calculating the unknown forces of reaction and interaction. The specific technique(s) of analysis is not material since statical determinacy is a guarantee that the unknown forces have unique values. Combinations of conditions of symmetry or antisymmetry, and/or special loading situations, may permit determination of certain unknown forces by inspection. Implicit in stating the values of unknown forces by inspection is visualization of an appropriate free-body diagram and the mental formulation and solution of the associated equations of equilibrium. Experience teaches that all too often the visualization and mental calculation approach to force analysis leads to incorrect results, not because the technique is invalid but because of the difficulty of maintaining complete, correct mental images. Assumptions concerning values for unknown forces are not necessary and should play no role in analysis. In all cases it is always possible—and should be easy—to verify results obtained by inspection and/or with the aid of assumptions by constructing free-body diagrams and substitution of purported results into equations of equilibrium. Examples in this book are solved using the universally recommended, time-tested technique of constructing complete free-body diagrams, writing the required equations of equilibrium, and then solving the equations.

1.8 PLANE-FRAMED STRUCTURES

Structures are, in fact, three-dimensional objects. However, the great majority of practical structures are adequately represented by two-dimensional planar models. Framed structures in which all members lie in a single plane are said to be *plane-framed*. The example and exercise problems are limited to plane-framed structures under coplanar loads; however, it should be noted that the principles discussed to this point apply to all structures.

Example 1.1

Determine the reactions at the supports and the forces of interaction at the b-end of member bc of the structure shown in Figure 1.2a. The essential free-body diagrams are shown in Figures 1.2b and c. For equilibrium of the entire frame:

$$0 = (+ccw)\Sigma M_C = \frac{50}{3}(50) + 35(30) - 45\left(\frac{2}{\sqrt{5}}A\right) - 20\left(-\frac{1}{\sqrt{5}}A\right) \Rightarrow A = 38.3 \text{ k}$$

$$0 = \uparrow\Sigma F = C_1 + \frac{2}{\sqrt{5}}A - 30 - \frac{3}{5}(50) \Rightarrow C_1 = 25.8 \text{ k}$$

$$0 = \rightarrow\Sigma F = -C_2 + \frac{1}{\sqrt{5}}A - \frac{4}{5}(50) \Rightarrow C_2 = -22.9 \text{ k}$$

8 Chapter 1 Analysis of Framed Structures

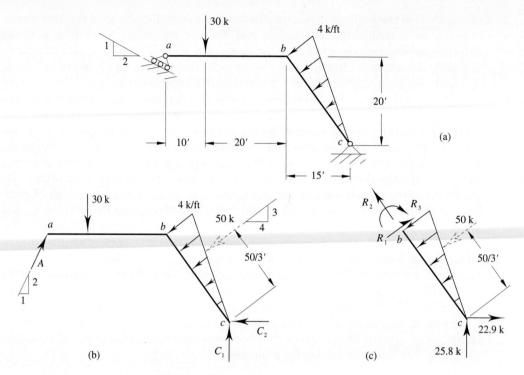

FIGURE 1.2 Three unknown reaction components.

For equilibrium of member bc:

$$0 = (+cw)\,\Sigma M_b = R_3 + \frac{25}{3}(50) - 15(25.8) - 20(22.9) \Rightarrow R_3 = 428 \text{ k-ft}$$

$$0 = \nearrow \Sigma F = R_1 - 50 + \frac{4}{5}(22.9) + \frac{3}{5}(25.8) \Rightarrow R_1 = 16.2 \text{ k}$$

$$0 = \nwarrow \Sigma F = R_2 + \frac{4}{5}(25.8) - \frac{3}{5}(22.9) \Rightarrow R_2 = -6.90 \text{ k}$$

Example 1.2

Determine the reactions for the frame shown in Figure 1.3a. By inspection the frame is fully restrained and it is statically determinate. The free-body diagram for the entire frame (Figure 1.3b) shows four unknown forces. The free-body diagram for member AB (Figure 1.3c) also shows four unknown forces. Between the two free-body diagrams there is a total of six unknown forces and exactly six independent equations of equilibrium can be written, three for each free-body diagram. For equilibrium of the entire frame:

$$0 = (+cw)\Sigma M_C = 12A_1 - 6A_2 + 6(30) - 6(36)$$
$$0 = \rightarrow \Sigma F = -A_2 - C_1 + 30$$
$$0 = \uparrow \Sigma F = A_1 + C_2 - 36$$

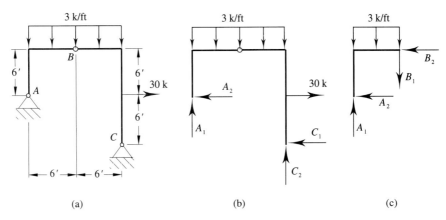

(a) (b) (c)

FIGURE 1.3 Four unknown reaction components.

Moment equilibrium of member AB requires that

$$0 = (+cw)\Sigma M_B = 6A_1 + 6A_2 - 3(18)$$

The two moment equations yield:

$$A_1 = 5\text{ k}, \quad A_2 = 4\text{ k}$$

The force equations then give:

$$C_1 = 26\text{ k}, \quad C_2 = 31\text{ k}$$

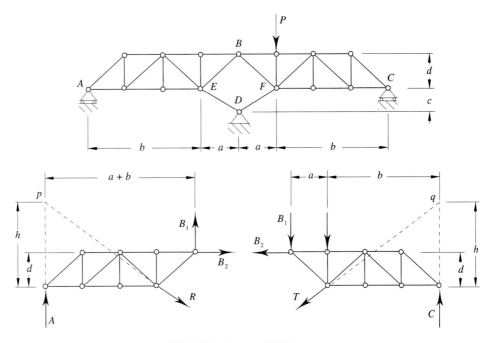

FIGURE 1.4 Symmetric Wichert truss.

Example 1.3

The structure shown in Figure 1.4 is an example of a symmetric Wichert truss. Any attempt to calculate the reactions and bar tensions manually by the commonly applied methods of sections and/or joints (these techniques are reviewed in Chapter 2) will fail because all sections contain more than three unknown forces and all joints contain more the two unknown forces. If there is a solution, it can be obtained with the aid of the two free-body diagrams as shown. Between the two free-body diagrams one can write six independent equations of equilibrium and solve them for the six unknown forces as shown. The remainder of the analysis offers no complications. A pair of equations to determine B_1 and B_2 are obtained from moment equilibrium of each half of the truss:

$$0 = (+ccw)\Sigma M_p \Rightarrow (b + a)B_1 + (h - d)B_2 = 0$$
$$0 = (+cw)\Sigma M_q \Rightarrow -(b + a)B_1 + (h - d)B_2 = bP$$

The point p is at the intersection of the lines of action of the reaction force at A and the force due to the tension R in bar DE. The point q is similarly located. The equations have a solution provided that $h \neq d$ or, equivalently, provided that $cb \neq ad$. If $h = d$, $cb = ad$, the truss is incompletely restrained and, hence, geometrically unstable.

KEY POINTS

1. The linear theory of structures is based on two fundamental assumptions that are justified by experience: (1) deformations and rotations are small, and (2) Hooke's law is satisfied. The validity of the principle of superposition is a consequence of the assumptions.
2. The stability and statical determinacy/indeterminacy properties of a structure are determined by the characteristics of the independent equations of equilibrium written with respect to the undeformed geometry of the structure.
3. The forces of reaction and interaction in a statically determinate structure are determined uniquely from the equations of equilibrium alone.
4. Removal of redundant mechanical restraints in a statically indeterminate structure produces a specific statically determinate cut-back or primary structure corresponding to the removed restraints.

PROBLEMS

1.1 Calculate the reactions.

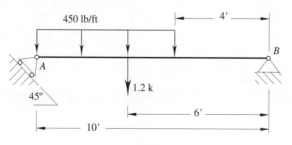

PROBLEM 1.1

1.2 Calculate the reactions.

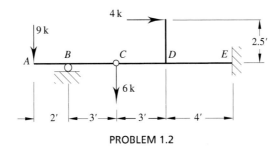

PROBLEM 1.2

1.3 Calculate the reactions. The support at A transmits axial force and bending moment but does not transmit shear.

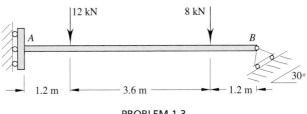

PROBLEM 1.3

1.4 Calculate the reactions.

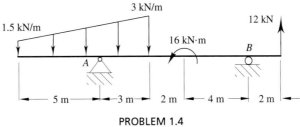

PROBLEM 1.4

1.5 Calculate the reactions.

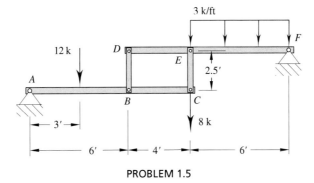

PROBLEM 1.5

1.6 The uniform beam weighs 300 kg/m. Calculate the reactions. Take $g = 9.81$ m/s^2.

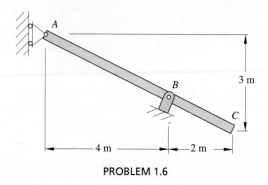

PROBLEM 1.6

1.7 Calculate the reactions and the forces of interaction on the B-end of member BCD.

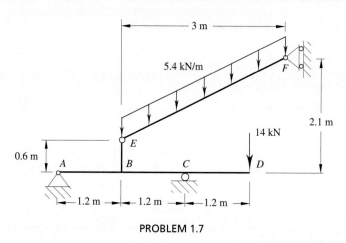

PROBLEM 1.7

1.8 Calculate the reactions and the forces of interaction on the B-end of member BD.

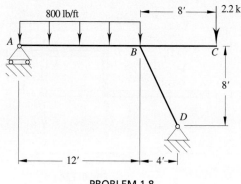

PROBLEM 1.8

1.9 Calculate the reactions and the forces of interaction on the B-end of member BC.

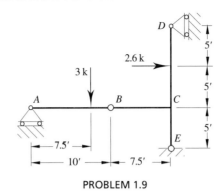

PROBLEM 1.9

1.10 Calculate the reactions and the forces of interaction on the B-end of member BC.

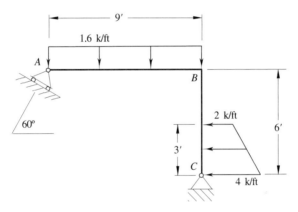

PROBLEM 1.10

1.11 Calculate the reaction at F and the forces of interaction on the E-end of member DE.

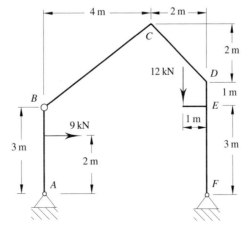

PROBLEM 1.11

1.12 Calculate the reactions at A and E and the forces of interaction on the D-end of member DE.

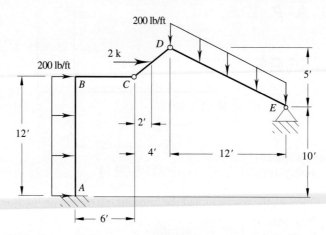

PROBLEM 1.12

CHAPTER 2

Trusses

A truss is a special type of structure made up of long, thin, straight members connected together only at their ends. In a space truss, the connections are idealized as frictionless ball and socket joints. Connections in a plane truss are modeled as frictionless pins. Ordinarily, except for the dead loads of the individual members, the loads on a truss are applied only at the joints. The present discussion is limited to the force analysis of plane trusses that are loaded only at the joints. Some commonly used plane bridge and roof trusses are shown in Figure 2.1.

Pin-connected joints may be found in some very old trusses. In modern truss construction, however, the joints are riveted, bolted, or welded. Practice and sophisticated computer analyses show that for the purpose of a force analysis such joints may be treated satisfactorily as being pin-connected provided that: (1) the members are long and thin, and (2) the centroidal axis of each member coincides with the corresponding reference line for that member on the line diagram of the truss.

A truss is classified as being *simple*, *compound*, or *complex* according to its construction. The construction of a simple truss begins with the formation of a triangle with three members and three pins. Thereafter, the truss is expanded by adding members two at a time. The two members are conceived as first being pin-connected to each other and then the free ends of the joined members are connected to two different

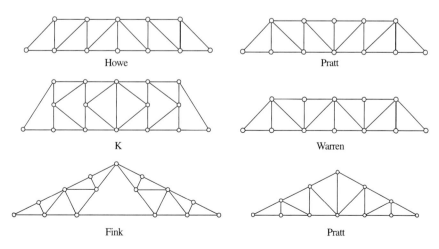

FIGURE 2.1 Common trusses.

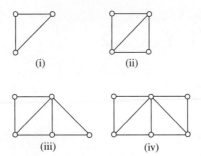

FIGURE 2.2 Simple truss construction.

existing joints. The two new members must be arranged so that their reference lines are not collinear. The construction of a simple truss is illustrated in Figure 2.2. The Howe, Pratt, K, and Warren trusses shown in Figure 2.1 are examples of simple trusses.

Compound trusses are formed by joining simple trusses. The Fink truss shown in Figure 2.1 is an example of a compound truss. The Wichert truss of Example 1.3 in Chapter 1 is another example of a compound truss.

A truss that cannot be classified as being simple or compound is said to be a complex truss. The truss shown in Figure 1.1 is an example of a complex truss.

Analytically, simple and compound trusses are characterized by the fact that they are readily amenable to direct piecemeal force analysis by a combination of the methods of sections and joints. The force analysis of a complex truss demands more than just the methods of joints and sections.

2.1 STABILITY AND DETERMINACY

All of the remarks made on these topics in Chapter 1 still apply. The outstanding feature of a pin-connected truss in relation to stability and determinacy lies in the ease with which the number of independent equations of equilibrium and the number of unknown forces can be determined. Assume the truss is supported only by some combination of pins and/or rollers (or their equivalents). Let:

b = number of bars (members)
j = number of joints (pins)
r = number of independent components of forces of reaction at supports

There are three independent equations of equilibrium for each member and for each joint there are two independent equations of equilibrium. Therefore, the number of independent equations of equilibrium is

$$N_e = n = 3b + 2j$$

In general, because the members are pin-connected at their ends, each member has two unknown components of forces of interaction at its ends. These are the forces of interaction exerted on the ends of the member by the associated pins. Therefore, the number of unknown forces is

$$N_u = m = 4b + r$$

and

$$N_e - N_u = n - m = 2j - b - r$$

From the discussion in Chapter 1, if $N_e > N_u$, the truss is geometrically unstable.
For a fully restrained truss, from the discussion in Chapter 1,

$$N_u - N_e = d = b + r - 2j$$

is the degree of indeterminacy. In this chapter attention is focussed on statically determinate trusses; i.e., trusses for which $d = 0$, or for which

$$b + r = 2j$$

Of course, the fact that $b + r = 2j$ does not guarantee stability.

2.2 FORCES OF INTERACTION/INTERNAL FORCE RESULTANTS

The discussion is limited to trusses that are loaded only at the joints. Consider a free-body diagram of a typical bar in such a truss as shown in Figure 2.3. The three independent equations of equilibrium for the member yield the three results

$$P_a = 0, \quad P_b = 0, \quad F_a = F_b$$

The results are expected because the bar is a *two-force member*. F_a and $F_b = F_a$ represent the scalar parts of the forces of interaction exerted by the pins at a and b on the bar (equal and opposite forces are exerted by the bar on the respective pins). These

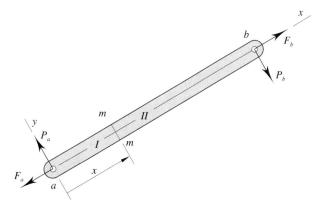

FIGURE 2.3 Straight two-force member.

forces and the components of forces of reaction at the supports constitute the remaining unknowns to be determined.

The forces of interaction and the resultant of the internal forces over a plane cross section are related. The internal force resultants are considered next. As shown in Figure 2.3, let the *x*-coordinate axis be taken positive from *a* toward *b*. The *x* axis overlies the reference axis of the bar which is understood to be scribed on the bar. The reference axis deforms as the bar deforms, whereas the *x* axis always remains straight. The *y*-coordinate axis is perpendicular to the *x* axis and directed so that $x \otimes y \Rightarrow z$ is a coordinate axis directed upward out of the plane of the paper. At any location *x* draw line *mm* parallel to the *y* axis. The line represents a plane cross section viewed on edge. The section divides the bar into two segments, *I* and *II*, as shown. The segments provide each other with mutual support at *mm*. Mutual distributed forces of interaction are generated at the cross section. The distribution of the forces of interaction is unknown, but a representation can be given of their resultant evaluated with respect to the point of intersection of the reference axis for the bar and the plane cross section. The representation defines a so-called sign convention for the internal force resultants. The representation is set out with the aid of a pair of free body diagrams as shown in Figure 2.4.

In Figure 2.4, *M*, *V*, and *N* are scalars to be applied as multipliers to unit vectors as shown to obtain the related forces. The orientations of the unit vectors are fixed with respect to the coordinate axes. Newton's third law is taken into account by reversal of the unit vectors on segments *I* and *II* at the cross section. *M* is called the bending moment, *V* is called the shear, and *N* is called the "axial force" or the tension. In general, *N*, *V*, and *M* vary with the location *x*; however, it is noted that here,

$$0 = \Sigma F_y \Rightarrow V(x) = 0$$
$$0 = \Sigma F_x = N - F \Rightarrow N(x) = F = \text{a constant}$$

Thus, for a truss loaded only at the joints, the shear and the bending moment are zero over the entire length of the bar, and the tension *N* is a constant and equal to the scalar part of the force of interaction at the ends of the bar. Because of this the problem of the force analysis of a truss is understood to be that of calculating the reactions and the bar tensions.

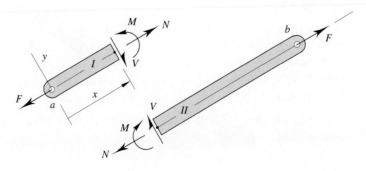

FIGURE 2.4 Stress resultants.

2.3 FORCE ANALYSIS OF STATICALLY-DETERMINATE TRUSSES

The general procedure applicable to all trusses is to draw a free-body diagram for each joint, demand that the resultant of the forces be zero, and solve the resulting set of $2j$ equations for the $2j$ unknown force components. It turns out that for the overwhelming majority of trusses of practical interest the analysis can be carried out efficiently in piecemeal fashion using a combination of the methods of joints and sections. The methods simply make use of two facts:

1. if a structure is in equilibrium, all parts must be in equilibrium;
2. for a statically-determinate structure there is exactly one set of values for the unknown forces such that equilibrium will prevail.

2.4 METHOD OF JOINTS

The basic idea here is to find a sequence of joints such that for each joint in the sequence the related free-body diagram contains no more than two unknown forces (and these forces are not collinear). Because the force system is concurrent, one need only write and solve two independent equations of equilibrium for the unknown forces. The method of joints is easily mastered but has two serious drawbacks: (1) an error committed in the analysis of the first joint in a sequence is, in general, propagated to result in additional errors; (2) often the method is very inefficient when the forces in only a few members are required.

2.5 METHOD OF SECTIONS

Here the idea is to find a sequence of finite parts of the truss that contain at least two joints and no more than three unknown forces (and these forces do not have parallel lines of action). The force system is coplanar and not concurrent. For such a force system three independent equations of equilibrium can be written and solved for the unknown forces. The method of sections is effective in sidestepping the shortcomings of the method of joints.

2.6 CONVENTIONS AND PRACTICE

The design of a member when the member is in compression is radically different from the case when the member is in tension. Accordingly, it is imperative that the "bar forces" or "tensions" be identified so that it is clear whether a member is in tension or compression. Two conventions for designation of a bar force are:

1. use C to indicate compression and use T to indicate tension;
2. use a negative sign to indicate compression and use no sign or a plus sign to indicate tension.

Good engineering practice is to summarize the results of a truss analysis by listing calculated quantities on a line diagram of the truss as the results are generated.

Chapter 2 Trusses

In this way the analyst can see what has been done and what remains to be done, and can pick up errors much more quickly than if the summary is not made or is postponed.

Example 2.1

Calculate the reactions and bar forces for the truss shown in Figure 2.5. The truss is amenable to analysis by the method of joints by selecting joints in the sequence a, b, e, f, c, g. The free-body diagrams for joints $a, b, e,$ and f are shown in Figure 2.5.

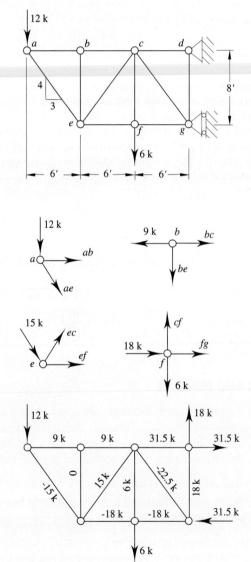

FIGURE 2.5 Method of joints.

Joint a:
$$(4/5)ae + 12 = 0 \Rightarrow ae = -15 \text{ k}$$
$$(3/5)ae + ab = 0 \Rightarrow ab = 9 \text{ k}$$

Joint b: By inspection, $be = 0, bc = 9 \text{ k}$.

Joint e:
$$(4/5)15 - (4/5)ec = 0 \Rightarrow ec = 15 \text{ k}$$
$$(3/5)15 + (3/5)ec + ef = 0 \Rightarrow ef = -18 \text{ k}$$

Continuing in this way through the remaining joints in the sequence given above leads to the results shown on the line diagram of the truss.

Example 2.2

The method of sections is particularly useful in the calculation of bar forces for selected bars in a truss. Consider the determination of the force in bar bc and in bar ec of the truss in the previous example. The truss is reproduced in Figure 2.6.

To obtain the desired results, look for a section that contains at least one of the unknown forces and contains no more than three unknown forces. The desired section is obtained by "cutting" bars bc, ec, and ef and constructing a free-body diagram of the portion of the truss to the left of the section. This eliminates the need for calculating reactions. Equilibrium of the free-body requires that

$$0 = \uparrow \Sigma F = (4/5)ec - 12 \Rightarrow ec = 15 \text{ k}$$
$$0 = (+cw)\Sigma M_e = 8bc - 6(12) \Rightarrow bc = 9 \text{ k}$$

As expected (and as they must), the results coincide with those obtained by the method of joints.

Example 2.3

This example serves to illustrate that, in general, it is most efficient to use the methods of sections and joints together. The problem is to calculate the tension in bars ab, bd, and eb of the K-truss shown in Figure 2.7.

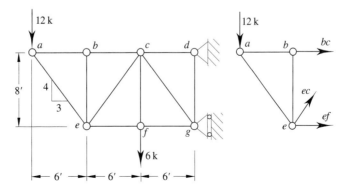

FIGURE 2.6 Method of sections.

22 Chapter 2 Trusses

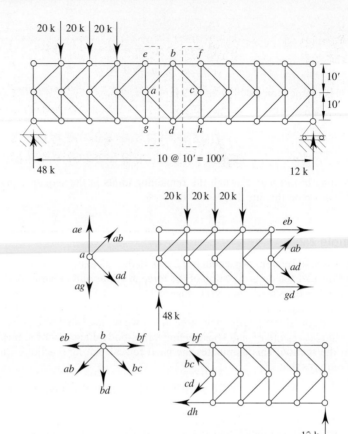

FIGURE 2.7 K-truss.

The reactions are as shown. The force in ab is obtained from equilibrium of joint a and of the section shown for the left half of the truss. From the free-body diagram of joint a,

$$0 = \rightarrow \Sigma F \Rightarrow ab + ad = 0$$

For the left half of the truss,

$$0 = \uparrow \Sigma F \Rightarrow (\sqrt{2}/2)ab - (\sqrt{2}/2)ad = 12$$

from which $ab = 8.49$ k. From equilibrium of joint c and the right half of the truss, $bc = -8.49$ k. From the free-body diagram of joint b,

$$0 = \downarrow \Sigma F \Rightarrow (\sqrt{2}/2)ab + (\sqrt{2}/2)bc + bd \Rightarrow bd = 0$$

Since the horizontal components of the bar forces in ab and in ad are equal and opposite, equilibrium of the left half of the truss requires that

$$0 = (+cw)\Sigma M_g = 20eb + 40(48) - 10(20) - 20(20) - 30(20)$$

from which $eb = -36$ k. Observe that eb could have been obtained from equilibrium of the left portion of the truss after cutting bars $eb, ea, ag,$ and gd.

Example 2.4

Calculate the forces in all members of the Fink truss shown in Figure 2.8. All angles are 30, 60, or 90 degrees. The reactions have been calculated and are as shown. Observe that significant portions of the analysis can be carried out without knowledge of the reactions. Consider the free-body diagram of joint d. Although there are three unknown forces, two of these are collinear and as a consequence force equilibrium in the direction "1" immediately yields the value $dp = -1$ k.

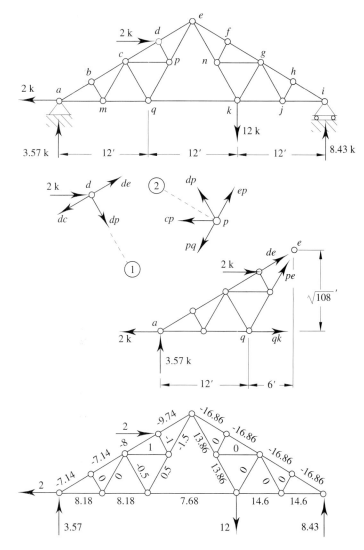

FIGURE 2.8 Fink truss.

A similar situation exists at joint p. Force equilibrium in the direction "2" yields $cp = 1$ k. Joints b and m may now be analyzed in the same way to obtain $bm = 0$, $mc = 0$. Next, with cm and cp known, force equilibrium of joint c gives $cq = -0.5$ k. A free-diagram of joint q leads immediately to the result $qp = 0.5$ k.

Returning to the free-body diagram of joint p (with dp, cp, and pq known), one finds $ep = 1.5$ k. The tensions in de and qk and verification of the result obtained for pe follow from equilibrium of the section obtained by cutting bars de, pe, and qk. Summing moments in the clockwise sense about joints a, q, and e, respectively, yields:

$$0 = 4.5\sqrt{3}(2) - 12(\sqrt{3}/2)pe \Rightarrow pe = 1.5 \text{ k}$$
$$0 = 12(3.57) + 4.5\sqrt{3}(2) + 12(0.5de) \Rightarrow de = -9.74 \text{ k}$$
$$0 = 18(3.57) + 6\sqrt{3}(2) - 1.5\sqrt{3}(2) - 6\sqrt{3}qk \Rightarrow qk = 7.68 \text{ k}$$

The right-hand portion of the truss may be treated in a similar manner. The results of the analysis are tabulated on a line diagram of the truss as shown in Figure 2.8, in which all forces are in kips.

Example 2.5

For the truss shown in Figure 2.9, calculate the bar tensions for $a = 2$ ft, $b = c = 3$ ft, $d = 4$ ft, $P = 2$ k, and $Q = 4$ k. This is a complex truss and, as such, it is not amenable to solution by the methods of sections and joints. The fundamental and most general method of solution is to write

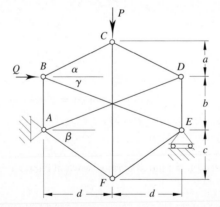

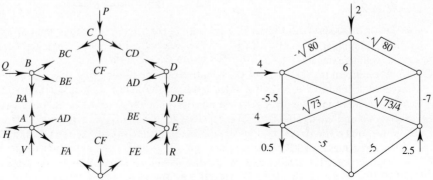

FIGURE 2.9 Complex truss.

and solve all of the joint equilibrium equations simultaneously. An alternate method is that developed by Henneberg (explained in the text by Timoshenko and Young) which involves the use of a "substitute" member and superposition. Another scheme, which is used here, is to solve for the forces in all members and for one of the loads, say, Q, in terms of one of the bar forces and the other load, P. Then, P and Q are put equal to their assigned values to complete the solution. This procedure shows that the truss is unstable if $a = c$ (or $\alpha = \beta$). The analysis is set out in general terms to the point where the claim is established. Because of the symmetry and for simplicity, the solution is formulated in terms of the tension CF, P, and Q.

By inspection of the free-body diagrams for joints C and F, $BC = CD$, and $FA = FE$. These results are used in the analysis which follows. For equilibrium of joints C and B,

$$0 = \downarrow \Sigma F_C = CF + 2BC \sin \alpha + P \Rightarrow BC = -\frac{P + CF}{2 \sin \alpha}$$

$$0 = \rightarrow \Sigma F_B = Q + BC \cos \alpha + BE \cos \gamma$$

from which

$$BE \cos \gamma = -Q + \frac{(P + CF) \cot \alpha}{2}$$

For equilibrium of joints F and E,

$$0 = \uparrow \Sigma F_F = CF + 2FE \sin \beta \Rightarrow FE = \frac{-CF}{2 \sin \beta}$$

$$0 = \leftarrow \Sigma F_E = FE \cos \beta + BE \cos \gamma$$

Therefore,

$$BE \cos \gamma = \frac{CF \cot \beta}{2}$$

Combining the two equations for $BE \cos \gamma$ yields

$$CF(\cot \beta - \cot \alpha) = P \cot \alpha - 2Q$$

It is now clear that equilibrium can prevail for arbitrary load only if $\alpha \neq \beta$, which requires $a \neq c$. For $a = 2$ ft, $b = c = 3$ ft, and $d = 4$ ft, $\cot \alpha = 2$, $\cot \beta = 4/3$, and with $P = 2$ k and $Q = 4$ k, $CF = 6$ k. Calculation of the remaining unknown forces offers no difficulty. The results are summarized on the line diagram shown in Figure 2.9, in which all forces are in kips.

Example 2.6

Analyze the truss shown in Figure 2.10. Because of the transverse loading, bars ab and ed are not two-force members. As can be seen from the free-body diagrams, these members have transverse as well as axial components of forces of interaction at their ends.

The tension is constant over the length of ed. The tension in ab has one value between a and the 16 k load and a different value between the load and b. In addition, the shear and bending moment in these members are non-zero. To complete the force analysis of the truss, members ab and ed are removed and replaced by the forces of interaction they exert on the respective joints. The structure that remains is in equilibrium and all members of the structure are two-force members. The bar tensions are easily found by application of the methods of joints and sections. The results, in kips, are displayed on a line diagram of the truss as shown in Figure 2.10.

26 Chapter 2 Trusses

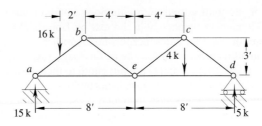

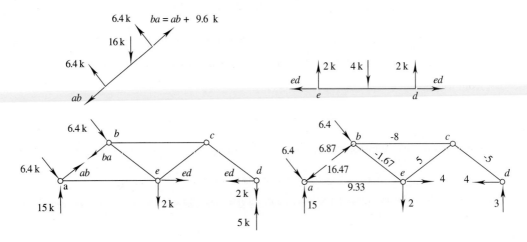

FIGURE 2.10 Members with loads between ends.

KEY POINTS

1. The most general procedure for the force analysis of any statically determinate truss consists of writing and then solving simultaneously the joint equilibrium equations.

2. The great majority of practical trusses are amenable to a piecemeal force analysis using the methods of joints and sections.

3. The idea of the method of joints is to find a sequence of joints such that for each joint in the sequence the related free-body diagram contains no more than two unknown, non-collinear forces, which can be determined by writing and solving two equations of equilibrium.

4. The idea of the method of sections is to find a sequence of sections such that for each section in the sequence the related free-body diagram contains no more than three unknown, non-parallel forces, which can be determined by writing and solving three equations of equilibrium.

5. It is essential to identify the axial force in a member as being a compression or a tension because the design of a compression member is radically different from the design of a tension member.

PROBLEMS

2.1 Calculate the force in each member. Indicate tension or compression. Display all results on a line diagram of the truss.

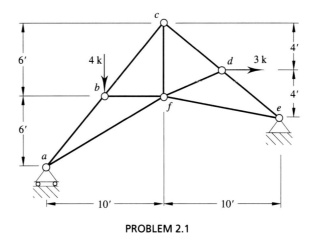

PROBLEM 2.1

2.2 Calculate the force in each member. Indicate tension or compression. Display all results on a line diagram of the truss.

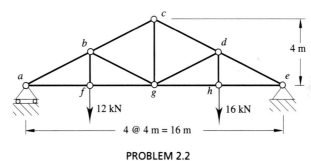

PROBLEM 2.2

2.3 Calculate the force in each member. Indicate tension or compression. Display all results on a line diagram of the truss.

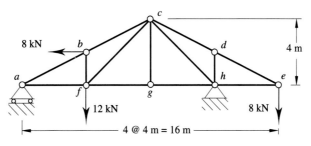

PROBLEM 2.3

2.4 Calculate the force in each member. Indicate tension or compression. Display all results on a line diagram of the truss.

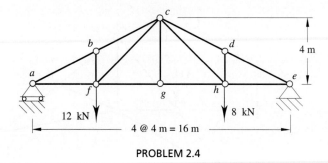

PROBLEM 2.4

2.5 Calculate the force in each member. Indicate tension or compression. Display all results on a line diagram of the truss. All triangles are isoceles triangles.

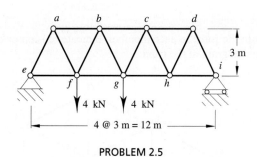

PROBLEM 2.5

2.6 Calculate the tension in members bc, bg, and gc.

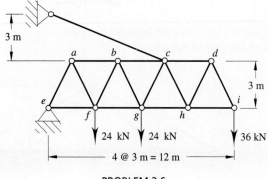

PROBLEM 2.6

2.7 Calculate the tension in members dh, ch, and gh.

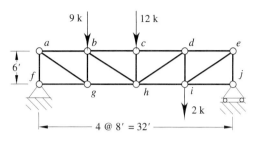

PROBLEM 2.7

2.8 Calculate the tension in members dg, fg, and bg.

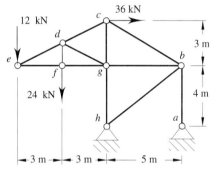

PROBLEM 2.8

2.9 Calculate the tension in members ce, cf, and ef.

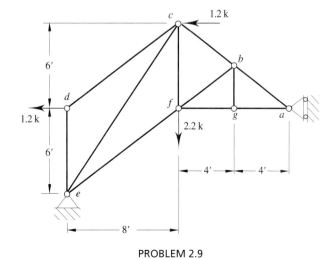

PROBLEM 2.9

2.10 Calculate the tension in members fe, gf, and ch.

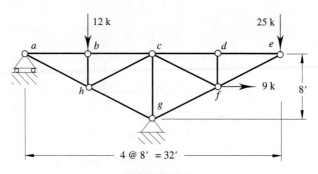

PROBLEM 2.10

2.11 Calculate the tension in members ab, bc, and ad.

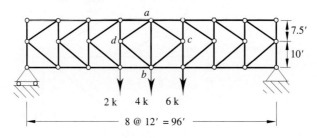

PROBLEM 2.11

2.12 Calculate the tension in members bc and cf.

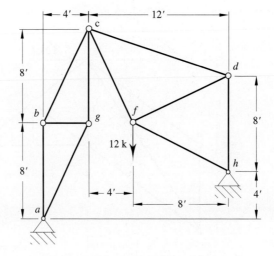

PROBLEM 2.12

2.13 Calculate the tension in members *ab*, *ac*, and *bc*.

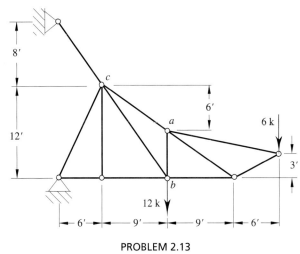

PROBLEM 2.13

2.14 Calculate the tension in members *ab*, *ad*, and *bc*.

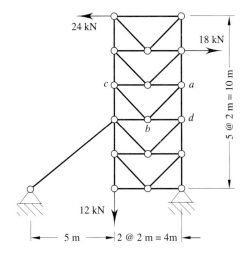

PROBLEM 2.14

2.15 Calculate the tension in members *ab*, *ac*, and *ad*.

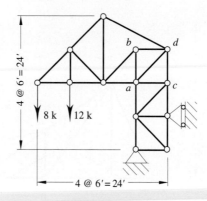

PROBLEM 2.15

2.16 Calculate the tension in members *ab* and *bc*.

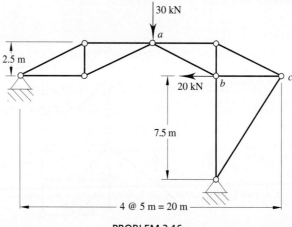

PROBLEM 2.16

2.17 The radial bars are of length L. Interior angles of the triangles are either 30° or 75°. Calculate the tension in members ab, bc, and ac.

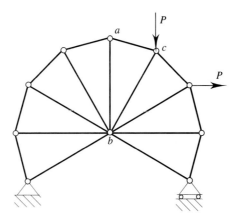

PROBLEM 2.17

2.18 Calculate the tension in members ab and ac.

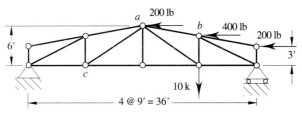

PROBLEM 2.18

2.19 All triangles are equilateral triangles. Calculate the tension in members ab, ac, and bc.

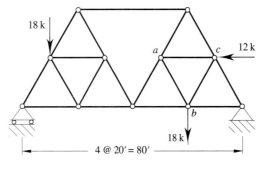

PROBLEM 2.19

2.20 Calculate the tension in members *ab*, *ac*, and *bc*.

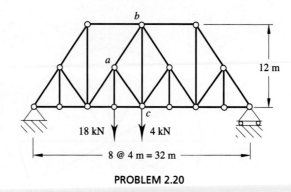

PROBLEM 2.20

2.21 Calculate the tension in member *ab*.

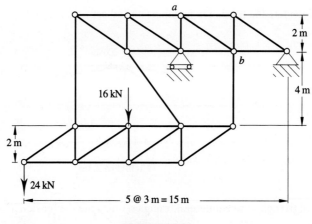

PROBLEM 2.21

2.22 Calculate the tension in member *ab*.

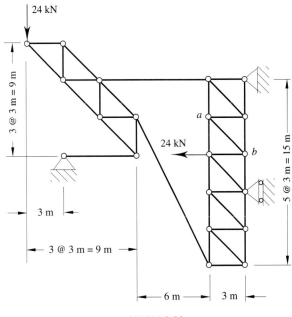

PROBLEM 2.22

2.23 Calculate the tension in each member of the symmetric Wichert truss.

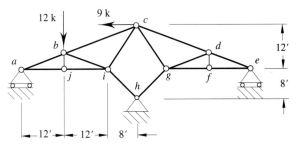

PROBLEM 2.23

2.24 Calculate the axial and transverse components of the forces of interaction at the ends of all members of the truss.

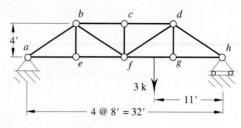

PROBLEM 2.24

2.25 Calculate the axial and transverse components of the forces of interaction at the ends of all members of the truss for which all angles are either 45 degrees or 90 degrees.

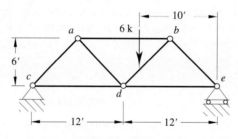

PROBLEM 2.25

CHAPTER 3

Internal Force Resultants

In the case of a long, thin member such as is encountered in a framed structure, the important stresses or internal forces occur on plane cross sections. For the purpose of design, with the aid of the methods of the mechanics of materials, the stresses on a cross section can be evaluated with sufficient accuracy when the resultant of the stresses is known. For the plane-framed structures and loadings under consideration, the resultant lies in the plane of the frame. It is the resultant evaluated with respect to the centroid of the cross section that must be known to evaluate the stresses by the methods of strength of materials. The locus of the centroids of the cross sections of a member defines the reference line for the member. For a straight member, the reference line becomes the reference axis.

Given the loading between the ends of a member and knowing the forces of reaction or interaction exerted on the ends of the member, one can determine the resultant of the stresses from the requirement of equilibrium. In general, the resultant consists of a force and a couple. The force is resolved into an axial component, called the tension or axial force, and a transverse component called the shear. The moment of the couple is called the bending moment. A conventional representation is given to the components of the resultant. The representation is understood to define a sign convention. The sign convention employed here is the same as that used in the discussion of trusses. For reinforcement of basic ideas, the convention is reviewed here. The discussion is limited to straight members under a general combination of transverse and axial loads.

To set out the convention, consider a free-body diagram of a typical straight member as shown in Figure 3.1. The member could have any orientation in the plane, but is shown in the horizontal position for convenience. The forces at the ends of the member are treated as known quantities and the loads between the ends are given.

In Figure 3.1, a and b are the end points of the reference axis of the member, a physical axis in the member. The coordinate axes, which are fixed in space, are chosen with the origin at a and with the x-axis directed from a toward b. The y-axis is chosen so that a z-axis, with z in the direction of $x \otimes y$, points upward out of the plane of the page. Functions $w(x)$ and $p(x)$ define the intensities of the distributed transverse and axial loads (forces per unit of length along the axis of the member). $w(x)$ and $p(x)$ are scalar multipliers to be applied to unit vectors in the directions of the associated arrows to obtain the respective loads. Line m–m represents the plane cross section at location x as seen on edge. The cross section separates the member into two segments: Segment I consists of the material between end a and the cross section, and Segment II consists of the material between the cross section and end b. The representation of the internal

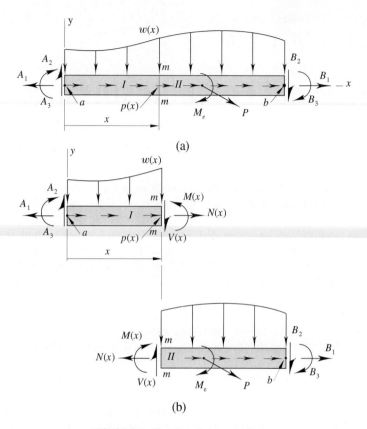

FIGURE 3.1 Loads and stress resultants.

force resultant on a cross section is shown in Figure 3.1b. $M = M(x)$, $N = N(x)$, and $V = V(x)$ are scalar multipliers to be applied to unit vectors in the directions of the respective arrows to obtain the corresponding force quantities. Newton's third law or, alternatively, the requirements of equilibrium for each segment considered separately and for the member as a whole is taken into account by taking the unit vectors on the cross section for Segment *II* to be opposite to the corresponding vectors on Segment *I*. The only unknown quantities shown on the free-body diagrams are M, V, and N. Because the member is in equilibrium, because there are three unknown quantities, and because three independent equations of equilibrium can be written, M, V, and N can be determined simply by writing and solving the equations of equilibrium.

3.1 INTERNAL AND EXTERNAL FORCE FUNCTIONS AND DIAGRAMS

$V = V(x)$, $M = M(x)$, and $N = N(x)$ are the shear, bending moment, and axial force functions for the member. Plots of these functions against location in the member are called the shear, bending moment, and axial force diagrams, respectively. $w(x)$ and $p(x)$ are the transverse and axial distributed load functions. Plots of $w(x)$ and $p(x)$ over the length of the member are called the transverse and axial load diagrams, respectively. Differential and integral relationships of considerable practical as well as theoretical importance connect the internal and external load functions. Generally, the internal and external load functions and/or their derivatives are only segmentally continuous; that is to say, it is only in segments that each and all of the functions has a unique analytical description. End points of the segments are readily identified as points at which a concentrated external force and/or couple is applied, or points at which the analytical descriptions of the distributed load functions change. The points of discontinuity are obvious when the internal force resultants are determined with the aid of free-body diagrams and direct application of the equations of equilibrium.

3.2 SHEAR AND BENDING MOMENT FUNCTIONS AND DIAGRAMS FOR BEAMS

When the axial forces are either zero or negligibly small, the member is referred to as a beam. To illustrate basic concepts, evaluation of the shear and bending moment at specific points, the construction of shear and bending moment functions, and construction of the related diagrams for particular situations are considered next. Subsequently, generalizations of the relationships that connect the various internal-external load functions are derived.

Example 3.1

Calculate the shear and bending moment at sections a, b, and c of the beam shown in Figure 3.2 on page 40. The sketch immediately above the beam is included as a graphical reminder of the positive senses for M and V according to the established sign convention.

Free-body diagrams selected for the calculation of M and V at sections a, b, and c are as shown.

Section a:

$$0 = \uparrow \Sigma F = 5 - V \Rightarrow V = 5 \text{ k}$$
$$0 = (+ccw)\Sigma M. = M - 1(5) \Rightarrow M = 5 \text{ k-ft}$$

Section b:

$$0 = \uparrow \Sigma F = 5 - 7 - V \Rightarrow V = -2 \text{ k}$$
$$0 = (+ccw)\Sigma M. = M + 2(7) - 4(5) \Rightarrow M = 6 \text{ k-ft}$$

Section c:

$$0 = \uparrow \Sigma F = V + 2 \Rightarrow V = -2 \text{ k}$$
$$0 = (+cw)\Sigma M. = M - 1(2) \Rightarrow M = 2 \text{ k-ft}$$

Example 3.2

Derive expressions for the shear and bending moment functions and plot the shear and bending moment diagrams for the beam shown in Figure 3.3 on page 41. The sketch immediately above the beam is included as a graphical reminder of the positive senses for M and V according to the established sign convention. The end points of the three segments in which $V(x)$ and $M(x)$ are continuous are located at $x = 0, 2, 6,$ and 8 feet. Free-body diagrams for the segments are as shown.

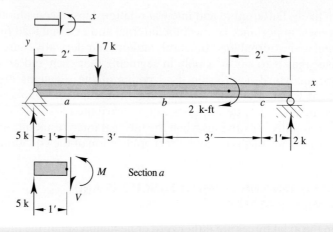

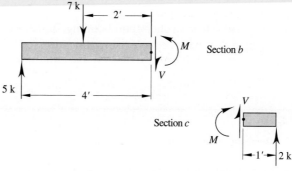

FIGURE 3.2 Free-body diagrams.

For $0 < x < 2$:
$$0 = \uparrow \Sigma F = 5 - V \Rightarrow V = 5$$
$$0 = (+ccw)\Sigma M_. = M - 5x \Rightarrow M = 5x$$

For $2 < x < 6$:
$$0 = \uparrow \Sigma F = 5 - 7 - V \Rightarrow V = -2$$
$$0 = (+ccw)\Sigma M_. = M - 5x + 7(x - 2) \Rightarrow M = 14 - 2x$$

For $6 < x < 8$:
$$0 = \uparrow \Sigma F = 2 + V \Rightarrow V = -2$$
$$0 = (+cw)\Sigma M_. = M - 2(8 - x) \Rightarrow M = 16 - 2x$$

It is observed that in each interval,

$$M' = \frac{dM}{dx} = V$$

In words, in each interval the derivative of the bending moment function is equal to the shear function.

Section 3.2 Shear and Bending Moment Functions and Diagrams for Beams

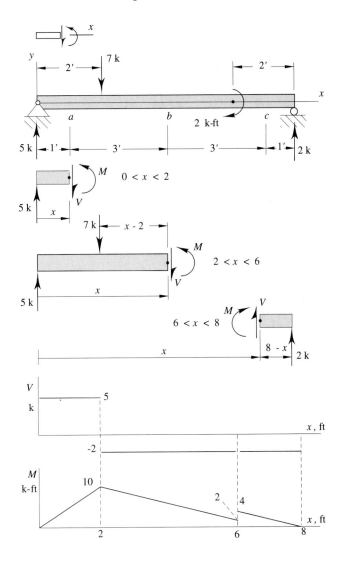

FIGURE 3.3 Discontinuities in shear and bending moment.

The shear and bending moment diagrams are as shown in Figure 3.3. Observe that the shear diagram has a discontinuity of 7 k at the point of application of the concentrated 7-k load. Also observe that the bending moment diagram has a discontinuity of 2 k-ft at the point of application of the applied concentrated couple of 2 k-ft.

Example 3.3

Derive expressions for the shear and bending moment functions and plot the shear and bending moment diagrams for the beam shown in Figure 3.4. The sketch immediately above the beam is included as a graphical reminder of the positive senses for M and V. The end points of the

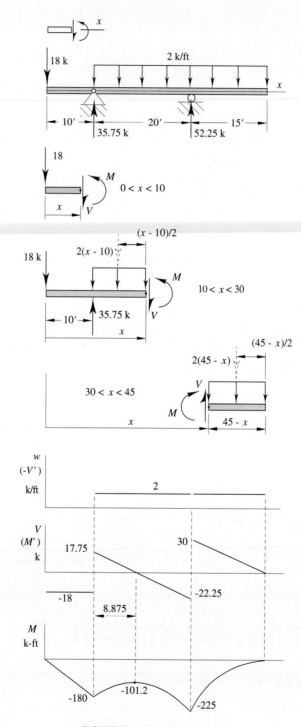

FIGURE 3.4 Distributed loads.

segments in which $V(x)$ and $M(x)$ are continuous with continuous derivatives are located at $x = 0, 10, 30,$ and 45. Free-body diagrams for each of the intervals are selected as shown.

For $0 < x < 10$:
$$0 = \downarrow \Sigma F = V + 18 \Rightarrow V = -18$$
$$0 = (+ccw)\Sigma M. = M + 18x \Rightarrow M = -18x$$

For $10 < x < 30$:
$$0 = \downarrow \Sigma F = V + 18 - 35.75 + 2(x - 10) \Rightarrow V = 17.75 - 2(x - 10)$$
$$0 = (+ccw)\Sigma M. = M + 18x - 35.75(x - 10) + (x - 10)^2$$
$$\Rightarrow M = -18x + 35.75(x - 10) - (x - 10)^2$$

For $30 < x < 45$:
$$0 = \uparrow \Sigma F = V - 2(45 - x) \Rightarrow V = 2(45 - x)$$
$$0 = (+cw)\Sigma M. = M + (45 - x)^2 \Rightarrow M = -(45 - x)^2$$

It is observed that in each interval,

$$M' = \frac{dM}{dx} = V \quad \text{and} \quad V' = \frac{dV}{dx} = -w$$

Thus, in addition to the derivative of the bending moment function being equal to the shear function, the derivative of the shear function is equal to the negative of the load function. These relations are not just fortuitous. It will be shown that the relations hold at all points where the derivatives exist.

A distributed load diagram as well as the shear and bending moment diagrams are shown in Figure 3.4.

3.3 RELATIONSHIPS BETWEEN INTERNAL AND EXTERNAL LOADS

The discontinuity and derivative relationships (and, hence, implied integral relationships) noted in the specific examples are valid for general loading of a member. The general relationships are established here for axial loading as well as for transverse loading. The relationships are of theoretical interest and of considerable practical interest because of their utility in the construction of internal force diagrams. The procedure for general loading differs from that in the specific examples only in how the resultants of the distributed loads are evaluated and how the moments of the distributed loads are evaluated. As in the specific examples, all results follow from an application of the conditions of equilibrium to free-body diagrams followed by application of the rules of the calculus.

Consider a portion of a member that is in equilibrium under the action of general loading as shown in Figure 3.5. In addition to the distributed transverse and axial loads the member is subjected to a concentrated force and a concentrated couple at location x_d.

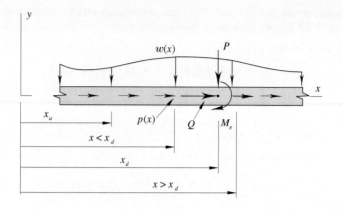

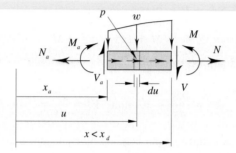

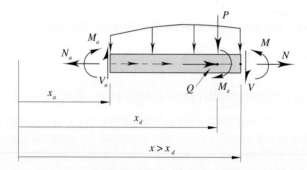

FIGURE 3.5 Generalized loading.

Consider equilibrium of the free-body diagram of that portion of the member contained between x_a and x_d.

$$0 = \rightarrow \Sigma F = N(x) - N_a + \int_{x_a}^{x} p(u)\,du$$

$$0 = \downarrow \Sigma F = V(x) - V_a + \int_{x_a}^{x} w(u)\,du$$

$$0 = (+ccw)\Sigma M_. = M(x) - M_a - V_a(x - x_a) + \int_{x_a}^{x} (x - u)w(u)\,du$$

Section 3.3 Relationships Between Internal and External Loads

Thus, for $x_a < x < x_d$,

$$N(x) = N_a - \int_{x_a}^{x} p(u)\, du$$

$$V(x) = V_a - \int_{x_a}^{x} w(u)\, du$$

$$M(x) = M_a + V_a(x - x_a) - \int_{x_a}^{x} (x - u)w(u)\, du$$

The expression for $M(x)$ can be expanded to read

$$M(x) = M_a + V_a(x - x_a) - x\int_{x_a}^{x} w(u)\, du + \int_{x_a}^{x} uw(u)\, du$$

By inspection of the free-body diagrams, it is clear that for $x_d < x$,

$$N(x) = N_a - \int_{x_a}^{x} p(u)\, du - Q$$

$$V(x) = V_a - \int_{x_a}^{x} w(u)\, du - P$$

$$M(x) = M_a + V_a(x - x_a) - P(x - x_d) - \int_{x_a}^{x} (x - u)w(u)\, du + M_e$$

The discontinuity in the tension N at x_d is equal to the difference in the tension at x_d+ and at x_d-. The discontinuities in the shear and the bending moment are evaluated in the same way. Therefore, the discontinuity relationships are:

$$N(x_d+) - N(x_d-) = -Q, \quad V(x_d+) - V(x_d-) = -P,$$
$$M(x_d+) - M(x_d-) = M_e$$

Differentiation of the expressions for $N(x)$ in either interval yields

$$\frac{dN}{dx} = N' = -p(x)$$

Differentiation of the expression for $V(x)$ in either interval gives

$$\frac{dV}{dx} = V' = -w(x)$$

Differentiation of the expanded form of the expression for $M(x)$ for $x_a < x < x_d$ yields

$$\frac{dM}{dx} = M' = V_a - \left(xw(x) + \int_{x_a}^{x} w(u)\, du\right) + xw(x) \equiv V(x)$$

The same operation yields the same result for $x > x_d$. Integration of the derivative relationships now gives

$$N(x) = N_a - \int_{x_a}^{x<x_d} p(x)\, dx, \quad V(x) = V_a - \int_{x_a}^{x<x_d} w(x)\, dx,$$

$$M(x) = M_a + \int_{x_a}^{x<x_d} V(x)\, dx$$

46 Chapter 3 Internal Force Resultants

The foregoing relations can be summed up as follows for any interval in which the functions are continuous and the associated derivatives exist (i.e., any interval in which there are no concentrated forces or couples and in which $w(x)$ and $p(x)$ are continuous):

1. The derivative of the shear function at location x is equal to the negative of the load function $w(x)$. Graphically this means that the slope of the shear diagram at location x has value equal to the negative of the ordinate $w(x)$ on the load diagram.
2. The derivative of the bending moment function at location x is equal to the shear function $V(x)$. Graphically this means that the slope of the bending moment diagram at location x has value equal to ordinate $V(x)$ on the shear diagram.
3. The increment in the shear between any two locations in which there are no concentrated applied loads is numerically equal to the negative of the area under the load diagram between the two locations.
4. The increment in the bending moment between any two locations in which there are no concentrated applied couples is numerically equal to the area under the shear diagram between the two locations.

Example 3.4

A beam is represented by its reference line as shown in Figure 3.6. Construct the shear and bending moment diagrams for the beam with the aid of the relationships among the internal and external loads. The sketch immediately above the beam is included as a graphical reminder of the positive senses for M and V according to the established sign convention.

The equation of the load function is

$$w(x) = (1/3)x$$

By inspection, the area under the load diagram from 0 to x is

$$A(x) = (1/6)x^2$$

Therefore,

$$V(x) = 6.5 - (1/6)x^2$$

The shear diagram shown in Figure 3.6 was drawn by plotting $V(0+)$, $V(12-)$, the abscissa $x = 6.24'$ for which $V = 0$, and sketching in the curve with slopes as indicated by the negative of the ordinates on the load diagram. The area under the shear diagram from 0 to x is obtained by integration of the shear function. Since $M(0+) = 0$, for $0 < x < 6$,

$$M(x) = 6.5x - (1/18)x^3$$

Because of the applied couple of moment 18 k-ft, for $6 < x < 12$,

$$M(x) = 18 + 6.5x - (1/18)x^3$$

The bending moment diagram shown in Figure 3.6 was drawn by plotting the critical points as shown and sketching in the curve with slopes as indicated by the ordinates on the shear diagram.

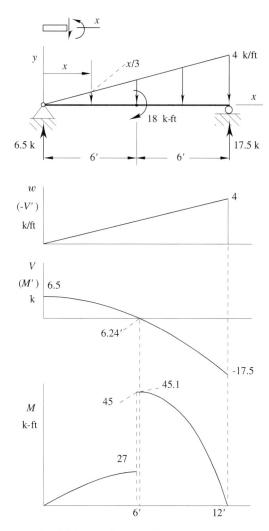

FIGURE 3.6   Application of internal/external load relations.

3.4   SHEAR AND BENDING MOMENT DIAGRAMS FOR FRAMES

The construction of shear and bending moment diagrams for a frame offers no difficulty if it is kept in mind that the construction consists in drawing the shear and bending moment diagrams for each member considered as a separate entity. For each member, the (x, y) coordinate axes are understood to be *local* coordinate axes appropriate for the member, i.e., the local x-axis is directed along the physical axis of the member under consideration. The starting point in the construction is the determination of the forces of reaction and/or interaction at the ends of each member. The ideas are best brought out by means of a specific example.

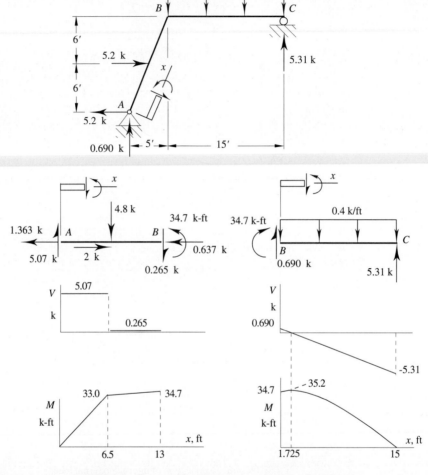

FIGURE 3.7 Shear and bending moment diagrams for a frame.

Example 3.5

Construct the shear and bending moment diagrams for the rigid frame shown in Figure 3.7. The required diagrams consist of nothing more than the diagrams for members AB and BC considered individually. Equilibrium of the free-body diagram obtained by cutting member AB at the B-end gives the axial, transverse, and moment forces of interaction at the B-end as shown in Figure 3.7. In Figure 3.7 the free-body diagram of member AB is shown with its reference axis rotated to the horizontal position. The resultant of the horizontal and vertical components of reaction at A has been resolved into an axial component of 1.363 k and a transverse component of 5.07 k. Because only concentrated forces act on member AB the shear and bending moment diagrams are straight-line segments as shown. Since the loading on member BC is uniform, the shear diagram is a straight line and the bending moment diagram is a parabola as shown. Only the critical points have been calculated. The known geometry of the curves was used to complete the diagrams.

KEY POINTS

1. Internal force resultants in a beam are vectors that are given a conventionalized representation that defines a sign convention for the tension, shear, and bending moment functions, which are scalars.
2. The sign convention or mode of representation of the internal force resultants must be known to interpret correctly the shear and bending moment functions and their respective plots (the shear and bending moment diagrams, respectively).
3. The differential relations among the load, shear and bending moment functions is sign convention dependent.
4. The basic elements in the "bottom line" for the construction of the internal force functions and their differential and integral relationships are: (a) knowledge of how the forces are represented, (b) free-body diagrams, and (c) equations of equilibrium.

PROBLEMS

3.1 Use free-body diagrams and equations of equilibrium to derive expressions for the shear and bending moment functions. Then plot the shear and bending moment diagrams.

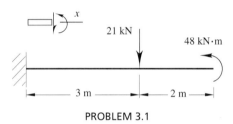

PROBLEM 3.1

3.2 Use free-body diagrams and equations of equilibrium to derive expressions for the shear and bending moment functions. Then plot the shear and bending moment diagrams.

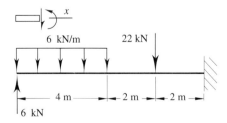

PROBLEM 3.2

3.3 Use free-body diagrams and equations of equilibrium to derive expressions for the shear and bending moment functions. Then plot the shear and bending moment diagrams.

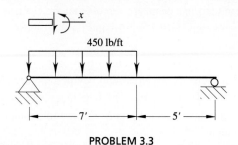

PROBLEM 3.3

3.4 Use free-body diagrams and equations of equilibrium to derive expressions for the shear and bending moment functions. Then plot the shear and bending moment diagrams.

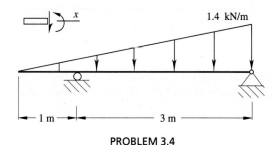

PROBLEM 3.4

3.5 Use free-body diagrams and equations of equilibrium to derive expressions for the shear and bending moment functions. Then plot the shear and bending moment diagrams.

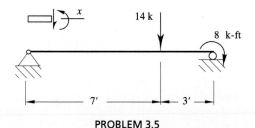

PROBLEM 3.5

3.6 Use free-body diagrams and equations of equilibrium to derive expressions for the shear and bending moment functions. Then plot the shear and bending moment diagrams. The connection at the right support can transmit bending moment and axial force but no shear.

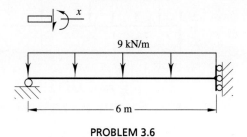

PROBLEM 3.6

3.7 Construct the shear and bending moment diagrams.

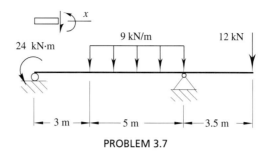

PROBLEM 3.7

3.8 Construct the shear and bending moment diagrams.

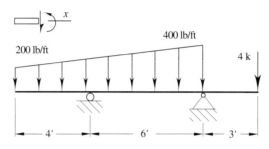

PROBLEM 3.8

3.9 Construct the shear and bending moment diagrams. The foundation pressure varies linearly.

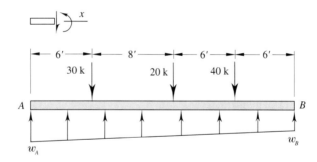

PROBLEM 3.9

3.10 Construct the shear and bending moment diagrams.

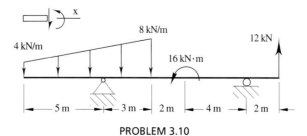

PROBLEM 3.10

3.11 Construct the shear and bending moment diagrams for beam $ABCD$.

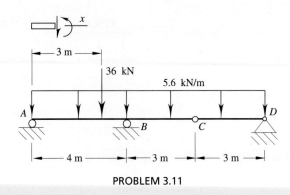

PROBLEM 3.11

3.12 Construct the shear and bending moment diagrams for beam $ABCD$. Members BE and ED are rigid.

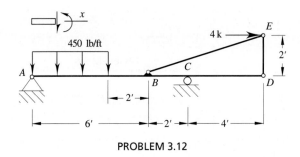

PROBLEM 3.12

3.13 Construct the shear and bending moment diagrams for beam $ABCD$.

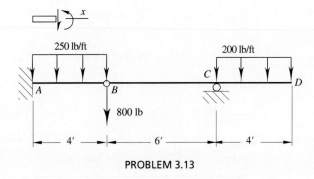

PROBLEM 3.13

3.14 Construct the shear and bending moment diagrams for beam *ABCDE*.

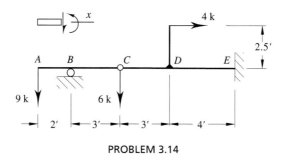

PROBLEM 3.14

3.15 Construct the shear and bending moment diagrams for beam *ABCDE*.

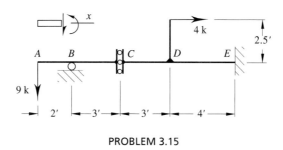

PROBLEM 3.15

3.16 The uniform beam *ABC* weighs 450 lb/ft. Construct the shear and bending moment diagrams.

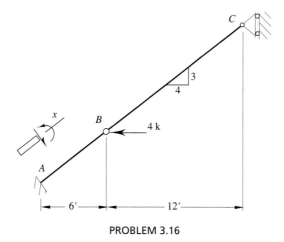

PROBLEM 3.16

3.17 The uniform beam *ABC* weighs 400 kg/m. Construct the shear and bending moment diagrams.

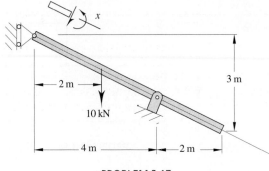

PROBLEM 3.17

3.18 For each member of the frame identify local coordinates and sign conventions for shear and bending moment and then construct the shear and bending moment diagrams.

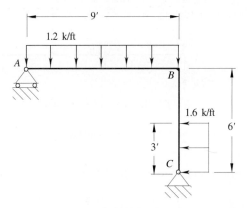

PROBLEM 3.18

3.19 For each member of the frame identify local coordinates and sign conventions for shear and bending moment and then construct the shear and bending moment diagrams.

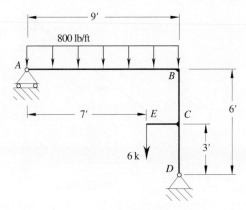

PROBLEM 3.19

3.20 For each member of the frame identify local coordinates and sign conventions for shear and bending moment and then construct the shear and bending moment diagrams.

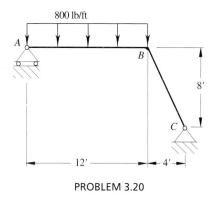

PROBLEM 3.20

3.21 For each member of the frame identify local coordinates and sign conventions for shear and bending moment and then construct the shear and bending moment diagrams. The connection at *C* develops axial force and bending moment but no shear.

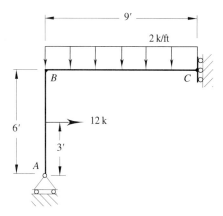

PROBLEM 3.21

3.22 For each member of the frame identify local coordinates and sign conventions for shear and bending moment and then construct the shear and bending moment diagrams.

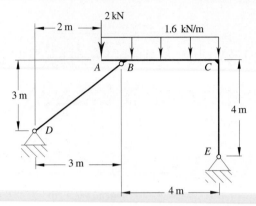

PROBLEM 3.22

3.23 For each member of the frame identify local coordinates and sign conventions for shear and bending moment and then construct the shear and bending moment diagrams.

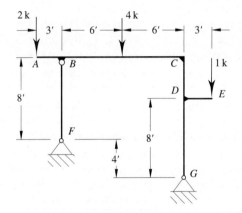

PROBLEM 3.23

3.24 For each member of the frame identify local coordinates and sign conventions for shear and bending moment and then construct the shear and bending moment diagrams.

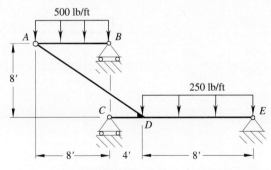

PROBLEM 3.24

CHAPTER 4

Moveable Loads/Influence Functions and Lines

Highway and railroad bridges are the outstanding examples of structures that may be subjected to loadings of variable placement. Each placement of a loading will produce a different effect on the structure. To provide a safe design it is essential to know how the loads should be placed to produce extreme values for such quantities as reactions, bending moments at critical or assigned locations, shears at critical or assigned locations, or any other quantity the extreme values of which may be critical.

To develop basic ideas related to moveable loads, consider the simplified situation shown in Figure 4.1. The loads P_1, P_2, P_3 and the spacings a_1, a_2 have fixed values. Z is a position coordinate that defines the location of the load train. For $Z < 0$ and $Z > L + a_1 + a_2$ the loads have no effect on the structure. The loads produce nonzero effects only for $0 < Z < L + a_1 + a_2$. For any assigned value of Z, determination of the reactions and the internal forces at any specific location requires only a straightforward application of statics. Determination of the extreme values of any one of the reactions or internal forces is, however, much more demanding. In principle, the extreme values of B, for example, are found by using statics to express B analytically in terms of the loads and Z. In other words, in principle, one would use statics to construct the function $B(Z)$ and then analyze $B(Z)$ to find its extreme values. Even for the simple structure shown, the process of constructing $B(Z)$ would be so difficult as to be considered a practical impossibility. The problem becomes all the more difficult as

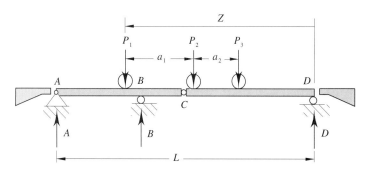

FIGURE 4.1 Moveable loads.

the complexity of the structure and the number of quantities to be evaluated are increased. Fortunately, the structures under consideration are such that the principle of superposition applies. The problem is made as simple as it can be made with the aid of the principle of superposition and the concepts of the *influence function* and the *influence line*.

4.1 INFLUENCE FUNCTIONS/INFLUENCE LINES

It is noted at the very outset that the concept of the influence function makes sense only when the structure under consideration is such that the principle of superposition applies. Thus, the notion of influence function is meaningless in the analysis of a cable structure under loads that are capable of inducing significant changes in the geometry of the system. Reconsider the structure of Figure 4.1, but subjected only to a load Q applied at location s as shown in Figure 4.2. Let p denote any point in the structure and let F_p denote the value of some quantity of interest at the point. The quantity could be a reaction, a bending moment, a shear, a deflection, or any other quantity of interest. For example, p may be taken to be the point B, and F taken to be the reaction at B, the bending moment over the support at B, or the shear just to the right (or just to the left) of the support. The value of F_p depends on the location, s, at which Q acts and the value of Q, i.e., $F_p = F_p(s, Q)$. By the principle of superposition, the value of F_p is Q times the value of F_p when Q is a unit load, i.e., when $Q = 1$. Therefore, by the principle of superposition,

$$F_p = Qf_p(s)$$

By definition $f_p(s)$ is the influence function for quantity F at location p. A plot of $f_p(s)$ against s is the influence line for quantity F at location p. By inspection, $f_p(s) = 0$ for $s < 0$ and for $s > L$. With $Q = 1$, the above relation becomes $F_p = f_p(s)$. Thus, to determine the influence function all one need do is use the principles of statics to evaluate F_p for $Q = 1$.

In practical applications the influence line is of primary importance. For most statically-determinate structures, the influence lines for reactions, forces of interaction, and internal forces are piecewise linear, i.e., linear in segments. Initially, this will be established algebraically for specific examples. Subsequently, it will be shown (in

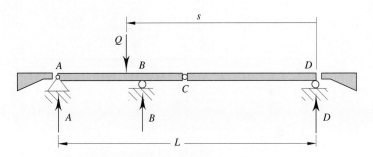

FIGURE 4.2 Loading for definition of influence function.

Chapter 5) how the shape of an influence line for force in a statically determinate structure can be established with the aid of the principle of virtual work. Clearly, the total effort involved in constructing an influence line is significantly reduced if one has prior knowledge of the shape of the influence line.

Example 4.1

Construct the influence functions and the influence lines for the reaction at the left support and the shear and bending moment at section B for the beam shown in Figure 4.3. The essential free-body diagrams for the analysis are as shown in Figure 4.3. Two free-body diagrams are necessary to calculate the shear and moment at B; one for $0 < s < 8$ when the unit load is not on the segment shown, and one for $8 < s < 12$ when the unit load is on the segment.

For equilibrium of the complete beam,

$$0 = (+ccw)\Sigma M_C = 1(s) - 12A \Rightarrow A = s/12$$

For equilibrium of segment AB with $0 < s < 8$,

$$0 = \downarrow \Sigma F = V_B - A \Rightarrow V_B = s/12$$
$$0 = (+ccw)\Sigma_B = M_B - 4A \Rightarrow M_B = s/3$$

For equilibrium of segment AB with $8 < s < 12$,

$$0 = \downarrow \Sigma F = V_B + 1 - A \Rightarrow V_B = (s/12) - 1$$
$$0 = (+ccw)\Sigma M_B = M_B + (s - 8) - 4A \Rightarrow M_B = 8 - 2s/3$$

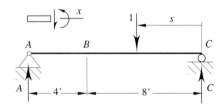

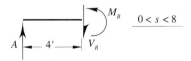

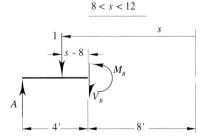

FIGURE 4.3 Simple beam example.

60 Chapter 4 Moveable Loads/Influence Functions and Lines

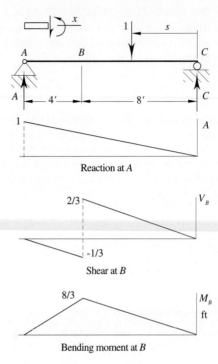

FIGURE 4.4 Influence lines.

The equations show clearly that the influence functions are segmentally linear in s. Therefore, the plots are just straight line segments as shown in Figure 4.4. Clearly, from the formulations of the equations of equilibrium alone (i.e., without actual solution of the equations) one could have ascertained the piecewise linearity and saved time in making the plots by restricting calculations to the end-point values which occur at $0+, 8-, 8+$, and $12-$. It is to be observed that the "unit" load is dimensionless. It follows that A and V_B are dimensionless and M_B has the dimension of length.

Example 4.2

Construct the influence lines for the reactions at a, b, and c, the bending moment at b and d, and the shear at e for the structure shown in Figure 4.5. It should be noted that section e is just to the right of the pin at e.

The essential free-body diagrams for the calculation of the desired quantities are as shown in Figure 4.5. By careful inspection of the free-body diagrams, it is determined that each of the influence lines is a piecewise linear function of s. The practical meaning of this is that all of the essential ordinates on the influence lines can be obtained by evaluating each of the quantities for $s = 0+, 10-, 10+, 20-, 20+, 40-, 40+$, and $60-$. The piecewise linearity of the influence lines can be verified by writing the equations of equilibrium for each of the free bodies for the indicated ranges of s. Results obtained from the equations of equilibrium are given below. The influence lines are shown in Figure 4.6.

Section 4.1 Influence Functions/Influence Lines

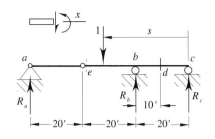

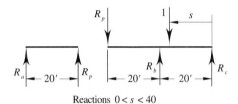

Reactions $0 < s < 40$

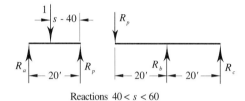

Reactions $40 < s < 60$

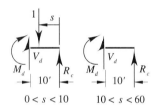

Bending moment at d

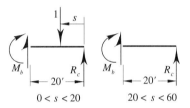

Bending moment at b

Shear at e

FIGURE 4.5 Free-body diagrams.

Reactions:

$$R_a = R_p = 0, \quad R_b = s/20, \quad R_c = 1 - s/20 \qquad 0 < s < 40$$
$$R_a = (s - 40)/20, \quad R_p = 1 - (s - 40)/20,$$
$$R_b = 2[1 - (s - 40)/20)], \quad R_c = -1 + (s - 40)/20 \qquad 40 < s < 60$$

Bending moment at d:

$$M_d = 10[1 - (s/20)] - [1(10 - s)] \qquad 0 < s < 10$$
$$M_d = 10[1 - (s/20)] \qquad 10 < s < 40$$
$$M_d = -10[1 - (s - 40)/20] \qquad 40 < s < 60$$

Bending moment at b:

$$M_b = 20[1 - s/20] - [1(20 - s)] = 0 \qquad 0 < s < 20$$
$$M_b = 20[1 - s/20] \qquad 20 < s < 40$$
$$M_b = -20[1 - (s - 40)/20] \qquad 40 < s < 60$$

Shear at e:

$$V_e = 0 \qquad 0 < s < 40$$
$$V_e = [(s - 40)/20] - 1 \qquad 40 < s < 60$$

Chapter 4 Moveable Loads/Influence Functions and Lines

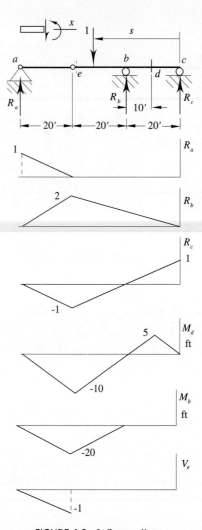

FIGURE 4.6 Influence lines.

Example 4.3

Construct the influence lines for the shear and the bending moment at sections e and f of the frame shown in Figure 4.7. The influence line for either A or B is needed to obtain the desired results. Moment equilibrium for the entire frame shows that the influence line for A must be as shown in Figure 4.7b. Two free-body diagrams are required to construct the influence lines for

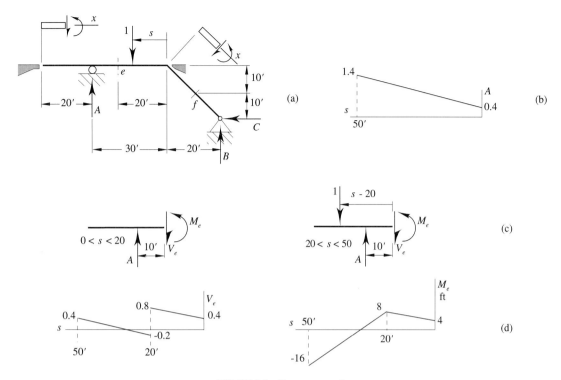

FIGURE 4.7 Frame example.

shear and bending moment at e. Equilibrium of the free-bodies shown in Figure 4.7c leads to the influence lines as shown in Figure 4.7d.

One additional free-body diagram is need to construct the influence lines for the shear and bending moment at f. The free-body diagrams along with the influence lines are shown in Figure 4.8. It is to be noted that separate local coordinates are chosen for each member. The coordinates and associated sign convention are illustrated on the sketches.

4.2 APPLICATION OF INFLUENCE LINE CONCEPT

Loads on a structure are classified as dead loads or live loads. The values and placements of dead loads are fixed. Typically, the dead load is nothing more than the gravity load of the structure itself. Loads whose magnitudes and/or placements are variable are called live loads. Because of the invariable character of dead loading, a one-time analysis is all that is needed to determine its effect on a structure. Generally, multiple analyses are needed to determine the effect of live loads. The combined effect of dead load and live load is determined by application of the principle of superposition. Influence functions (or lines) together with the principle of superposition can be used to investigate the effects of complex live loading, and combinations of dead and live loading. To illustrate the basic ideas involved, consider a simply supported beam subject to

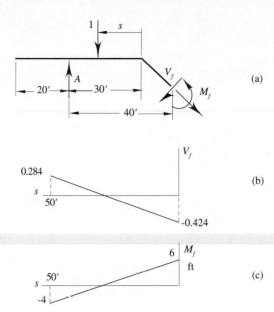

FIGURE 4.8 Frame example (continued).

one placement of a train of live loads as shown in Figure 4.9. Suppose it is required to determine the shear and bending moment at section B. The influence lines for shear and bending moment at B are available and will be used for the calculation. By the principle of superposition the net shear at B is the shear due to the 10-k load plus the shear due to the 12-k load plus the shear due to the 14-k load. Again, by the principle of superposition the effect of the 10-k load is 10 k times the effect of a unit load in the same position. From the influence line, the shear induced by the unit load at the same position as the 10-k load is $-1/6$. Thus, the shear at B due to the 10-k load is $-10(1/6)$ k. The same reasoning for the 12-k and 14-k loads together with the principle of superposition gives a net shear at B of

$$V_B = 10(-1/6) + 12(1/2) + 14(1/3) = 9 \text{ k}$$

Similar reasoning applied to the bending moment at B gives

$$M_B = 10(4/3) + 12(2) + 14(4/3) = 56 \text{ k-ft}$$

4.3 GENERALIZED APPLICATION

In general, loading consists of distributed as well as concentrated loads and the influence function is not necessarily linear in segments. The upper part of Figure 4.10 represents the loaded surface of a structure. The lower part of the figure represents the influence line for some quantity F at some point p in the structure. The net value of F at p, F_p, is the sum of the values due to each of the concentrated forces plus the value due to the distributed load. Due to the concentrated forces

$$F_p = Q_1 f_p(s_1) + Q_2 f_p(s_2) + Q_3 f_p(s_3) \rightarrow F_p = \Sigma Q_i f_p(s_i)$$

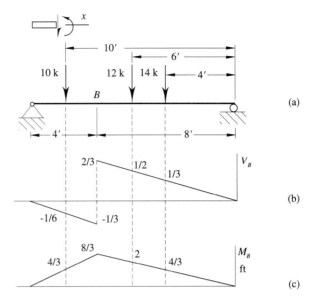

FIGURE 4.9 Application of the influence line concept.

The effect of the distributed load is evaluated in steps. First, the length L of the loaded surface is subdivided into elements each of elemental length ds. The resultant of the distributed load on an element is $dR = w(s)\,ds$ which is treated as a concentrated force at position s. The net value of F_p due to the distributed load is approximately the same as that obtained by evaluating the sum over all elements of the terms $dR\,f_p = [w(s)\,ds]\,f_p(s)$. In the limit as $ds \to 0$ the result is exact and given by the integral

$$F_p = \int_0^L w(s) f_p(s)\,ds$$

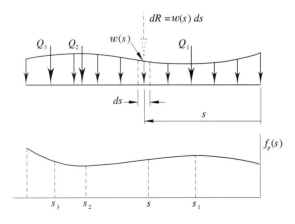

FIGURE 4.10 Generalized loading.

Chapter 4 Moveable Loads/Influence Functions and Lines

In the common case of a uniform load over a portion of the loaded surface, $w(s)$ is of constant intensity w_0, in which case w_0 can be taken out from under the integral sign. The resulting integral formulation can then be interpreted as asserting that the value of F_p due to w_0 is equal to the product of w_0 and the area under the influence line over the length of the segment on which w_0 acts. The net value of F_p for all loads is given by

$$F_p = \Sigma Q_i f_p(s_i) + \int_0^L w(s) f_p(s)\, ds$$

Example 4.4

The simply supported beam shown in Figure 4.11 carries a uniform dead load of 0.5 k/ft. The live load, the placement of which is arbitrary, consists of a 20-k concentrated force and a uniform distributed load of 0.6 k/ft. Calculate the maximum and minimum (in the algebraic sense) shear that can be induced at section B by these loads. The influence line for the shear at B has been determined and is as shown. The problem is solved with the aid of the principle of superposition and the properties of the influence line. The distributed dead load is constant and extends over the full length of the beam as shown in Figure 4.11c.

Accordingly, the shear due to the dead load is the product of the dead load and the net area under the influence line:

$$V_{Bdl} = 0.5[(1/2)(-1/3)4 + (1/2)(2/3)8] = 1.0\ \text{k}$$

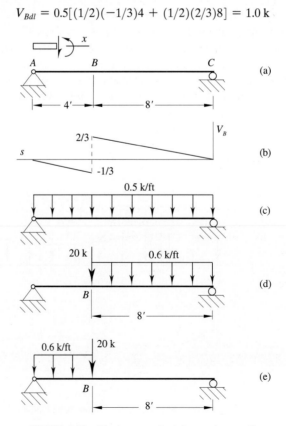

FIGURE 4.11 Maximum and minimum shear at B.

As shown in Figure 4.11d, to produce the maximum live shear, place the 20-k load just to the right of B and place the live distributed load over segment BC where all of the ordinates on the influence line are positive. The maximum live shear is the superposition of the shear due to the concentrated load and the distributed load and is

$$V_{Bll} = 20(2/3) + 0.6(1/2)(2/3)8 = 14.93 \text{ k}$$

Superimposing the dead load and live load shears give a maximum positive shear of $V_B = 1 + 14.93 = 15.93$ k. Placement of the live load to induce the minimum live shear is shown in Figure 4.11e. The minimum live shear is

$$V_{Bll} = -20(1/3) - 0.6(1/2)(1/3)4 = -7.07 \text{ k}$$

The minimum shear at B is again found by superposition of the dead load and live load shears. The minimum shear is $V_B = 1 - 7.07 = -6.07$ k.

4.4 FLOOR GIRDERS

A schematic diagram of a flooring system is shown in Figure 4.12. The stringers are understood to be simply supported by the floor beams which in turn are simply supported by the girder. The arrangement is such that only concentrated live loading is transferred to the girder at the panel points.

The mechanism of load transfer and a method of force identification are shown in Figure 4.13. Also shown are shear and bending moment diagrams for the girder. Because the shear is constant between panel points, shears are referred to as panel shears. It is evident from the bending moment diagram that regardless of the loading on the stringers, the extreme values for the bending moment must occur at a panel point. Influence lines for panel shears and for bending moments at the panel points play a very important role in the analysis of floor girders for multiple cases of live load. The constructions require only direct applications of the principles of statics, attention to the details of the stringer arrangement, and patience.

Example 4.5

Construct the influence lines for the bending moment at B and the shear in panel BC of the floor girder shown in Figure 4.14. From the free-body diagram of segment AB of the girder it is clear that R_A, A, and B are essential to the calculation of the required quantities. From equilibrium of the free-body diagram of the system as a whole, the influence line for R_A is determined to be one

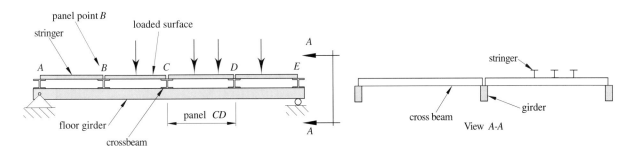

FIGURE 4.12 Floor girder system.

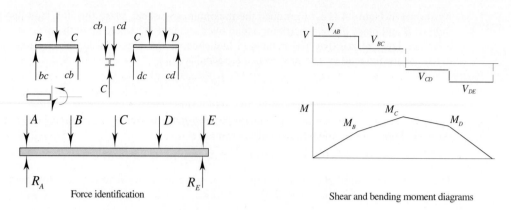

FIGURE 4.13 Floor girder details.

straight line as shown. Because stringers AB and BC are simply supported, it is readily determined that $A = 0$ for $0 < s < 30$, and that A increases linearly from 0 to 1 for $30 < s < 40$. The reaction B is 0 unless the unit load is on stringer AB or stringer BC. For $20 < s < 30$, B increases linearly from 0 to 1, and for $30 < s < 40$ B decreases linearly from 1 to 0. Accordingly, the influence lines for A and B are as shown. These results are easily verified by construction of the free-body diagrams for stringers AB and BC and for cross beams A, B, and C. Force and moment equilibrium of segment AB yield, for all s,

$$V_{BC} = R_A - A - B, \quad M_B = 10(R_A - A)$$

Because each of the quantities R_A, A, and B vary linearly for $0 < s < 20$, $20 < s < 30$, and $30 < s < 40$, so do V_{BC} and M_B. Therefore, the influence lines for V_{BC} and M_B may be constructed by carrying out the specific calculations for the ordinates for $s = 0+, 20-, 20+, 30-,$

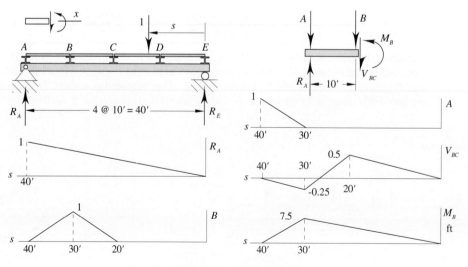

FIGURE 4.14 Influence lines.

30+, and 40−, and constructing the straight line segments for each range of s. The influence lines turn out to be as shown in Figure 4.14.

In this example the influence lines are continuous for $0 < s < 40$. A different stringer-floor beam arrangement may, however, result in influence lines with discontinuities. For example, replacing the three 10-foot-long stringers AB, BC, and CD by two 15-foot-long stringers ABF and FCD produces the system shown in Figure 4.15, in which BF is a 5-foot overhang for ABF and FC is a 5-foot overhang for FCD. The effect of the overhangs is to produce discontinuous influence lines for the cross-beam reactions A and B, the panel shear V_{BC}, and the bending moment M_B as shown in Figure 4.15. (Note that the overhangs also induce discontinuities in the influence line for other quantities such as M_C, for example.)

4.5 BRIDGE TRUSSES

Flooring systems for bridge trusses are similar in arrangement to those for which girders provide the main support. Stringers transfer the floor loads to cross beams which in turn transfer the loads to the truss at the joints of the truss. A schematic diagram of such a system as it would be treated for the purpose of analysis is shown in Figure 4.16. The expanded view of a joint shows how the members are connected. The members are positioned so that their centroidal axes intersect at a common point which is understood to define the theoretical location of the pin connection. The arrangement ensures that all live loads will be transferred to the truss at the joints. As was noted earlier, the joints are not, in fact, pin-connected. Accordingly, the joints are capable of developing bending moments. Bending action is minimized by positioning the members so that their centroidal axes intersect at a common point. Bending that does occur

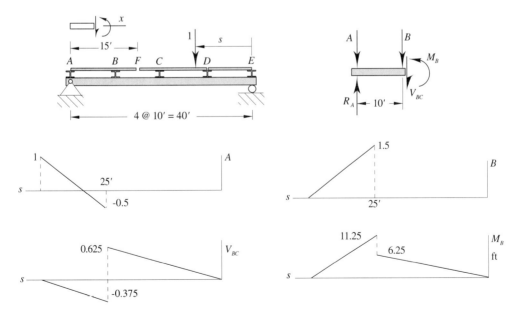

FIGURE 4.15 Overhanging stringers.

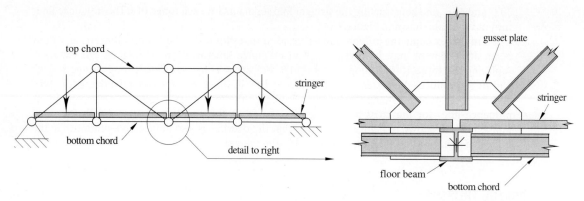

FIGURE 4.16 Bridge truss.

is due to axial deformations. The (secondary) stress due to bending is small relative to the (primary) stress due to direct tension or compression.

Example 4.6

Construct the influence lines for the tensions in bars $cb, cf, cg, ch,$ and hg of the truss shown in Figure 4.17. Each stringer is 20 feet long and is simply supported at its ends. Free-body diagrams (not shown) of the stringers can be used to establish that the forces transmitted to the pins on the bottom chord are piece-wise linear functions of s in the intervals defined by the ends of the stringers, i.e., the forces are linear in the intervals $0 < s < 20, 20 < s < 40, 40 < s < 60, 60 < s < 80$. It follows from the free-body diagrams that each of the unknown forces is also piece-wise linear in the same intervals. A free-body diagram of joint g is omitted because by inspection $cg = g$ for all load positions. Thus, a simple (but not necessarily most efficient) procedure for constructing the influence lines is to calculate the required quantities at the ends of the intervals (i.e., $0+$, $20-, 20+, 40-, 40+, 60-, 60+,$ and $80-$) and connect the end points by straight lines. Calculated

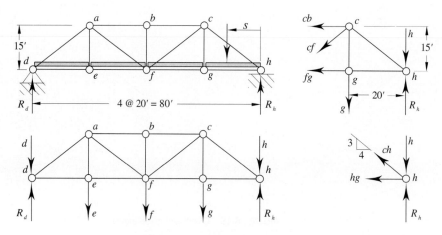

FIGURE 4.17 Bridge truss example.

Section 4.5 Bridge Trusses

TABLE 4.1 Influence coefficients

80−	60+	60−	40+	40−	20+	20−	0+	s
0	1/4	1/4	1/2	1/2	3/4	3/4	1	R_h
0	0	0	0	0	0	0	1	h
0	0	0	0	0	1	1	0	g
0	−2/3	−2/3	−4/3	−4/3	−2/3	−2/3	0	cb
0	5/12	5/12	5/6	5/6	−5/12	−5/12	0	cf
0	0	0	0	0	1	1	0	cg
0	−5/12	−5/12	−5/6	−5/6	−5/4	−5/4	0	ch
0	1/3	1/3	2/3	2/3	1	1	0	hg

values of the ordinates on the influence lines at the ends of each interval are listed in Table 4.1. The influence lines are shown in Figure 4.18.

The stringer arrangements play an important role in determining the characteristics of the influence lines for some members in a truss. To illustrate this point, reconsider the example truss with the stringer arrangement shown in Figure 4.19. The four 20-foot-long stringers de, ef, fg, and gh of the truss in Figure 4.17 have been replaced by stringer de of length 20′ and stringers ek and kh each of 30′ length. Stringers ek and kh have a 10-foot-long overhang. The ranges of s to be considered now are: $0 < s < 30$, $30 < s < 60$, and $60 < s < 80$. Because the influence lines are known to be linear in each of the segments, the calculations are carried out only for $s = 0+, 30−, 30+, 60−, 60+,$ and $80−$. The influence line for the tension in cf is as

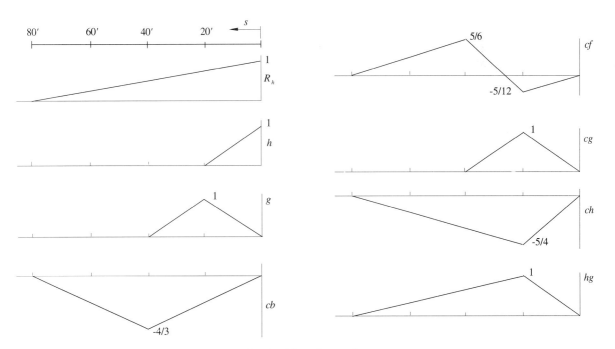

FIGURE 4.18 Influence lines.

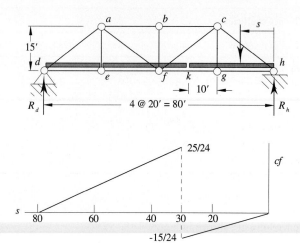

FIGURE 4.19 Tension in bar cf.

shown in Figure 4.19. The influence line for cf for the modified stringer arrangement should be compared to the influence line for the same member with the original stringer arrangement.

4.6 EXTREME EFFECTS DUE TO FIXED LOADS OF VARIABLE PLACEMENT

The simplified situation illustrated in Figure 4.20 is used to establish the basic features of the class of problems to be considered. The objective is to determine the extreme values of bending moment that can be produced at support B by the given load train as it passes over the beam from right to left and then to solve the same problem when the order of the loads is reversed (i.e., when the loads pass from right to left with the 5-k load in the trailing position).

It is obvious that the minimum moment (in the algebraic sense) will occur when Z, which defines the location of the lead load is at least 30′; however, to get a picture of

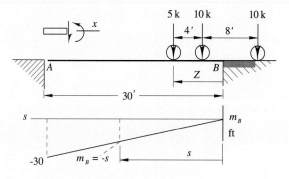

FIGURE 4.20 Load train.

Section 4.6   Extreme Effects Due to Fixed Loads of Variable Placement

how the bending moment varies with Z, and to answer the question of extreme values with certainty, M_B is determined as a function of Z for all Z in the range $0 < Z$. This is most easily accomplished with the aid of the principle of superposition and the properties of influence lines. The influence line for the bending moment at B for $0 < s < 30$ is as shown in Figure 4.20. The equation of the influence line is $m_B = 0$ for $s < 0$ and for $s > 30$, and $m_B = -s$ for $0 < s < 30$. By the principle of superposition the contribution of any load to the total bending moment at B is the product of the load and the ordinate on the influence line at the position of the load. Also, by the principle of superposition the total bending moment is the sum of the bending moments due to each of the loads considered separately. With Z and s measured from B, and with $m_B(s)$ as the equation of the influence line, the net moment at B is given by

$$-M_B = 5m_B(Z) + 10m_B(Z - 4) + 10m_B(Z - 12)$$

In using the above formulation, it is essential to recall that $m_B(s) = 0$ for $s < 0$, $m_B(s) = -s$ for $0 < s < 30$, and $m_B(s) = 0$ for $s > 30$. Because $m_B(s)$ is linear in segments so is $M_B(Z)$; the end points of the segments for $M_B(Z)$ being determined according to the locations of each of the loads. Thus, on the basis of the foregoing discussion, $M_B(Z)$ is found to be given by:

$$
\begin{aligned}
Z < 0 \qquad & M_B = 0 \\
0 < Z < 4 \qquad & -M_B = 5Z \\
4 < Z < 12 \qquad & -M_B = 5Z + 10(Z - 4) \\
12 < Z < 30 \qquad & -M_B = 5Z + 10(Z - 4) + 10(Z - 12) \\
30 < Z < 34 \qquad & -M_B = \phantom{5Z + {}}10(Z - 4) + 10(Z - 12) \\
34 < Z < 42 \qquad & -M_B = \phantom{5Z + 10(Z - 4) + {}}10(Z - 12) \\
42 < Z \qquad & -M_B = 0
\end{aligned}
$$

When the loads pass over the beam in reverse order the corresponding relations are:

$$
\begin{aligned}
Z < 0 \qquad & -M_B = 0 \\
0 < Z < 8 \qquad & -M_B = 10Z \\
8 < Z < 12 \qquad & -M_B = 10Z + 10(Z - 8) \\
12 < Z < 30 \qquad & -M_B = 10Z + 10(Z - 8) + 5(Z - 12) \\
30 < Z < 38 \qquad & -M_B = \phantom{10Z + {}}10(Z - 8) + 5(Z - 12) \\
38 < Z < 42 \qquad & -M_B = \phantom{10Z + 10(Z - 8) + {}}5(Z - 12) \\
42 < Z \qquad & -M_B = 0
\end{aligned}
$$

Plots of the functions defined by the above equations are shown in Figure 4.21.

The procedure for writing out the expressions that define how a quantity such as M_B varies with the load position is simple enough, but not very practical in application when the influence function contains discontinuities and there are several loads in the load train. The problem with implementation of the procedure is related to the growth in the number of discontinuities in the function or the slope of the general function

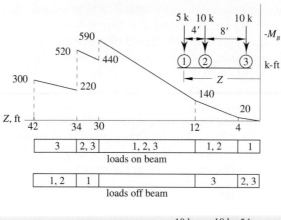
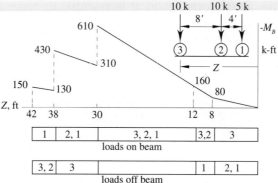

FIGURE 4.21 Bending moment at B.

$F_p(Z)$ that corresponds to M_B. Fortunately, there is no real need for the equations. The characteristics of the loading together with the piecewise linearity of the influence function guarantee that: (a) $F_p(Z)$ will be piecewise linear, and (b) the extreme values of $F_p(Z)$ will occur when one of the loads is adjacent to an end point of a segment of the influence line. The properties of $F_p(Z)$ are the basis for two approaches to solving the problem of finding the extreme values.

Property (b) of $F_p(Z)$ is the basis of the first approach, which employs engineering judgment and trial-and-error calculations. Judgment founded on experience is used to select values of Z that are realistic candidates for the placement of the loads for which $F_p(Z)$ will be an extreme. Clearly, in each case Z must be such that one of the loads is adjacent to an end point of a segment of the influence line. The corresponding values of $F_p(Z)$ are calculated and the extreme value of the calculated quantities is taken to be an extreme value of $F_p(Z)$. This line of attack is appropriate when there are two or at most three loads and the influence line is relatively simple.

The second approach is nothing more than a numerical procedure for systematic calculation of the values of $F_p(Z)$ at the end points of the intervals. A generalized load train and a generalized piecewise linear influence function for a quantity F are shown

Section 4.6   Extreme Effects Due to Fixed Loads of Variable Placement   75

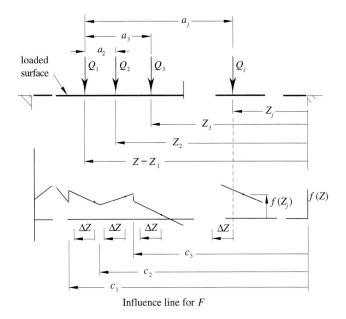

FIGURE 4.22   Generalized load train and influence function.

in Figure 4.22 in which $a_1 = 0, a_2, a_3, \ldots, a_j, \ldots, a_n$ are the constant distances of the loads from the lead load. The position of the lead load and, therefore, the position of every load is defined by the distance Z measured from the beginning of the loaded surface. $Z = Z_1$ defines the current positions of the loads. Because ordinates on the influence line are undefined at points of discontinuity, it is understood that Z is such that none of the loads is at a point of discontinuity in $f(s)$. In addition, it is assumed that Z is such that none of the loads is situated at a point at which a discontinuity occurs in the slope of $f(s)$. By the principle of superposition, the current value of F is given by

$$F = F(Z) = \Sigma Q_j f(Z_j)$$

The distances $(c_1 - Z), (c_2 - Z), \ldots, (c_n - Z)$ represent the maximum forward movements that can be given to $Q_1, Q_2, \ldots, Q_n$, respectively, before the associated load is adjacent to a point of discontinuity in either the ordinate or the slope of the influence line. Let ΔZ be the minimum of the foregoing quantities. Then, the associated increment in F, ΔF, is given by

$$\Delta F = \Delta Z \Sigma Q_j m_j$$

where m_j is the slope of the segment of the influence line under Q_j. The foregoing increment in Z causes at least one of the loads to be adjacent to a point of discontinuity in $f(s)$ or in the slope of $f(s)$. If $\Delta Z = 0$ one or more of the loads is adjacent to a point of discontinuity in the influence line or in the slope of the influence line. In either case the discontinuity is crossed by taking $\Delta Z = 0+$ and the increment in F is given by

$$\Delta F = \Sigma Q_j \{f(Z_j+) - f(Z_f-)\}$$

By starting with $Z = 0+$ and selecting increments ΔZ according to the above relations one can determine the end points and the end-point values of $F(Z)$. The extreme values of F can then be determined by inspection. Of course, the procedure need not be started at $Z = 0+$; it may be started at any point and extended to cover only a range of values of Z for which judgment indicates an extreme value of F will occur.

Example 4.7

Determine the variation in the shear and the bending moment at section b as the load train passes over the beam shown in Figure 4.23. The influence lines for the shear and bending moment at the section are as shown. Observe that for $0 < s < 48'$ the influence line for shear has a discontinuity at $s = 24'$, while the slope of the line is continuous at a constant value of $(1/36)$/ft. The influence line for the bending moment is continuous, but with a discontinuity in slope at $s = 24'$. The slope is $1/3$ for $s < 24'$ and $-2/3$ for $s > 24'$. Before going on to the calculation it is noted that ordinarily one would be interested only in the extreme values of the bending moment and shear at the section. In such cases, one may construct an abbreviated table of values which includes results only for the ranges of values of the load placement variable Z in which an extreme is likely to occur. It should be kept in mind that, in general, the extreme effects produced by the loads taken in the given order will not be the same as the extreme effects produced when the ordering of the loads is reversed. Tables 4.2 and 4.3 contain the results for the shear and bending moment calculations, respectively.

In Tables 4.2 and 4.3 the entry in column 1 is the current position of the load train which coincides with the position of Q_1, the entry in column 5 on any line is the current posi-

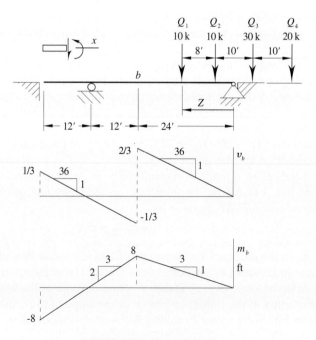

FIGURE 4.23 Example loading.

Section 4.6 Extreme Effects Due to Fixed Loads of Variable Placement

TABLE 4.2 Shear calculation

	Z ft (1)	ΔV_b kip (2)	V_b kip (3)	ΔZ ft (4)	Q_1@ ft (5)	Q_2@ ft (6)	Q_3@ ft (7)	Q_4@ ft (8)	ΔZ_1 ft (9)	ΔZ_2 ft (10)	ΔZ_3 ft (11)	ΔZ_4 ft (12)
(1)	0+		0	8	0+	−8	−18	−28	24	**8**	18	28
(2)	8−	2.222	2.222	0	8−	0	−10	−20	16	**0**	10	20
(3)	8+	0	2.222	10	8+	0	−10	−20	16	24	**10**	20
(4)	18−	5.555	7.777	0	18−	10	0	−10	6	14	**0**	10
(5)	18+	0	7.777	6	18+	10	0	−10	**6**	14	24	10
(6)	24−	8.333	16.111	0	24−	16	6	−4	**0**	8	18	4
(7)	24+	−10	6.111	4	24+	16	6	−4	24	8	18	**4**
(8)	28−	5.555	11.666	0	28−	20	10	0	20	4	14	**0**
(9)	28+	0	11.666	4	28+	20	10	0	20	**4**	14	24
(10)	32−	7.777	19.444	0	32−	24	14	4	16	**0**	10	20
(11)	32+	−10	9.444	10	32+	24	14	4	16	24	**10**	20
(12)	42−	19.444	28.888	0	42−	34	24	14	6	14	**0**	10
(13)	42+	−30	−1.111	6	42+	34	24	14	**6**	14	24	10
(14)	48−	11.666	10.555	0	48−	40	30	20	**0**	8	18	4
(15)	48+	−3.333	7.222	4	48+	40	30	20	*	8	18	**4**
(16)	52−	6.666	13.888	0	52−	44	34	24	*	4	14	**0**
(17)	52+	−20	−6.111	4	52+	44	34	24	*	**4**	14	24
(18)	56−	6.666	0.555	0	56−	48	38	28	*	**0**	10	20
(19)	56+	−3.333	−2.777	10	56+	48	38	28	*	*	**10**	20
(20)	66−	13.888	11.111	0	66−	58	48	38	*	*	**0**	10

tion of load Q_1, the entry in column 6 is the current position of Q_2, the entry in column 7 is the current position of Q_3, and the entry in column 8 is the current position of Q_4. The entry in column 9 of any line is the maximum forward movement that can be assigned to load Q_1 without having the load cross a point of discontinuity in the ordinate or slope of the associated influence line. Similarly, the entries in columns 10 through 12 on any line are the maximum forward movements that can be assigned to loads Q_2, Q_3, and Q_4, respectively, without having the load cross a point of discontinuity in the ordinate or slope of the influence line. An asterisk in any of the columns numbered 9 through 12 indicates that the corresponding load has passed off the beam to the left and its forward movement is no longer a factor in the calculations. The maximum forward movement that can be assigned to the load train without having any of the loads cross a point of discontinuity in the ordinate or slope of the influence line is the minimum of the values of the incremental movements displayed in columns 9 through 12. This minimum is shown in boldface type and becomes the incremental movement, ΔZ, of the load train, which is listed in column 4. In Table 4.2 (Table 4.3) the value of the shear (bending moment) corresponding to the current placement of the load train is shown in column 3. The increment in the shear (bending moment) due to the forward movement ΔZ is calculated using the previously developed formulas and is listed in column 2. The entry in column 3 on any line is the sum of the entry in column 2 of the same line and the entry in column 3 of the previous line.

To illustrate specific calculations, consider the shear at section b with the load train in the placement $Z = 18'+$, which corresponds to Line 5 in Table 4.2. Here, Q_1, Q_2, and Q_3 are on the segment of the beam for which the respective slopes of the influence line are $m_1 = m_2 = m_3 = 1/36$. (Note that, in general, the loads will be on different segments of the influence line and the corresponding slopes will not be equal.) The slope of the segment of

Chapter 4 Moveable Loads/Influence Functions and Lines

TABLE 4.3 Bending moment calculation

	Z ft (1)	ΔM_b k-ft (2)	M_b k-ft (3)	ΔZ ft (4)	Q_1@ ft (5)	Q_2@ ft (6)	Q_3@ ft (7)	Q_4@ ft (8)	ΔZ_1 ft (9)	ΔZ_2 ft (10)	ΔZ_3 ft (11)	ΔZ_4 ft (12)
(1)	0+		0	8	0+	−8	−18	−28	24	**8**	18	28
(2)	8−	26.666	26.666	0	8−	0−	−10	−20	16	**0**	10	20
(3)	8+	0	26.666	10	8+	0+	−10	−20	16	24	**10**	20
(4)	18−	66.666	93.333	0	18−	10	0	−10	6	14	**0**	10
(5)	18+	0	93.333	6	18+	10	0	−10	**6**	14	24	10
(6)	24−	100	193.33	0	24−	16	6	−4	**0**	8	18	4
(7)	24+	0	193.33	4	24+	16	6	−4	24	8	18	**4**
(8)	28−	26.666	220.0	0	28−	20	10	0	20	4	14	**0**
(9)	28+	0	220.0	4	28+	20	10	0	20	**4**	14	24
(10)	32−	53.333	273.33	0	32−	24	14	4	16	**0**	10	20
(11)	32+	0	273.33	10	32+	24	14	4	16	24	**10**	20
(12)	42−	33.333	306.66	0	42−	34	24−	14	6	14	**0**	10
(13)	42+	0	306.66	6	42+	34	24	14	**6**	14	24	10
(14)	48−	160	146.66	0	48−	40	30	20	**0**	8	18	14
(15)	48+	80	226.6	4	48+	40	30	20	*	8	18	**4**
(16)	52−	−80	146.6	0	52	44	34	24	*	**4**	10	**0**
(17)	52+	0	146.6	4	52	44	34	24	*	**4**	14	24
(18)	56−	−160	−13.33	0	56−	48	38	28	*	**0**	10	20
(19)	56+	80	66.67	10	56+	48	38	28	*	*	**10**	20
(20)	66−	−333.3	−266.6	0	66−	58−	48	38	*	*	**0**	10
(21)	66+	240.0	−26.66	10	66+	58	48	38	*	*	*	**10**
(22)	76−	−133.3	−160	0	76−	68	58	48	*	*	*	**0**

the influence line on which Q_4 is located ($s < 0$) is $m_4 = 0$. The maximum forward movement that can be given to the load train with no load crossing a discontinuity is $6'-$. This movement will place Q_1 to the right of and adjacent to the discontinuity in the influence line for shear at section b. The corresponding increment in the shear is

$$\Delta V_b = \Delta Z \, \Sigma Q_j m_j = 6(10 + 10 + 30)(1/36) = 8.333 \text{ k}$$

Thus, with $Z = 24'-$ the shear at section b is $V_b = 7.777 + 8.333 = 16.111$ k as shown in column 3 of line 6 in Table 4.2. With Q_1 adjacent to the discontinuity at section b, $\Delta Z = 0$ in the next step. Because only Q_1 crosses a point of discontinuity in the ordinate of the influence line, the increment in the shear is

$$\Delta V_b = \Sigma Q_j \{f(Z_j+) - f(Z_j-)\} = 10[(-1/3) - (2/3)] = -10 \text{ k}$$

After the movement Q_1 is at $Z = 24'+$, just to the left of the discontinuity in the influence line for shear at Section b and the corresponding shear is $V_b = 16.111 - 10 = 6.111$ k as shown in column 3, line 7 of Table 4.2.

To complete the demonstration calculations, consider the increment in the bending moment as the load train is advanced from $Z = 24'+$. From Line 7 in Table 4.3, $\Delta Z = 4'$. The slope of the segment of the influence line for the bending moment (Figure 4.23) corresponding to the position of Q_1 is $m_1 = -2/3$ and the slope of the segment corresponding to the positions of Q_2 and Q_3 is $m_2 = m_3 = 1/3$. For Q_4 which is located at $Z = -4'$, $m_4 = 0$. Thus, the increment

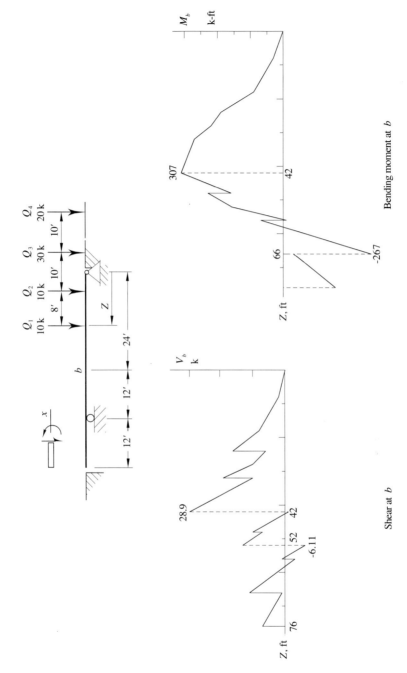

FIGURE 4.24 Variation of V_b and M_b with position of the load train.

in the bending moment is

$$\Delta M_b = \Delta Z \sum Q_j m_j = 4[10(-2/3) + 40(1/3)] = 26.666 \text{ k-ft}$$

which is shown in column 2, line 8 of Table 4.3.

Inspection of Table 4.2 shows that the maximum and minimum values for the shear at section b are 28.9 k and -6.11 k and occur with load Q_1 positioned at $Z = 42'-$ and $Z = 52'+$, respectively. The extreme values for the bending moment at section b are (from Table 4.3) 307 k-ft and -267 k-ft, respectively. Plots of the shear function $V_b(Z)$ defined by the entries in Table 4.2 and the bending moment function $M_b(Z)$ defined by the entries in Table 4.3 are shown in Figure 4.24.

The plots in Figure 4.24 and the extensive details given in Tables 4.2 and 4.3 are provided to assist the reader in getting a feel for the problem of analyzing the effects of moveable loads. Clearly, the plots are not required, and the tables may be abbreviated in several ways.

KEY POINTS

1. Influence lines, which are plots of influence functions, are extremely useful in evaluating the extreme effects due to combined loads and/or loads of variable placement on structures for which the principle of superposition applies.

2. By the principle of superposition the value of specific quantity F measured at a particular point p, F_p, due to concentrated load Q at location s, is given by $F_p = Q f_p(s)$. By definition $f_p(s)$ is the influence function for F at p. It follows that $f_p(s)$ is the value of F_p when the unit load $Q = 1$ is placed at location s.

3. The basic procedure for constructing the influence function $f_p(s)$ follows from the definition; it consists in placing a unit load at location s and calculating the value of f at p using basic principles of mechanics (see Key Point 2.).

4. In many applications analysis can be used to determine the geometric properties of an influence line so that only a relatively few ordinates need be calculated.

PROBLEMS

4.1–4.5 General instructions: Use free-body diagrams and equations of equilibrium to derive the required equations.

4.1 Derive the equations for the influence functions and then plot the influence lines for the reactions at A and B, the shear at C, and the bending moment at C.

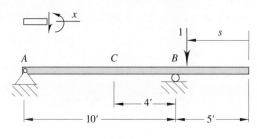

PROBLEM 4.1

4.2 Derive the equations for the influence functions and then plot the influence lines for the reaction at A, the bending moment at A, and the shear at B.

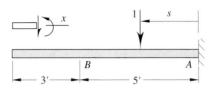

PROBLEM 4.2

4.3 Derive the equations for the influence functions and then plot the influence lines for the reaction at B and the bending moment at A.

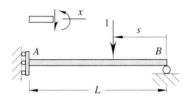

PROBLEM 4.3

4.4 Derive the equations for the influence functions and then plot the influence lines for the reaction at A, the bending moment at A, and the shear at C.

4.5 Derive the equations for the influence functions and then plot the influence lines for the shear just to the left of B, the shear just to the right of B, and the bending moment at B.

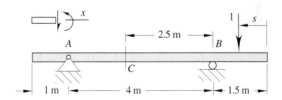

PROBLEMS 4.4 and 4.5

4.6 Construct the influence lines for the reaction at A and the shear at D.

4.7 Construct the influence lines for the reaction at B and the bending moment at D.

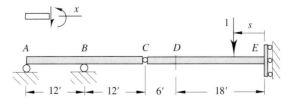

PROBLEMS 4.6 and 4.7

4.8 Construct the influence lines for the bending moment at A and the shear just to the left of C.

4.9 Construct the influence lines for the shear at A and the bending moment at C.

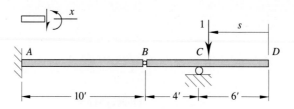

PROBLEMS 4.8 and 4.9

4.10 Construct the influence lines for the shear at A and the bending moment at B.

4.11 Construct the influence lines for the shear at B and the bending moment at A.

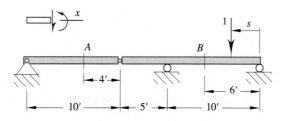

PROBLEMS 4.10 and 4.11

4.12 Construct the influence lines for the bending moment at A and D.

4.13 Construct the influence lines for the shear at A and the shear just to the left of D.

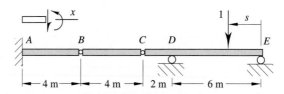

PROBLEMS 4.12 and 4.13

4.14 Construct the influence lines for the reaction at A and the shear at C.

4.15 Construct the influence lines for the reaction at D and the bending moment at C.

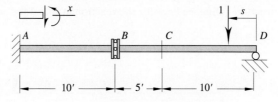

PROBLEMS 4.14 and 4.15

4.16 Construct the influence lines for the moment of the reaction couple at D and the bending moment at the B-end of BC.

4.17 Choose local coordinates and appropriate sign conventions and construct the influence lines for the shear and the bending moment at the *B*-end of member *BD*.

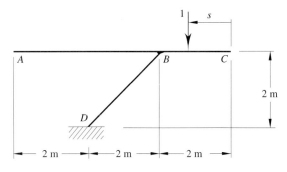

PROBLEMS 4.16 and 4.17

4.18 Construct the influence lines for the shear and the bending moment at the *C*-end of *CD*.

4.19 Choose local coordinates and appropriate sign conventions and construct the influence lines for the shear and the bending moment at the *C*-end of member *AC*.

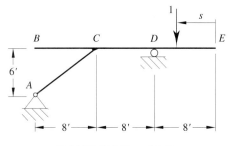

PROBLEMS 4.18 and 4.19

4.20 Construct the influence lines for the horizontal reaction at *A* and the shear and bending moment at *E*.

4.21 Choose local coordinates and appropriate sign conventions and construct the influence lines for the shear and the bending moment at the *C*-end of member *CD*.

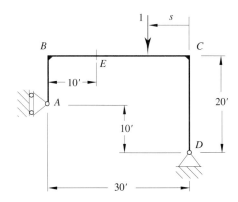

PROBLEMS 4.20 and 4.21

4.22 Construct the influence lines for the horizontal reaction at A and the shear and bending moment at the B-end of BC.

4.23 Construct the influence lines for the vertical reaction at E and the shear and bending moment at the D-end of CD.

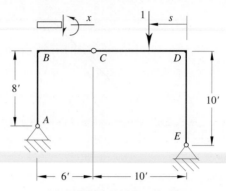

PROBLEMS 4.22 and 4.23

4.24 Calculate the extreme values of the shear and the bending moment at section D due to a dead load of 900 lb/ft over the full length of the beam and a concentrated dead load of 4,000 lb at $s = 7'$ together with all possible combinations of a distributed live load of 450 lb/ft and a 6,000-lb concentrated live load.

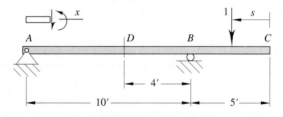

PROBLEM 4.24

4.25 Calculate the extreme values of the bending moment at A and the shear just to the left of B due to a dead load of 600 lb/ft over the full length of the structure and a concentrated dead load of 9,000 lb at $s = 10'$ together with all possible combinations of a distributed live load of 400 lb/ft and a 3,000-lb concentrated live load.

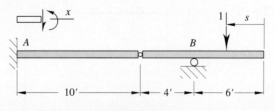

PROBLEM 4.25

4.26 Calculate the extreme values of the shear and the bending moment at section A due to a dead load of 6 kN/m over the full length of the structure and a 20 kN concentrated load at $s = 3$ m together with all possible combinations of a distributed live load of 9 kN/m and a 20 kN concentrated live load.

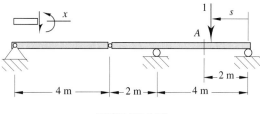

PROBLEM 4.26

4.27 Construct the influence lines for the shear in panel AB and the bending moment at B.
4.28 Construct the influence lines for the shear in panel BC and the bending moment at C.

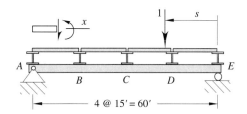

PROBLEMS 4.27 and 4.28

4.29 Construct the influence lines for the shear in panel AB and the bending moment at B.
4.30 Construct the influence lines for the shear in panel DE and the bending moment at D.

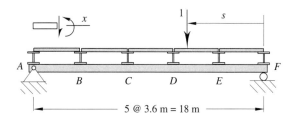

PROBLEMS 4.29 and 4.30

4.31 Construct the influence lines for the shear in panel BC and the bending moment at B.

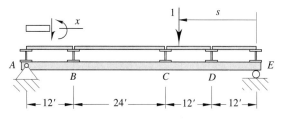

PROBLEM 4.31

4.32 Construct the influence lines for the shear in panel DE and the bending moment at D.

4.33 Construct the influence lines for the shear in panel EF and the bending moment at E.

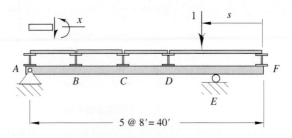

PROBLEMS 4.32 and 4.33

4.34 Construct the influence lines for the shear in panel BC and the bending moment at B.

4.35 Construct the influence lines for the shear in panel CD and the bending moment C.

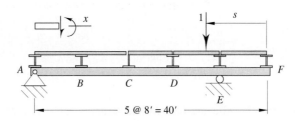

PROBLEMS 4.34 and 4.35

4.36 Construct the influence lines for the shear in panel BC and the bending moment at B.

4.37 Construct the influence lines for the shear in panel CD and the bending moment at D.

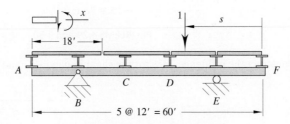

PROBLEMS 4.36 and 4.37

4.38 Construct the influence lines for the shear in panel EF and the bending moment at E.

4.39 Construct the influence lines for the shear in panel *DE* and the bending moment at *D*.

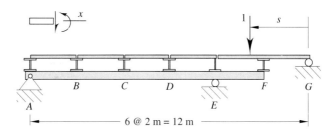

PROBLEMS 4.38 and 4.39

4.40 Construct the influence lines for the shear in panel *CD* and the bending moment at *D*.
4.41 Construct the influence lines for the shear in panel *DE* and the bending moment at *B*.

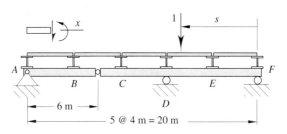

PROBLEMS 4.40 and 4.41

4.42 Construct the influence lines for the shear in panel *CD* and the bending moment at *B*. Note: $0 < s < 8$.
4.43 Construct the influence lines for the shear in panel *BC* and the bending moment at *D*. Note: $0 < s < 8$.

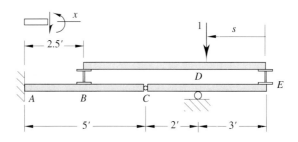

PROBLEMS 4.42 and 4.43

4.44 Construct the influence lines for the shear in panel *CD* and the bending moment at *A*.

4.45 Construct the influence lines for the shear in panel AB and the bending moment at D.

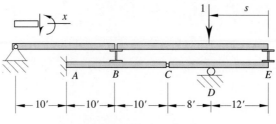

PROBLEMS 4.44 and 4.45

4.46 Calculate the extreme values for the shear in panel BC and the bending moment at C due to a dead load of 10 kN/m over the full length of the loaded surface and a 30 kN concentrated dead load at $s = 10$ m together with all possible combinations of a distributed live load of 12 kN/m and a concentrated live load of 40 kN.

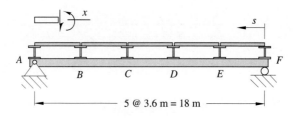

PROBLEM 4.46

4.47 Calculate the extreme values for the shear in panel DE and the bending moment at E due to a uniform dead load of 600 lb/ft over the full length of the loaded surface and a concentrated dead load of 12 k at $s = 25'$ together with all possible combinations of a uniform live load of 450 lb/ft and a 36-k concentrated live load.

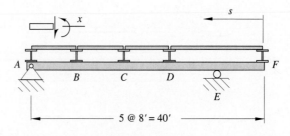

PROBLEM 4.47

4.48 Calculate the extreme values for the shear in panel BC and the bending moment at E due to a uniform dead load of 620 lb/ft over the full length of the loaded surface and a concentrated dead load of 16 k at $s = 28'$ together with all possible combinations of a uniform live load of 480 lb/ft and a 24-k concentrated live load.

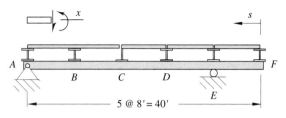

PROBLEM 4.48

4.49 Construct the influence lines for the tension in bars a, b, and c.

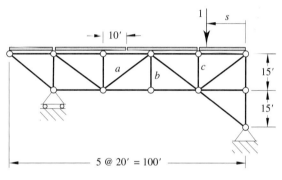

PROBLEM 4.49

4.50 Construct the influence lines for the tension in bars a, b, and c.

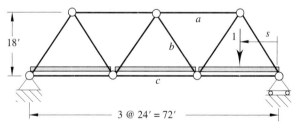

PROBLEM 4.50

4.51 Construct the influence lines for the tension in bars $a, b,$ and c.

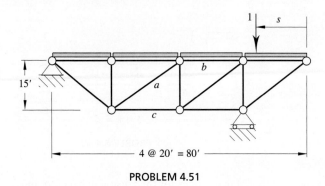

PROBLEM 4.51

4.52 Construct the influence lines for the tension in bars $a, b,$ and c.

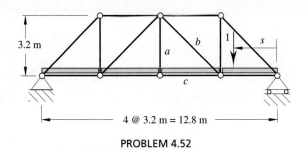

PROBLEM 4.52

4.53 Construct the influence lines for the tension in bars $a, b,$ and c.

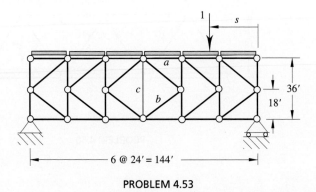

PROBLEM 4.53

4.54 Calculate the extreme values for the shear in panel BC and the bending moment at C due to the right to left passage of the load train.

4.55 Calculate the extreme values for the shear in panel BC and the bending moment at C due to the right to left passage of the load train, with the loads taken in reverse order.

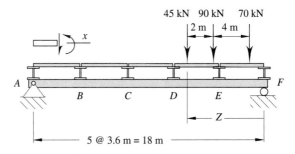

PROBLEMS 4.54 and 4.55

4.56 Calculate the extreme values for the shear in panel BC and the bending moment at C due to the right to left passage of the load train.

4.57 Calculate the extreme values for the shear in panel BC and the bending moment at C due to the right to left passage of the load train, with the loads taken in reverse order.

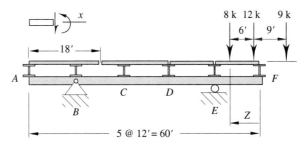

PROBLEMS 4.56 and 4.57

4.58 Calculate the extreme values for the shear in panel BC and the bending moment at C due to the right to left passage of the load train.

4.59 Calculate the extreme values for the shear in panel *BC* and the bending moment at *C* due to the right to left passage of the load train, with the loads taken in reverse order.

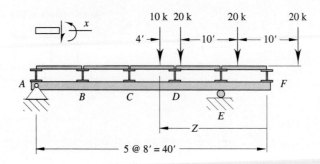

PROBLEMS 4.58 and 4.59

4.60 Calculate the extreme values for the tension in bars *a* and *b* due to right to left passage of the load train.

4.61 Calculate the extreme values for the tension in bars *a* and *b* due to right to left passage of the load train, with the loads taken in reverse order.

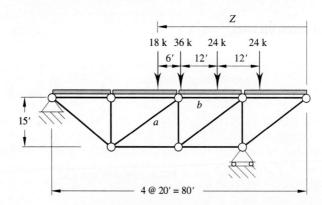

PROBLEMS 4.60 and 4.61

CHAPTER 5

Energy Methods (Part 1)

5.1 INTRODUCTION

Energy methods of mechanics, either alone or in conjunction with vector methods, can be used to advantage in the solution of a wide variety of problems in structural analysis. Elements of energy principles are introduced in basic courses in physics, statics, dynamics, and mechanics of deformable bodies. All too often this introduction is fragmented and inadequate. Accordingly, the present discussion focuses on a few principles and important related concepts and their specific formulations for applications to the analysis of plane-framed structures that make up the main subject matter of introductory courses in the analysis of statically determinate and statically indeterminate structures. Applications in this chapter are centered on problems in rigid-body statics. Application of energy methods to the displacement analysis of deformable bodies is taken up later.

5.2 BASIC CONCEPTS

Certain concepts from elementary mechanics are essential to the development of energy principles. A *particle* is understood to be a body of dimensions so small that it may be treated as occupying a point in space. The concept of rotation plays no role in the mechanics of a particle. A solid body of finite dimensions may be thought of as a mechanical system made up of particles that are held into a cohesive mass by means of mutual forces of interaction that are equal in magnitude, opposite in sense and collinear. (This is the Newtonian model of a body.) The simultaneous positions of all particles that make up a body is understood to define the *configuration*, *position*, or *placement* of the body. A change in position of any particle in a body defines a displacement of the body. Accordingly, any change in the configuration of a body defines a displacement of the body.

A mechanical system in which the positions of the particles cannot be assigned independently is understood to be a *constrained mechanical system*. The conceptual rigid body is the classical example of a constrained mechanical system. The idealized beam for which plane cross sections remain plane is an example of a constrained system of central importance in structural analysis.

Mathematically the configuration of a mechanical system is defined by means of independent variables called *position variables*, *position coordinates*, or *generalized coordinates*. There are many ways in which position coordinates may be chosen;

however, often the geometry of the system suggests an appropriate choice for these variables. The *degrees of freedom* of a mechanical system is defined to be equal to the number of independent position variables that are required to define completely the configuration of the system. A single particle free to move in space has three degrees of freedom, as does a plane rigid body constrained to move in its own plane. A plane rigid body constrained to pivot about an axis normal to its plane has one degree of freedom. Displacements of a constrained mechanical system are defined by changes in the values of the position coordinates. Accordingly, a set of changes or increments in the position variables is understood to be a displacement.

The mechanical state, i.e., state of stress, strain, etc., throughout an *elastic structure* is defined completely by its configuration. Thus, determination of the configuration of a structure is of considerable importance. Even when considered within the framework of certain standard assumptions, a structure is an enormously complex infinite degree-of-freedom mechanical system. Nevertheless, often an exact analysis can be arranged in such a way that the structure may be treated as an n degree-of-freedom system. In general, however, one must settle for a simplified analysis in which the configuration (and, hence, the mechanical state) of a structure is determined only approximately. In the simplified analysis reasonable assumptions concerning how members deform are introduced in such a way as to derive an n degree-of-freedom mathematical model of the structure that is relatively easy to analyze. Clearly, this is done on a regular basis in conventional analysis; however, that it is done is not always explicitly admitted. Obviously, the quality of the results depend on how closely the model can be made to approximate the given structure.

The mathematical significance of the statement that a structure is modeled as an n degree-of-freedom system may be summed up as follows: Let

$$\mathbf{r} = \mathbf{r}(z_1, z_2, \ldots, z_n)$$

be the position of a particle in an n degree-of-freedom system with position variables

$$z_1, z_2, \ldots, z_n$$

The particle displacement due to the displacement defined by

$$\Delta z_1, \Delta z_2, \ldots, \Delta z_n$$

is given by

$$\mathbf{u} = \mathbf{r}(z_1 + \Delta z_1, z_2 + \Delta z_2, \ldots, z_n + \Delta z_n) - \mathbf{r}(z_1, z_2, \ldots, z_n)$$

For infinitesimally small displacements,

$$\Delta z_1 \to \delta z_1, \Delta z_2 \to \delta z_2, \ldots, \Delta z_n \to \delta z_n, \text{ and } \mathbf{u} \to \delta \mathbf{u}$$

and the corresponding particle displacement to the first order is

$$\delta \mathbf{u} = \frac{\partial \mathbf{r}}{\partial z_1} \delta z_1 + \frac{\partial \mathbf{r}}{\partial z_2} \delta z_2 + \cdots + \frac{\partial \mathbf{r}}{\partial z_n} \delta z_n = \sum \frac{\partial \mathbf{r}}{\partial z_k} \delta z_k$$

5.3 VIRTUAL DISPLACEMENT/VIRTUAL WORK

A *virtual displacement* is any possible (admissible) infinitesimal displacement considered without reference to cause. Virtual displacements, by definition, must be consistent with all mechanical constraints that have been imposed on the system.

In the practical application of energy principles, there is a need for a method of identifying virtual displacements and the first order infinitesimal changes or *variations* in quantities which are functions of virtual displacements. The accepted custom is to take whatever symbol is used to identify a quantity and to apply the lower case Greek letter δ as a prefix. Thus, if u denotes displacement, δu denotes virtual displacement; if φ denotes rotation, $\delta\varphi$ denotes virtual rotation; if ε denotes strain, $\delta\varepsilon$ denotes virtual strain.

Three nominally different procedures are available for obtaining virtual strain and virtual rotation in terms of virtual displacements. First, one may employ the basic defining relation to obtain the exact increment in a quantity and then linearize the relation. Second (essentially the same as the first), one may use the basic defining relation and the differential calculus. Third, one may employ the basic defining relation and small-angle geometry.

The work done by a force on a virtual displacement is defined to be the *virtual work* of the force and is denoted δW. From the definition of work and the condition that the virtual displacement is infinitesimally small,

$$\delta W = \mathbf{F} \cdot \delta \mathbf{u}$$

In general, the force varies with the position of the particle on which it acts. In evaluating the virtual work, it is understood that the force is evaluated at the position from which the virtual displacement is measured. The virtual work of a mechanical system is defined to be equal to the total virtual work done by all the forces that act on all the particles in the system. Thus,

$$\delta W = \sum \mathbf{F}_k \cdot \delta \mathbf{u}_k$$

where the summation is over all particles in the system.

For a constrained system, the configuration of which is defined by position variables,

$$z_1, z_2, \ldots, z_n$$

it is understood that all physical virtual quantities, e.g., δu, $\delta\varphi$, $\delta\varepsilon$, and so on, are to be expressed in terms of the infinitesimal changes

$$\delta z_1, \delta z_2, \ldots, \delta z_n$$

that define the virtual displacement. From the definition of position and displacement, and the calculus, it follows that with

$$\mathbf{r} = \mathbf{r}(z_1, z_2, \ldots, z_n)$$

being the physical position of a particle, the physical virtual displacement of the particle is given by

$$\delta \mathbf{r} = \delta \mathbf{u} = \frac{\partial \mathbf{r}}{\partial z_1} \delta z_1 + \frac{\partial \mathbf{r}}{\partial z_2} \delta z_2 + \cdots + \frac{\partial \mathbf{r}}{\partial z_n} \delta z_n$$

From the definition of virtual work and the formulation for the virtual displacement, it follows that for the constrained mechanical system the virtual work can be expressed in the form

$$\delta W = Q_1 \delta z_1 + Q_2 \delta z_2 + \cdots + Q_n \delta z_n = \Sigma Q_k \delta z_k$$

The truth of the claim is established as follows by direct substitution of the expression for $\delta \mathbf{u}_k$ into the expression for the virtual work:

$$\delta W = \mathbf{F}_1 \cdot \left(\frac{\partial \mathbf{r}_1}{\partial z_1} \delta z_1 + \frac{\partial \mathbf{r}_1}{\partial z_2} \delta z_2 + \cdots + \frac{\partial \mathbf{r}_1}{\partial z_n} \delta z_n \right) +$$

$$\mathbf{F}_2 \cdot \left(\frac{\partial \mathbf{r}_2}{\partial z_1} \delta z_1 + \frac{\partial \mathbf{r}_2}{\partial z_2} \delta z_2 + \cdots + \frac{\partial \mathbf{r}_2}{\partial z_n} \delta z_n \right) +$$

$$\cdots$$

$$\mathbf{F}_m \cdot \left(\frac{\partial \mathbf{r}_m}{\partial z_1} \delta z_1 + \frac{\partial \mathbf{r}_m}{\partial z_2} \delta z_2 + \cdots + \frac{\partial \mathbf{r}_m}{\partial z_n} \delta z_n \right)$$

Summing by columns yields

$$\delta W = Q_1 \delta z_1 + Q_2 \delta z_2 + \cdots + Q_n \delta z_n = \Sigma Q_k \delta z_k$$

where

$$Q_1 = \mathbf{F}_1 \cdot \frac{\partial \mathbf{r}_1}{\partial z_1} + \mathbf{F}_2 \cdot \frac{\partial \mathbf{r}_2}{\partial z_1} + \cdots + \mathbf{F}_m \cdot \frac{\partial \mathbf{r}_m}{\partial z_1}$$

$$Q_2 = \mathbf{F}_1 \cdot \frac{\partial \mathbf{r}_1}{\partial z_2} + \mathbf{F}_2 \cdot \frac{\partial \mathbf{r}_2}{\partial z_2} + \cdots + \mathbf{F}_m \cdot \frac{\partial \mathbf{r}_m}{\partial z_2}$$

$$\cdots$$

$$Q_n = \mathbf{F}_1 \cdot \frac{\partial \mathbf{r}_1}{\partial z_n} + \mathbf{F}_2 \cdot \frac{\partial \mathbf{r}_2}{\partial z_n} + \cdots + \mathbf{F}_m \cdot \frac{\partial \mathbf{r}_m}{\partial z_n}$$

in which m is the number of forces.

5.4 CONTACTING SURFACES AND FORCES

All contacting surfaces are idealized in that they are assumed to be perfectly smooth (i.e., frictionless) and to maintain contact. Only normal forces can be transmitted across such surfaces. Because there can be no relative displacement along the normal to a surface of contact and because the contact forces are equal in magnitude, opposite in sense, and collinear, the net work of the forces of contact is always zero. The contacting surfaces at pins, rollers, and other connections commonly used in structural applications are normally assumed to be of the frictionless type.

5.5 FORCES

The forces in a mechanical system consisting of interconnected bodies are separated into three groups:

1. Internal or mutual forces of interaction that act between particles belonging to the same body. In the Newtonian model of a body, these forces always occur in pairs that are equal, opposite, and collinear.
2. Forces of interaction along the contacting surfaces of different bodies.
3. The remaining forces, which are prescribed and are called the external forces.

5.6 RIGID DISPLACEMENTS

Rigid displacements play a central role in the solution of statics problems by the method of virtual work. Accordingly, it is important that one know the relationship between the displacements of the points in a body that undergoes a rigid displacement. Only plane displacements are considered herein. The statement that a body undergoes a rigid displacement means that in the displacement the distances between every pair of points in the body are unchanged. This in turn means that every line in the body undergoes the same rotation. The net displacement of any point can always be expressed as the composition of a rigid translation and a rigid rotation about a reference point. The relationship of the displacement of any point to the displacement of the reference point depends on the magnitude of the rotation. Although primary interest is in the relationships that are appropriate for (infinitesimally) small rotations, it is instructive to consider finite rotation first.

With reference to Figure 5.1, let a be the reference point and let b be any other point. The rotation θ is measured positive by the right hand rule about the unit vector $\mathbf{k}$ which is taken to point upward out of the plane of the paper. The unit vector $\boldsymbol{\mu}$ is

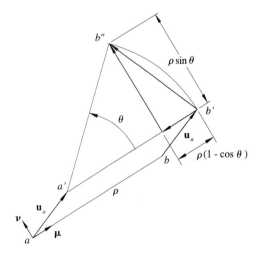

FIGURE 5.1 Rigid displacement.

directed from *a* toward *b* and the unit vector $\boldsymbol{v}$ is given by $\boldsymbol{v} = \mathbf{k} \times \boldsymbol{\mu}$. ρ is the distance from *a* to *b*. From the geometry shown,

$$\mathbf{u}_b = \mathbf{u}_a + \mathbf{b}'\mathbf{b}''$$

in which $\mathbf{b}'\mathbf{b}''$ is the displacement of *b* relative to *a* due to the rotation. Also from the geometry shown,

$$\mathbf{b}'\mathbf{b}'' = \rho[(\sin \theta)\boldsymbol{v} - (1 - \cos \theta)\boldsymbol{\mu}]$$

5.7 SMALL DISPLACEMENTS

For (infinitesimally) small displacements the rotations must be small. Therefore, with

$$|\theta| \ll 1, \quad \sin \theta \approx \theta, \quad \cos \theta \approx 1$$

and the expression for $\mathbf{u}_b$ reduces to

$$\mathbf{u}_b = \mathbf{u}_a + \rho \theta \boldsymbol{v}$$

Observe that for small rotations the displacement of point *b* due to the rotation about *a* is perpendicular to the radius ρ. Also observe that infinitesimally small rigid body displacements are related to each other in exactly the same way as velocities are related in a rigid body motion. The geometry of small rigid virtual displacements consistent with the formulation is shown in Figure 5.2. For displacements due to virtual translation $\delta\mathbf{u}_a$ and to virtual rotation $\delta\theta$,

$$\delta\mathbf{u}_b = \delta\mathbf{u}_a + \rho \, \delta\theta \, \boldsymbol{v}$$

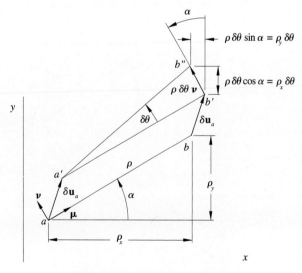

FIGURE 5.2 Rectangular components.

5.8 RECTANGULAR COMPONENTS

In most applications one has need of the rectangular components of the displacements due to small translation and small rotation. Let δu_x and δu_y be the components of displacement parallel to and positive in the directions of the x and y coordinate axes, respectively. Then, from the geometry shown in Figure 5.2,

$$\delta u_{bx} = \delta u_{ax} - \rho \, \delta\theta \sin\alpha = \delta u_{ax} - \rho_y \delta\theta$$
$$\delta u_{by} = \delta u_{ay} - \rho \, \delta\theta \cos\alpha = \delta u_{ay} - \rho_x \delta\theta$$

Clearly, ρ_x and ρ_y are just the projections of ρ onto the x and y axes, respectively. In most practical problems these projections are given and the formulations for the displacements are easy to implement.

5.9 INSTANTANEOUS CENTER

The most general small rigid-body displacement can be expressed in terms of a rotation about an axis through a point called the instantaneous center of the displacement. By definition, the instantaneous center is located at that point which moves with the body but undergoes no translation as a consequence of the displacement. Let c identify the instantaneous center. Then, with c as reference, the virtual displacements of points a and b are given by

$$\delta u_a = (\rho_a \delta\theta) v_a, \quad \delta u_b = (\rho_b \delta\theta) v_b$$

As shown in Figure 5.3a, if δu_a is not parallel to δu_b the instantaneous center is located at the point of intersection of the normal to δu_a drawn through a and the normal to δu_b drawn through b. Observe that to locate the instantaneous center in this case it is sufficient to know only the orientations of the displacement vectors at a and b.

If the displacements at a and b are parallel and not equal they must also be normal to the line connecting the two points. In this case the instantaneous center lies on the line connecting the points as shown in Figure 5.3b; however, to locate the center one

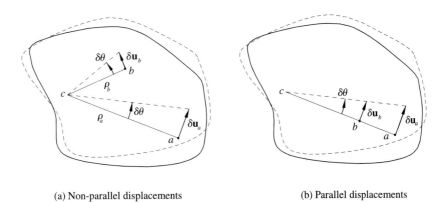

(a) Non-parallel displacements (b) Parallel displacements

FIGURE 5.3 Instantaneous center.

must know the ratio of the two displacements. It is to be noted that if *a* is constrained so that it cannot be displaced, then *a* must be the instantaneous center. If *a* and *b* are constrained so that neither can move, then the displacement of every point is zero. The application of the instantaneous center will be demonstrated in example problems.

5.10 PRINCIPLE OF VIRTUAL WORK

The principle of virtual work is taken as the fundamental principle of the energy method; it is taken as a first principle that does not require proof. The principle may be stated as follows:

> A mechanical system is in an equilibrium configuration if and only if the virtual work of all the forces is zero for arbitrary virtual displacements from that configuration.

In this statement the virtual work of all the forces is understood to mean the virtual work of all the forces that act on all the particles in the system. Also, it is to be understood that the system may consist of any number of interconnected bodies. Because the virtual work of forces along (frictionless) contacting surfaces is zero, the virtual work of all the forces is

$$\delta W = \delta W_e + \delta W_i$$

in which the subscripts e and i identify the virtual work of the external and internal forces, respectively. Clearly, the principle of virtual work may now be formulated as

$$\delta W = \delta W_e + \delta W_i = 0$$

By introducing the definition

$$\delta W_i \equiv -\delta W_\sigma$$

the principle may be given the alternate, somewhat more convenient, form

$$\delta W_e = \delta W_\sigma$$

Subsequently it will be shown that δW_σ represents the total virtual work of the stresses on the virtual strains induced by the virtual displacements.

For a rigid body or for a deformable body that undergoes a rigid virtual displacement, $\delta W_i = 0$. Accordingly, for a mechanical system consisting of any number of rigid bodies or any combination of rigid and deformable bodies, each of which undergoes a rigid virtual displacement, the virtual work formulation of the condition of equilibrium takes the special form

$$\delta W_e = 0$$

5.11 VIRTUAL WORK OF FORCES ON A RIGID VIRTUAL DISPLACEMENT

A theorem of considerable practical importance relates the virtual work of a system of forces on a rigid virtual displacement to the virtual work of the resultant of the forces. The theorem asserts that:

Section 5.11 Virtual Work of Forces on a Rigid Virtual Displacement

The virtual work of any system of forces on a rigid virtual displacement is equal to the virtual work of the resultant of that force system evaluated with respect to any point on the same virtual displacement.

The resultant of a force system with respect to a point consists of a force and a couple. The force is equal to the vector sum of all of the forces in the given system and the moment of the couple is equal to the sum of the moments of all the forces in the given system taken about the point. (Note: the resultant of a force system is itself a force system.) By definition the virtual work of any system of forces is equal to the sum of the virtual works of the individual forces in the system. From the foregoing properties it follows that the theorem is true for any system of forces if it is true for a system consisting of one force. Figure 5.4 contains the basic information needed to establish the theorem. The force system is shown in Figure 5.4a. Shown in Figure 5.4b is the resultant of the force system in Figure 5.4a evaluated with respect to point a. The virtual displacements are shown in Figure 5.4c. The virtual work of the forces in Figure 5.4a on the displacements in Figure 5.4c is

$$\delta W = \mathbf{F} \cdot \delta \mathbf{u}_b = \mathbf{F} \cdot (\delta \mathbf{u}_a + \rho\, \delta\theta\, \mathbf{v}) = \mathbf{F} \cdot \delta \mathbf{u}_a + \delta\theta(\rho \mathbf{F} \cdot \mathbf{v}) = \mathbf{F} \cdot \delta \mathbf{u}_a + M_a \delta\theta$$

Therefore, for any system of forces,

$$\delta W = \mathbf{R} \cdot \delta \mathbf{u}_a + M_a \delta\theta$$

where $\mathbf{R}$ is the vector sum of all the forces in the system and M_a is the sum of the moments of all the forces in the system about an axis through a. (Note that M_a and $\delta\theta$ are measured positive in the same sense about the axis.) Thus, the theorem is proved.

The theorem leads to some very important results. The force system shown in Figure 5.5a is representative of the forces of interaction that are conceived to act between pairs of particles in the Newtonian model of a body. The forces are equal in magnitude, opposite in sense and collinear. The vector sum of all the forces in the system and the sum of the moments of the forces in the system about any axis is zero. Therefore, the virtual work of the pair on a rigid virtual displacement is zero and from

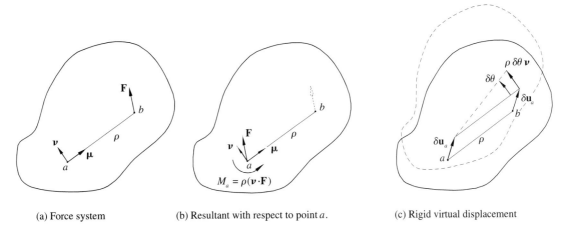

(a) Force system   (b) Resultant with respect to point a.   (c) Rigid virtual displacement

FIGURE 5.4 Virtual work of forces on a rigid virtual displacement.

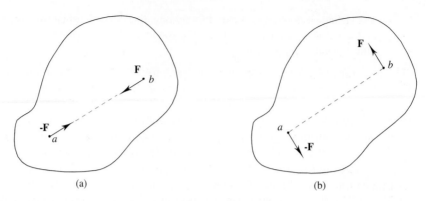

FIGURE 5.5 Force pairs.

this it follows that the virtual work of the internal forces is zero on any rigid virtual displacement. The equal and opposite forces shown in Figure 5.5b comprise a couple. The moment of the couple is the same about all axes. The virtual work of the pair of forces on a rigid virtual displacement is, from the theorem, equal to the product of the moment of the couple and the virtual rotation.

5.12 VECTOR REPRESENTATIONS AND WORK FORMULATIONS

In Figure 5.6, **F** is a force vector, $\delta\mathbf{u}$ and $\delta\mathbf{v}$ are virtual displacement vectors, and **i, j, λ, μ**, and **ν** are unit vectors. All other quantities are scalars. The top row (a) of Figure 5.6 illustrates what amounts to complete detailed representations of the force vector and the virtual displacement vectors. Almost invariably, detailed representations are too cumbersome to work with. Frequently, in practice, the simplified representations illustrated in the bottom row (b) of Figure 5.6 are employed instead. In the simplified representations, the scalar multipliers to be applied to the unit vectors are shown adjacent to them, and the unit vectors are simply shown without specific identification. At times, difficulty is encountered in formulating correct virtual work

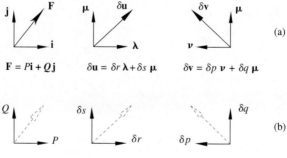

FIGURE 5.6 Vector representations.

expressions when simplified representations are used. The problem centers on connective signs and is eliminated by keeping in mind that work is evaluated with the aid of the dot product and that the arrows in the simplified representation really stand for unit vectors in the directions in which they are drawn. This is the equivalent of looking at part (b) of Figure 5.6, but thinking part (a) of the figure. From the simplified representations, the virtual work of **F** on δ**u** is given by $\delta W = P\,\delta r + Q\,\delta s$ and the virtual work of **F** on δ**v** is given by $\delta W = -P\,\delta p + Q\,\delta q$. These formulations are easy to verify by direct calculation using the complete representations illustrated in Figure 5.6a.

5.13 EXAMPLE PROBLEMS

The example problems that follow are set up with two purposes in mind. The first is to simply illustrate how the principle of virtual work is employed to solve equilibrium problems of one-degree-of-freedom systems. These are particular examples of incompletely constrained systems that are, therefore, geometrically or kinematically unstable. The second purpose is to illustrate the reasoning employed in applying the principle of virtual work to solve equilibrium problems in the case of fully restrained structures. Because a fully restrained structure made of rigid bodies has no degrees of freedom, it cannot be subjected to a virtual displacement and, therefore, how to apply the principle of virtual work to the force analysis of such a system requires some explanation. To achieve clarity in the sketches associated with the examples it is necessary to show virtual displacements to a highly exaggerated scale. All calculations, however, are (as they must be) carried out on the basis of small-angle geometry in which the sine of an angle is approximated by the angle (in radian measure) and the cosine of the angle is approximated by unity.

Example 5.1

The rigid bar shown in Figure 5.7 is supported by a pin at B. Determine the force R to hold the bar in equilibrium in the position shown. The bar has one degree of freedom. Because B is a fixed point, it is also the instantaneous center for displacement. To find R by the principle of virtual work, subject the bar to a virtual displacement as shown in Figure 5.7b. The virtual work is formulated as

$$\delta W = P\,\delta v - R\,\delta u = 0$$

Replacing δv by its equivalent in terms of δu gives

$$\delta W = \left(\frac{Pb}{L} - R\right)\delta u = 0$$

The virtual work must be zero with $\delta u \neq 0$, but this condition can be satisfied only if $R = Pb/L$, which is exactly the same result as one obtains from moment equilibrium. Part (c) of Figure 5.7 shows the bar with the left end supported by a roller. With the roller in place, the bar (which is rigid) is fully restrained and cannot be subjected to displacement of any kind. To calculate the reaction at the roller by the method of virtual work, one first observes that a property of the roller is that it develops whatever reactive force normal to the direction of roll is necessary to keep the bar in the position shown. In other words, the reaction force A provided by the roller performs exactly the same function as R in the case of equilibrium of the bar with the roller absent. The observation is the basis of the procedure for calculating unknown forces in fully restrained structures by virtual work. For a particular unknown component of force the procedure is to conceptually remove the physical element that develops the force and replace it by the component of force that it develops. In this way, the fully restrained structure is reduced to a single-degree-of-freedom system for which the unknown force component can be found by virtual work.

Example 5.2

The two-bar mechanism ABC is maintained in equilibrium by means of a pair of equal and opposite couples applied to the ends of the members at the interior pin B as shown in Figure 5.8. Determine the moment M of the couples for equilibrium in the position shown. C is the instantaneous center for BC.

The mechanism has one degree of freedom. The relationship between the rotations of AB and BC is obtained from the conditions of rigidity of the members and the condition of compatibility of displacement enforced by the pin at B. The compatibility condition is that the displacements of the B end of AB and the B end of BC be equal. From equality of the axial components of displacement and rigidity of the bars it is concluded that A can undergo no displacement and, therefore, A must be the instantaneous center of AB. Equality of the transverse components of displacement at the B-end of AB and the B-end of BC leads to the relationship between the rotations as shown in part (b) of the figure. By the principle of virtual work, for equilibrium

$$\delta W = -M\,\delta\varphi - M\,\delta\theta + P\,\delta v = 0$$

After replacing $\delta\varphi$ and δv with their equivalents in terms of $\delta\theta$, the work equation reduces to

$$\delta W = \left(\frac{-ML}{a} + Pb\right)\delta\theta = 0$$

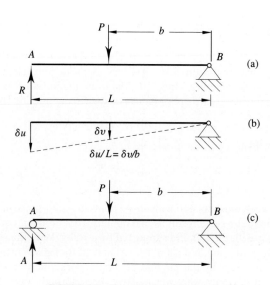

FIGURE 5.7   Reaction by virtual work.

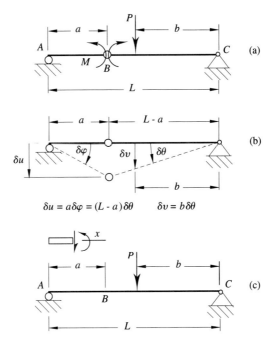

FIGURE 5.8 Bending moment by virtual work.

The demand that the work equation be satisfied for non-zero virtual displacements requires that the term in parentheses be zero. Satisfaction of the condition requires that

$$M = \frac{Pab}{L}$$

With respect to maintaining the system in equilibrium in the position shown, welding the members at the pin would produce exactly the same effect as the pair of couples. Welding would, however, convert the mechanism into a fully restrained beam (as shown in Figure 5.8c) in which weld material develops the equal and opposite couples required for equilibrium. Following the line of reasoning in the previous example, it is seen that to obtain the bending moment in the beam at section B by virtual work all one need do is (conceptually) remove the material that develops the bending moment at B and replace the material by the equal and opposite couples that it develops. In this way the beam becomes the original two-bar single-degree-of-freedom mechanism for which M can be found by virtual work. It is easy to verify that the bending moment in the beam at section B is as given above.

Example 5.3

The vertical stub ends of rigid bars AB and BC are rigidly attached to the members as shown in Figure 5.9. The rollers permit only relative displacements that are parallel to the direction of roll. The stub ends and the members are, therefore, required to remain at the same separation (essentially zero) and parallel when the mechanism is displaced. The problem is to find the equal and opposite forces required on the stub ends to keep the system in equilibrium in the position shown.

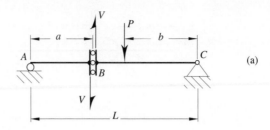

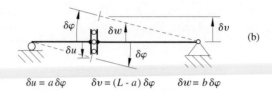

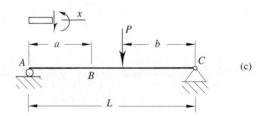

FIGURE 5.9 Shear by virtual work.

From the given conditions A is the instantaneous center of AB and C is the instantaneous center of BC. Since AB and BC must remain parallel, the virtual displacements, with the members rotated clockwise, must be as shown in Figure 5.9b. By the principle of virtual work, for equilibrium

$$\delta W = V\,\delta u + V\,\delta v - P\,\delta w = 0$$

By replacing each of the virtual displacements by its equivalent in terms of $\delta\varphi$, the equation is reduced to

$$\delta W = (VL - Pb)\,\delta\varphi = 0$$

from which

$$V = \frac{Pb}{L}$$

The result is exactly as expected; namely, V is equal to the shear in the simply supported beam obtained by welding the rollers in the mechanism at B. It is now clear how the shear at any location in the beam can be obtained by virtual work. The material at the section is replaced by

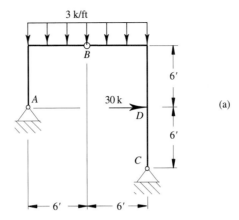

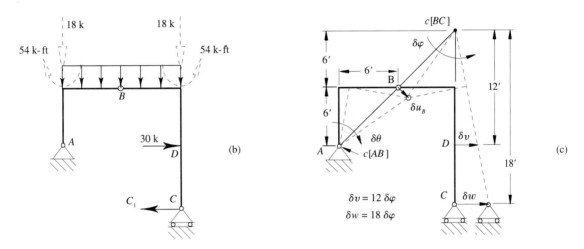

FIGURE 5.10 Frame reaction.

the roller device and the equal and opposite shear forces are applied to the roller device. In this way the beam has now been converted to a mechanism and the shear may be calculated by virtual work.

Example 5.4

Calculate the horizontal component of the reaction at support C of the frame shown in Figure 5.10a. Use the principle of virtual work.

The mechanism for the calculation is obtained by replacing the pin support at C by a horizontal roller as shown in Figure 5.10b. The force acting on the roller is C_1, the unknown reaction component to be determined. The resultant of the distributed load on part AB of the frame, evaluated with respect to the upper left corner of AB, consists of a force and a couple as shown by dashed lines. Similarly, the resultant of the distributed load on BC, evaluated with respect to the upper right corner of BC, is a force and a couple as shown. Virtual displacements of the frame are shown in Figure 5.10c. Point A is the instantaneous center of AB. The instantaneous center for BC is determined from the known direction of the displacement at B and at C. Because the displacement of the B-end of AB is perpendicular to the radius from A to B and because the displacement of the B-end of BC must be the same as that of the B-end of AB, it follows that the instantaneous center of BC, identified as $c[BC]$, must lie on the extension of the radius from A to B as shown. Since the displacement at C is horizontal, $c[BC]$ must also lie on a vertical line through C as shown. The horizontal component of displacement of the pin at B must have the same value whether B is considered to be the B-end of AB or the B-end of BC (a compatibility condition). In each case the horizontal component of displacement due to the rotation is equal to the product of the projection of the corresponding radius onto a vertical line (see the last paragraph of the discussion on rigid displacements). Therefore,

$$6 \, \delta\theta = 6 \, \delta\varphi \Rightarrow \delta\theta = \delta\varphi$$

By the principle of virtual work,

$$\delta W = 54 \, \delta\theta + 54 \, \delta\varphi + 30 \, \delta v - C_1 \, \delta w = 0$$

Substitution of the relations $\delta\theta = \delta\varphi$, $\delta v = 12 \, \delta\theta$, $\delta w = 18 \, \delta\theta$ gives

$$\delta W = [54 + 54 + 30(12) - 18C_1]\delta\theta = 0$$

from which $C_1 = 26$ kips. The result agrees with the result obtained for the frame by vector statics in Chapter 1.

5.14   INFLUENCE LINES

The principle of virtual work in the formulation $\delta W = \delta W_e = 0$ is a powerful aid in the construction of influence lines for statically determinate structures. The principle can be used to calculate ordinates on the influence line; however, its greatest utility is in establishing the geometry of the influence line. With knowledge of the geometry of the influence line all that remains to be done to establish the line is to calculate key ordinates. The fact that the principle of virtual work can be used to establish the geometry of the influence line comes out of a direct attempt to calculate any ordinate on the influence line by virtual work. The utility of the method depends on one's ability to analyze the geometry of small displacements. The ideas involved are best brought out by example.

Considerable detail is provided in the following examples to help the reader develop an ability to analyze small displacements and to develop confidence in the methodology. In practice one need include only the information that is essential to establishing the geometry of the influence line. As was pointed out above, once the geometry of the influence line is known, whatever technique appears to be most appropriate can be used to calculate the key ordinates.

Example 5.5

Construct the influence lines for the reaction at A and the shear and bending moment at B for the beam shown in Figure 5.11. Recall that the influence line for any quantity is a plot of the influence function that describes how the quantity varies with the location s of the unit load.

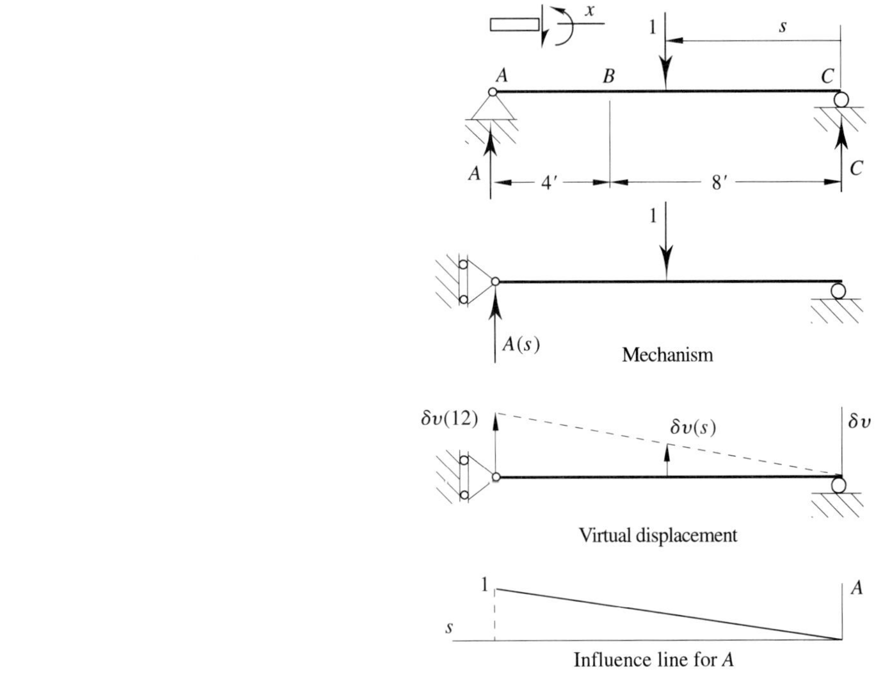

FIGURE 5.11 Reaction at A.

Because each of the quantities under consideration has a value that is required to satisfy the conditions of equilibrium, the influence function and the related line are obtained by application of the conditions of equilibrium. Here, the initial action is to make use of the principle of virtual work for the calculation.

Reaction at A The mechanism and associated virtual displacement for calculation of the reaction at A are as shown. The reaction at A and the virtual displacement at the point of application of the unit load are functions of s. To make this clear the quantities are identified as functions of s. By the principle of virtual work,

$$\delta W = A(s)\,\delta v(12) - (1)\,\delta v(s) = 0$$

from which

$$A(s) = \frac{\delta v(s)}{\delta v(12)}$$

Substitution of the equation for $\delta v(s)$ into the foregoing expression for $A(s)$, yields an explicit formula for $A(s)$, i.e., an explicit formula for the influence function for reaction A.

The substitution is not necessary for the purposes of constructing the influence line. What the last relation asserts is that the influence function $A(s)$ is proportional to the virtual displacement function $\delta v(s)$. Graphically this means that the geometry of the influence line is exactly the same as the geometry of the virtual displacement diagram. From this and the sketch of the virtual displacements it is clear that the influence line for A is a straight line that has ordinate zero at $s = 0$. To complete construction of the influence line only one additional ordinate is needed. By inspection, $A = 1$ when the unit load is at $s = 12$. Therefore, the influence line must be as shown below the virtual displacement diagram. This example illustrates the general idea of using the principle of virtual work to establish the geometry of an influence line for force. Once the geometry has been established, whatever calculational technique is most appropriate can be used to establish key ordinates.

Shear at B The mechanism and virtual displacement for the determination of the influence function (line) for $V_B(s)$ are as shown in Figure 5.12. By the principle of virtual work, for equilibrium,

$$\delta W = (4 + 8)\,\delta\varphi\, V_B(s) - (1)\,\delta v(s) = 0$$

Because $\delta v(s)$ is proportional to $\delta\varphi$, the foregoing relationship amounts to the statement that $V_B(s)$ is proportional to $\delta v(s)$. Therefore, the influence line for the shear at B has the same geometry as the virtual displacement function. The influence line is defined completely following calculation of the shear at B for $s = 8-$ and $8+$.

Bending Moment at B The mechanism and the associated virtual displacement for the construction of the influence line for the bending moment at B are shown in Figure 5.13. For equilibrium,

$$\delta W = (\delta\theta + \delta\varphi) M_B(s) - (1)\,\delta v(s) = 0$$

Because $\delta\varphi$ and $\delta v(s)$ are proportional to $\delta\theta$, $M_B(s)$ is proportional to $\delta v(s)$. Therefore, the influence line for the bending moment at B is as shown. The ordinate at $s = 8'$ was calculated by conventional vector statics.

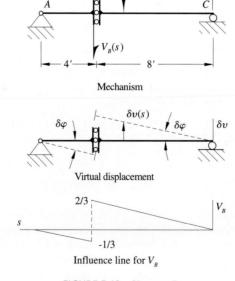

FIGURE 5.12 Shear at B.

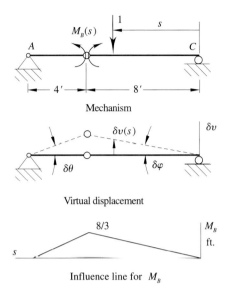

FIGURE 5.13 Bending moment at B.

Example 5.6

Construct the influence lines for the shear in panel BC and the bending moment at panel point B of the floor girder shown in Figure 5.14.

Shear in Panel BC The mechanism for the shear in panel BC is as shown. The rollers may be located anywhere in the panel. The location is of no consequence because the constraint provided by the rollers requires that the portions of the girder on either side of the rollers undergo the same rotation. As shown in the virtual displacement diagram, the virtual displacements are measured from the s axis, which coincides with the undisplaced position of the loaded surface, to the displaced position of the loaded surface. With the aid of the principle of virtual work, it follows by inspection that the influence line, which is proportional to the virtual displacement of the loaded surface, consists of two straight-line segments with a discontinuity at $s = 25'$. By inspection of the geometry of the displacements it is seen that all the data needed to construct the influence line is obtained by evaluating the shear in panel BC for $s = 25-$ and $s = 25+$. Statics gives the results shown. The influence line coincides with the results obtained in an earlier example.

Bending Moment at B The mechanism for the calculation of the bending moment at B is obtained by inserting a hinge at B as shown in Figure 5.15 on page 113. The associated virtual displacement is as shown. Again, by the principle of virtual work it is seen that the influence line consists of straight-line segments with a discontinuity at $s = 25'$. Evaluation of the bending moment at B for $s = 25-$ and for $s = 25+$ gives values for the end points of the line segments. The influence line as shown coincides with that obtained by vector methods in an earlier example.

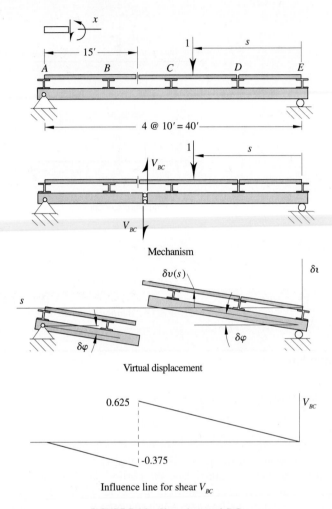

FIGURE 5.14 Shear in panel BC.

Example 5.7

Construct the influence lines for the tension in bars a and b of the truss shown in Figure 5.16.

Tension in Bar a The mechanism for the determination of the tension in bar a is as shown in Figure 5.16. In the virtual displacement, the shaded portions A and B displace as rigid bodies. Only key displacements are shown on the virtual displacement diagram. The relationship between the virtual rotations is determined from equality of the displacement of pin p considered as part of A and as part of B. The instantaneous center $c[B]$ is determined on the basis of the known directions of the displacements at point p and at the roller support. Equality of the vertical components of displacement of point p requires that $8\,\delta\theta = 24\,\delta\varphi$. The equilibrium condition is $\delta W = -N_a(24\,\delta\varphi) - (1)\,\delta v(s) = 0$, which asserts that the influence function $N_a(s)$ is

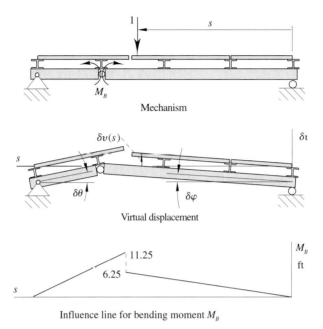

FIGURE 5.15 Bending moment at B.

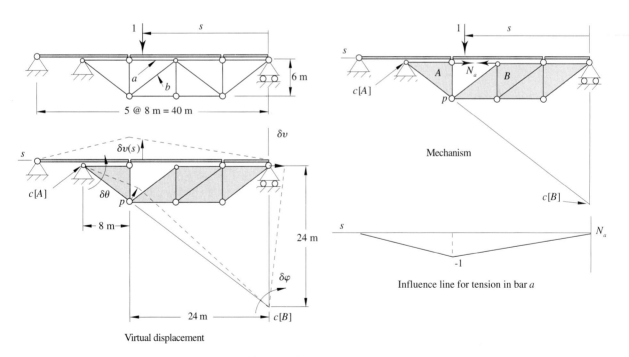

FIGURE 5.16 Tension in bar a.

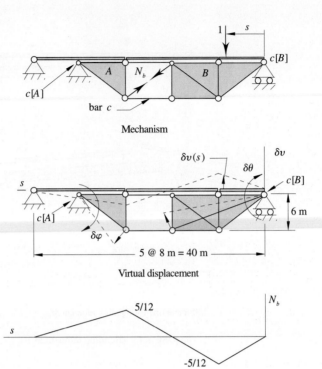

FIGURE 5.17 Tension in bar b.

proportional to the virtual displacement function $\delta v(s)$ and that the proportionality factor is a negative number. The influence line for the tension in bar a is as shown. Vector statics with the unit load placed at $s = 24$ m gives $N_a(24) = -1$. By inspection $\delta v(24) = 24\,\delta\varphi$. Substitution of this result into the work equation also gives $N_a(24) = -1$.

Tension in Bar b The mechanism for calculation of the tension in bar b is as shown in Figure 5.17. Shaded portions A and B of the truss displace as rigid bodies. Because the members of the top chord are rigid, there can be no displacement of the roller support. Therefore, the instantaneous center for B is at the roller. The relationship between the rotations is determined from the condition that bar c undergo no axial deformation. The virtual elongation of bar c is

$$\delta l = 6\,\delta\varphi - 6\,\delta\theta = 0$$

With forces and displacements resolved into rectangular components, the condition of equilibrium by virtual work is

$$\delta W = -0.8 N_b (6\,\delta\varphi) - 0.6 N_b (8\,\delta\varphi) - 0.6 N_b (16\,\delta\theta) - 1\,\delta v(s) = 0$$

Because $\delta\theta = \delta\varphi$, the equation reduces to $19.2 N_b\,\delta\varphi = -\delta v(s)$. Again, the work equation asserts that for equilibrium, $N_b(s)$ must be proportional to the virtual displacement function (with a negative proportionality factor). Therefore, the influence line for the tension in bar b is as shown in Figure 5.17. The results are easily verified by vector statics or by virtual work. For $s = 24$ m, $\delta v(24) = -8\,\delta\varphi$ and the work equation gives $N_b(24) = 5/12$.

KEY POINTS

1. A virtual displacement is an arbitrary, infinitesimally small, admissible displacement considered without reference to cause. Any displacement must be possible to be admissible. Admissible displacements must satisfy all conditions of constraint.
2. In an equilibrium problem, forces in the equilibrium position and virtual displacements measured from the equilibrium position are used to evaluate the virtual work.
3. To calculate a force of reaction or a pair of forces of interaction in a statically determinate structure by the method of virtual work: (a) remove the material that develops the unknown force or force pairs and replace the material by the force or force pairs it develops in the equilibrium position; (b) subject the derived single-degree-of-freedom mechanism to a virtual displacement and impose the condition $\delta W = 0$.
4. To determine the geometry of the influence line for a force of reaction or a pair of forces of interaction in a statically determinate structure, remove the material that develops the force or force pair and subject the derived single-degree-of-freedom mechanism to a virtual displacement. The influence line is then proportional to the virtual displacement of the loaded surface.

PROBLEMS

5.1–5.9 General instructions: Construct the single-degree-of-freedom mechanism appropriate to the determination of the assigned reaction or component of reaction, and then apply the principle of virtual work to calculate the assigned quantity.

5.1 Calculate the reaction at A and the vertical component of the reaction force at B. Follow the general instructions for Problems 5.1–5.9.

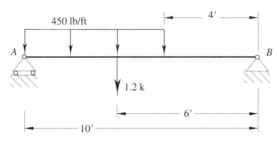

PROBLEM 5.1

5.2 Calculate the reaction at B and the vertical component of the reaction force at A. Follow the general instructions for Problems 5.1–5.9.

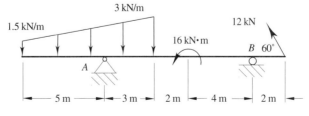

PROBLEM 5.2

5.3 Calculate the reaction at B, the moment of the reaction couple at E, the vertical component of the reaction force at E, and the horizontal component of the reaction force at E. Follow the general instructions for Problems 5.1–5.9.

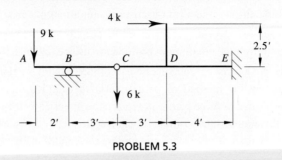

PROBLEM 5.3

5.4 Calculate the moment of the reaction couple at A and the reaction force at B. Follow the general instructions for Problems 5.1–5.9.

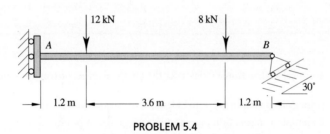

PROBLEM 5.4

5.5 Calculate the vertical reaction at A, the vertical reaction at F, and the horizontal reaction at F. Follow the general instructions for Problems 5.1–5.9.

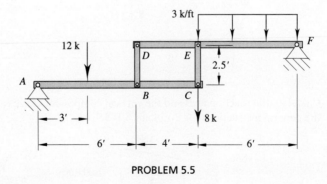

PROBLEM 5.5

5.6 Calculate the vertical reaction at A, the horizontal reaction at D, and the horizontal reaction at E. Follow the general instructions for Problems 5.1–5.9.

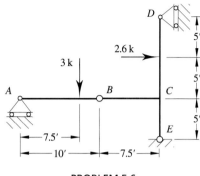

PROBLEM 5.6

5.7 The uniform member weighs 300 kg/m. Calculate the horizontal reaction at A, the horizonal reaction at B, and the vertical reaction at B. Follow the general instructions for Problems 5.1–5.9.

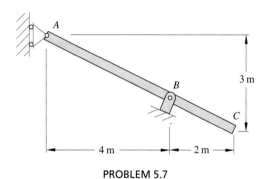

PROBLEM 5.7

5.8 Calculate the reaction at A, the horizonal reaction at C, and the vertical reaction at C. Follow the general instructions for Problems 5.1–5.9.

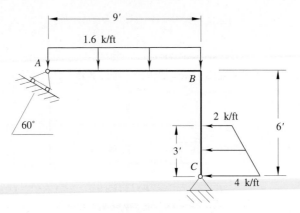

PROBLEM 5.8

5.9 Calculate the vertical reaction at A, the horizontal reaction at A, and the reaction at D. Follow the general instructions for Problems 5.1–5.9.

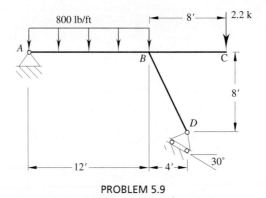

PROBLEM 5.9

5.10–5.37 General instructions: Construct the single-degree-of-freedom mechanism appropriate to the determination of the assigned reaction or force of interaction, use the principle of virtual work to establish the geometry of the influence line for the assigned quantity, and then calculate, by any means, the key ordinates on the influence line.

5.10 Construct the influence lines for the reaction at A, the shear at C, and the bending moment at C. Follow the general instructions for Problems 5.10–5.37.

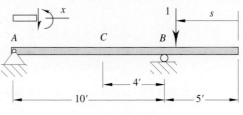

PROBLEM 5.10

5.11 Construct the influence lines for the reaction at A, the bending moment at B, and the shear just to the left of the support at B. Follow the general instructions for Problems 5.10–5.37.

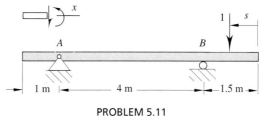

PROBLEM 5.11

5.12 Construct the influence lines for the reaction at A, the reaction at B, the shear at D, and the bending moment at E. Follow the general instructions for Problems 5.10–5.37.

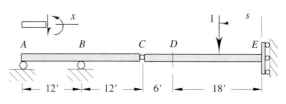

PROBLEM 5.12

5.13 Construct the influence lines for the reaction at A, the reaction at C, the bending moment at C, and the shear just to the left of C. Follow the general instructions for Problems 5.10–5.37.

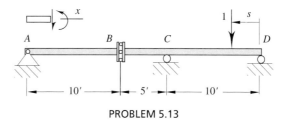

PROBLEM 5.13

5.14 Construct the influence lines for the shear at A and the bending moment at B. Follow the general instructions for Problems 5.10–5.37.

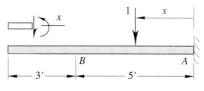

PROBLEM 5.14

5.15 Construct the influence lines for the bending moment at A, the shear at A, and the bending moment at B. Follow the general instructions for Problems 5.10–5.37.

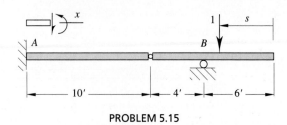

PROBLEM 5.15

5.16 Construct the influence lines for the reaction at A, the reaction at C, the shear at D, and the bending moment at D. Follow the general instructions for Problems 5.10–5.37.

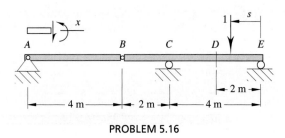

PROBLEM 5.16

5.17 Construct the influence lines for the bending moment at A and the shear just to the left of the support at B. Follow the general instructions for Problems 5.10–5.37.

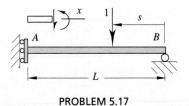

PROBLEM 5.17

5.18 Construct the influence lines for the reaction at A, the reaction at D, and the bending moment at D. Follow the general instructions for Problems 5.10–5.37.

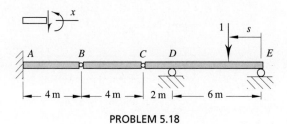

PROBLEM 5.18

5.19 Construct the influence lines for the shear in panel *BC* and the bending moment at *B*. Follow the general instructions for Problems 5.10–5.37.

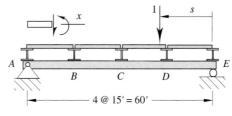

PROBLEM 5.19

5.20 Construct the influence lines for the shear in panel *BC* and the bending moment at *B*. Follow the general instructions for Problems 5.10–5.37.

5.21 Construct the influence lines for the shear in panel *CD* and the bending moment at *C*. Follow the general instructions for Problems 5.10–5.37.

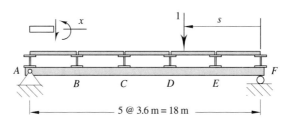

PROBLEMS 5.20 and 5.21

5.22 Construct the influence lines for the shear in panel *CD* and the bending moment at *E*. Follow the general instructions for Problems 5.10–5.37.

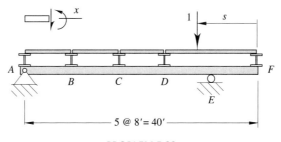

PROBLEM 5.22

5.23 Construct the influence lines for the shear in panel *BC* and the bending moment at *B*. Follow the general instructions for Problems 5.10–5.37.

5.24 Construct the influence lines for the shear in panel CD and the bending moment at C. Follow the general instructions for Problems 5.10–5.37.

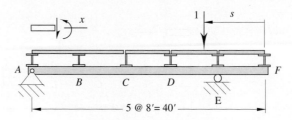

PROBLEMS 5.23 and 5.24

5.25 Construct the influence lines for the shear in panel BC and the bending moment at D. Follow the general instructions for Problems 5.10–5.37.

5.26 Construct the influence lines for the shear in panel CD and the bending moment at C. Follow the general instructions for Problems 5.10–5.37.

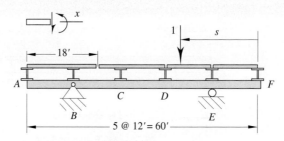

PROBLEMS 5.25 and 5.26

5.27 Construct the influence lines for the shear in panel EF and the bending moment at E. Follow the general instructions for Problems 5.10–5.37.

5.28 Construct the influence lines for the shear in panel DE and the bending moment at D. Follow the general instructions for Problems 5.10–5.37.

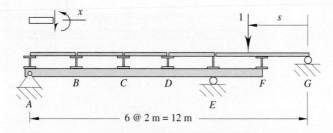

PROBLEMS 5.27 and 5.28

5.29 Construct the influence lines for the shear in panel CD and the bending moment at D. Follow the general instructions for Problems 5.10–5.37.

5.30 Construct the influence lines for the shear in panel *DE* and the bending moment at *B*. Follow the general instructions for Problems 5.10–5.37.

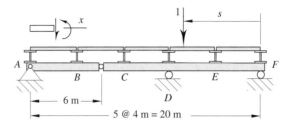

PROBLEMS 5.29 and 5.30

5.31 Construct the influence lines for the shear in panel *CD* and the bending moment at *D*. Follow the general instructions for Problems 5.10–5.37.

5.32 Construct the influence lines for the shear in panel *AB* and the bending moment at *A*. Follow the general instructions for Problems 5.10–5.37.

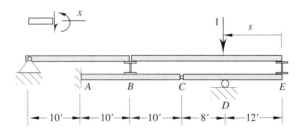

PROBLEMS 5.31 and 5.32

5.33 Construct the influence lines for the tension in bars *a*, *b*, and *c*. Follow the general instructions for Problems 5.10–5.37.

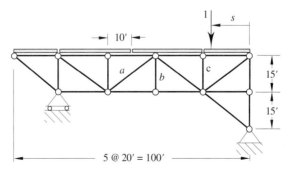

PROBLEM 5.33

5.34 Construct the influence lines for the tension in bars $a, b,$ and c. Follow the general instructions for Problems 5.10–5.37.

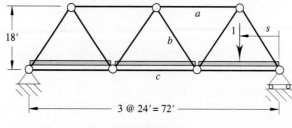

PROBLEM 5.34

5.35 Construct the influence lines for the tension in bars $a, b,$ and c. Follow the general instructions for Problems 5.10–5.37.

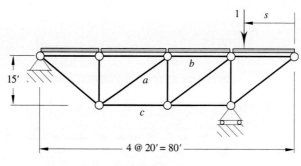

PROBLEM 5.35

5.36 Construct the influence lines for the tension in bars $a, b,$ and c. Follow the general instructions for Problems 5.10–5.37.

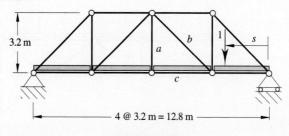

PROBLEM 5.36

5.37 Construct the influence lines for the tension in bars $a, b,$ and c. Follow the general instructions for Problems 5.10–5.37.

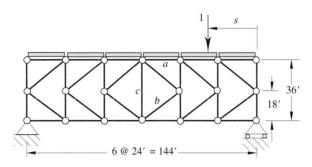

PROBLEM 5.37

CHAPTER 6

Displacement-Deformation Analysis of Structures

In general, displacement of a body results in a change in the geometry of the body, i.e., the body is deformed. The character and severity of the deformation is measured locally by the quantity called the strain. With the aid of a definition of strain, one can derive differential relationships that express the strain in terms of gradients (derivatives) of the displacements. Herein the relationships are called displacement-deformation equations.

There are two fundamental problems in the displacement-deformation analysis of a body:

1. Given the displacements, determine the deformations.
2. Given the deformations, determine the displacements.

Solution of either problem involves application of the calculus. Solution of the first problem requires use of the differential calculus while solution of the second problem involves the integral calculus.

For a plane-framed structure in which the members are long, thin, beam-like elements, the important deformations of a member are determined with sufficient accuracy by the axial and bending deformations of the reference line of the member. Consequently, to handle either problem for a framed structure, one needs the basic relations that express the deformations of the reference line in terms of displacements.

The developments that follow are limited to initially straight members. The ideas involved can be used to derive appropriate relations for plane-curved members. Deformations and rotations are unrestricted in the initial formulation. Subsequently, the relations are simplified for application to the important, practical case of small deformation and small rotation.

The fundamental assumption of beam theory in which shear deformation is negligible is that plane-cross sections remain plane and normal to the deformed reference axis of the beam. By virtue of this assumption the axial and bending deformations anywhere in the beam are defined completely by the local axial and bending deformation of the reference axis. The local axial deformation of the reference axis of the beam is measured by the strain ε and the local bending deformation is measured by the curvature κ. The basic relationships between displacement and deformation are derived from geometry and the definitions of ε and κ.

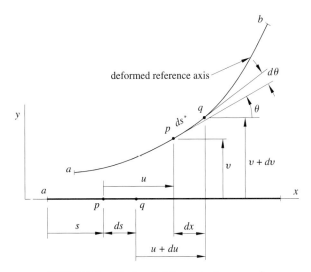

FIGURE 6.1 Displaced, deformed reference axis.

The geometry of the reference axis of a beam before and after deformation is shown in Figure 6.1, where s, which is measured along the undeformed reference axis, defines the location of any point on the reference axis of the undeformed beam; p and q are neighboring points; u and v are components of the displacement of p in the directions of the x and y coordinate axes, respectively; and θ is the angle between the tangent drawn at p and the x-axis. In the undeformed beam, the tangent drawn at p coincides with the x-axis and, therefore, θ is understood to be the rotation of the tangent drawn at p. Finally, ds and ds^* are the undeformed and deformed lengths, respectively, of the line element connecting p and q.

The components of displacement are understood to be functions of s, i.e., $u = u(s)$, $v = v(s)$ (in reality, $u = u(s; \lambda)$, $v = v(s; \lambda)$ where λ is a load parameter such that $u(s; 0) = v(s; 0) = 0$).

$$\varepsilon = \frac{ds^* - ds}{ds}, \quad \kappa = \frac{d\theta}{ds^*} = \frac{d\theta}{(1 + \varepsilon) ds}$$

From the geometry shown in Figure 6.1,

$$dv = ds^* \sin \theta = [(1 + \varepsilon) \, ds] \sin \theta$$
$$dx = ds^* \cos \theta = [(1 + \varepsilon) \, ds] \cos \theta$$
$$dx + u = ds + u + du \Rightarrow du = dx - ds$$

For the structures considered in this work the deformations and rotations are (infinitesimally) small, i.e., $|\varepsilon| \ll 1$, $|\kappa| \ll 1$, and $|\theta| \ll 1$. Accordingly,

$$\varepsilon \simeq \frac{du}{dx}, \quad \frac{ds}{dx} \simeq 1, \quad \theta \simeq \frac{dv}{dx}, \quad \kappa \simeq \frac{d\theta}{dx} \simeq \frac{d^2 v}{dx^2}$$

In practice, equalities are used in place of the approximations. Thus, practical applications are based on the relations

$$\varepsilon = \frac{du}{dx}, \quad \frac{ds}{dx} = 1, \quad \theta = \frac{dv}{dx}, \quad \kappa = \frac{d\theta}{dx} = \frac{d^2v}{dx^2}$$

Of the two basic problems in deformation and displacement analysis, it is almost always the second that is most difficult to solve; it is also the problem most frequently encountered in practice. What it entails is integration of a system of exact differential equations. For the usual straight, beam-like elements in plane-framed structures, for small deformation and small rotation, the integration to be performed is that of the differential equations

$$\frac{d^2v}{dx^2} \equiv v'' = \kappa = f(x)$$

$$\frac{du}{dx} \equiv u' = \varepsilon = g(x)$$

Rules for a particular scheme of integration of $y'' = f(x)$ are summed up in the moment-area theorems which are taken up below. In the application of the theorems, as with any method based on integration of the deformation-displacement relations, a displacement analysis problem (the second fundamental problem) is solved by what is termed a direct geometric approach. In this approach the problem of calculating displacements when deformations are given is confronted in its most basic form; namely, as a problem in geometry. The only truly different alternative to the direct geometric solution of the second problem is the energy method, which has been satisfactorily developed and applied only in the case of small displacements. The fundamental energy method is the method of virtual work. Development of the method amounts to a review of the principle of virtual work and then taking advantage of the principle. The direct geometric approach is treated in this chapter. Energy methods are developed in another chapter.

It follows from the linearity and the structure of the differential equations for u and v that the total displacement at any point in a structure can be obtained as the superposition of the displacement due to axial deformation (with no bending deformation) and the displacement due to bending (with no axial deformation). As a rule, in the cases of beams and frames, the loading and the proportions of the members are such that the contribution of axial deformation to displacement is either negligibly small or zero. Accordingly, in such cases it is customary to calculate only the displacements due to bending deformation.

6.1 TRUSSES

When the deformations and rotations are small, the displacements of the joints of a truss depend only on the axial deformations of the members; in other words, the joints of a truss undergo no displacement as a consequence of any bending deformation of its members. In this case, displacement calculations are based strictly on the geometry of displacements due to small deformations and small rotations and compatibility of displacements at the joints. The general procedure is analogous to the force analysis of a statically determinate truss by writing out all of the joint equations of equilibrium and then solving the equations simultaneously. Special methods, which correspond to a force analysis by the methods of sections and joints, have been developed but their implementation is awkward. The special analytical procedures have a graphical analog. The Williot-Mohr graphical method (see Timoshenko and Young) provides a very rapid and accurate approach to obtaining the displacements of all joints in simple and compound trusses. Application of the method presupposes adequate skills in graphics. It is an unfortunate fact that the de-emphasis of graphics in engineering education makes it unrealistic to attempt a presentation of the Williot-Mohr method here. (It is noted, however, that with adequate graphic software, it is relatively easy to implement the Williot-Mohr procedure.) In view of the foregoing discussion, only the energy method will be employed in the displacement analysis of trusses.

6.2 BEAMS

To illustrate the basic ideas of displacement analysis, assume a cantilever beam is deflected as shown in Figure 6.2 and that the displacements of the reference axis are given by

$$u(x) = 0$$
$$v(x) = \frac{A}{L^3}(3Lx^2 - x^3), \quad |A| \ll 1$$

The deflection function $v(x)$ satisfies the boundary conditions

$$v(0) = 0, \quad v'(0) = \theta(0) = \left\{\frac{dv}{dx}\right\}_{x=0} = 0$$

With $u(x) = 0$ and the deflection $v(x)$ as given, the only non-zero deformation is the bending deformation, which is measured by the curvature κ. Thus, the bending deformation is

$$\kappa = \frac{d^2v}{dx^2} = \frac{6A}{L^3}(L - x)$$

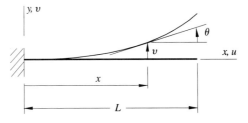

FIGURE 6.2 Cantilever beam, displacements given.

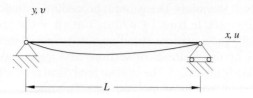

FIGURE 6.3 Simple beam, deformations given.

Now assume the same beam is given the same deformation but is simply supported as shown in Figure 6.3, and it is required to determine the deflection function $v(x)$. The boundary conditions on $v(x)$ are $v(0) = 0, v(L) = 0$. Two integrations of the expression for the curvature give

$$v'(x) = \theta(x) = \frac{6A}{L^3}\left(Lx - \frac{x^2}{2} + C_1\right), \quad v(x) = \frac{6A}{L^3}\left(L\frac{x^2}{2} - \frac{x^3}{6} + C_1 x + C_2\right)$$

The boundary conditions yield $C_2 = 0, C_1 = -L^2/3$. Therefore,

$$v(x) = \frac{6A}{L^3}\left(\frac{Lx^2}{2} - \frac{x^3}{6} - \frac{L^2 x}{3}\right)$$

For future reference, it is noted that $v(L/2) = -3A/8$.

In practice direct integration of differential equations to construct deflection functions $v(x)$ for the members in compound beams or frames is neither desirable nor necessary. In most practical problems it is the deflection at a relatively small number of designated points that is desired and this task can be accomplished without generating the deflection functions $v(x)$ for each member. The moment-area theorems developed by Mohr and also by Green can be employed to generate deflection functions; however, their greatest utility is in applications to problems in which deflections at only a few points are required. The theorems are a consequence of a particular method of integration of the differential equation $v''(x) = \kappa(x)$.

6.3 MOMENT-AREA THEOREMS

The moment-area theorems are applicable to the calculation of bending deflections $v(x)$ measured parallel to and positive in the direction of the y-coordinate axis of the local x-y coordinate axes associated with the member (these are the same local coordinates employed in the construction of shear and moment diagrams for the member). Geometry is employed to obtain the rectangular components of displacement corresponding to coordinate axes at any orientation to the local coordinate axes.

The moment-area theorems come directly from two integrations of the differential equation $v''(x) = \kappa(x)$, where $\kappa(x)$ is given. It is instructive, therefore, to consider ideas involved in calculating deflections by double integration before going on to the

Section 6.3 Moment-Area Theorems

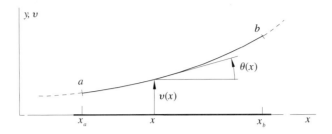

$\theta(x) = v'(x)$ and $v(x)$ continuous: $x_a \leq x \leq x_b$

$\theta(x) = v'(x) = \int \kappa(x)\, dx + C_1$

$v(x) = \int \theta(x)\, dx = \int \left[\int \kappa(x)\, dx \right] dx + C_1 x + C_2$

FIGURE 6.4 Deflection curve.

theorems. Let x_a and x_b be coordinates of the end points of segment ab of a beam that undergoes small bending deformation and small rotation as shown in Figure 6.4. Assume that in all cases where $x_a \leq x \leq x_b$, the deflection $v(x)$ and the rotation $\theta(x) = v'(x)$ are continuous, and $v''(x) = \kappa(x)$ is given.

A first integration of $v''(x) = \kappa(x)$ gives the rotation $\theta(x) = v'(x)$ to within a constant, C_1, of integration and a second integration determines the deflection $v(x)$ to within $C_1 x$ and a second constant, C_2, of integration. From differential equations it is known that to determine the constants of integration, i.e., to complete the determination of $v(x)$, there must be given exactly two appropriate independent conditions that are the equivalent of boundary conditions on v and/or θ at x_a and/or x_b. To be independent the boundary conditions must consist of two of the quantities: $v_a = v(x_a), \theta_a = v'(x_a)$, $v_b = v(x_b), \theta_b = v'(x_b)$ and one of the prescribed quantities must be v_a or v_b.

It has already been noted that the moment-area theorems are a consequence of a particular mode of integration of the given curvature function $\kappa(x)$ in the differential equation $v'' = \kappa(x)$. In application to the displacement analysis of beams and frames the utility of the theorems depends on: (a) the ability to evaluate areas and moments of areas, and (b) the ability to deal with small-angle geometry.

The theorems are derived with the aid of the calculus and a sketch on which the pertinent geometric quantities are shown in grossly exaggerated form to facilitate identification. The essential sketch and the related curvature diagram are shown in Figure 6.5, in which filled or solid circles and letter symbols identify points on the deformed reference axis. v_a is the deflection of point a and v_b is the deflection of point b. Tangents to the deformed reference axis drawn through the filled circles at points a and b are identified as t_a and t_b, respectively.

The angle of inclination of a tangent drawn at any point on the deformed reference axis, measured in radians from the x-axis and positive in the counterclockwise sense, is the rotation of the tangent from its position on the undeformed reference axis. Therefore, θ_a and θ_b are the rotations of the tangents drawn at a and at b, respectively.

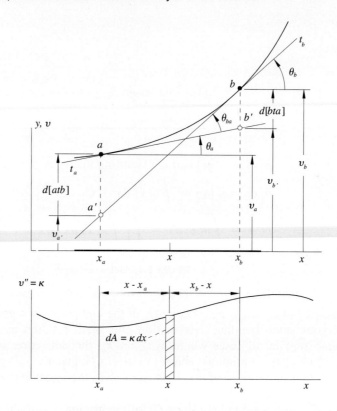

FIGURE 6.5 Deflection and curvature diagrams.

The counterclockwise rotation of the tangent drawn at b with respect to the counterclockwise rotation of the tangent drawn at a is denoted by θ_{ba}. The point of intersection of a tangent with a line drawn parallel to the v-axis is identified by an open circle and an appropriate primed letter. The v coordinate of a point on a tangent is the deflection of that point from the position the point occupied on the x-axis before deformation. a' in Figure 6.5 identifies that point on the tangent drawn at b that has x_a as its x coordinate. The deflection of a' is denoted $v_{a'}$. In the undeformed state the position of point a on the deflection curve and the position of point a' on the tangent drawn at b are coincident. The deflection of b from point b', denoted as $d[bta]$, is called the deflection of b from the tangent drawn at a, and the deflection of a from the tangent drawn at b is denoted by $d[atb]$. Motivation for the nomenclature and the symbolism follows directly from the geometry shown in Figure 6.5. From the figure,

$$v_b - v_{b'} = d[bta], \quad v_a - v_{a'} = d[atb], \quad \theta_{ba} = \theta_b - \theta_a$$

and, because all angles are small,

$$v_b = v_a + (x_b - x_a)\theta_a + d[bta]$$
$$v_a = v_b + (x_a - x_b)\theta_b + d[atb]$$

Formulas for the evaluation of θ_{ba}, $d[bta]$, and $d[atb]$ are derived subject to the conditions that $v(x)$ and $v'(x) = \theta(x)$ are continuous for $x_a \leq x \leq x_b$. It is assumed that $v''(x) = \theta'(x) = \kappa(x)$ is given. Integration of the expression for v'' gives

$$v'(x) = \theta(x) = \theta_a + \int_{x_a}^{x} \kappa(u)\, du$$

from which

$$\theta_{ba} = \theta_b - \theta_a = \int_{x_a}^{x_b} \kappa(x)\, dx$$

With the integral interpreted as an area (in the algebraic sense) the last relation represents the first moment-area theorem.

First Theorem:

> The rotation of the tangent drawn at b with respect to the tangent drawn at a is equal to the area under the curvature diagram between x_a and x_b.

Integration of the expression for $v'(x)$ from x_a to x_b gives

$$v_b = v_a + \theta_a(x_b - x_a) + \int_{x_a}^{x_b}\left[\int_{x_a}^{x} \kappa(u)\, du\right] dx$$

From the expression for v_b in terms of v_a, θ_a and $d[bta]$, it is clear that

$$v_b - v_{b'} = d[bta] = \int_{x_a}^{x_b}\left[\int_{x_a}^{x} \kappa(u)\, du\right] dx$$

The integral can be rewritten to give it a geometric interpretation. Integration by parts gives

$$\int_{x_a}^{x_b}\left[\int_{x_a}^{x} \kappa(u)\, du\right] dx = \left[x \int_{x_a}^{x} \kappa(u)\, du\right]_{x_a}^{x_b} - \int_{x_a}^{x_b} x\kappa(x)\, dx$$

$$= x_b \int_{x_a}^{x_b} \kappa(x)\, dx - \int_{x_a}^{x_b} x\kappa(x)\, dx$$

$$= \int_{x_a}^{x_b} [x_b - x]\kappa(x)\, dx$$

Therefore,

$$d[bta] = \int_{x_a}^{x_b} [x_b - x]\kappa(x)\, dx$$

The geometric interpretation of the integral is summed up in the second moment-area theorem.

Second Theorem:

> The deflection of b from the tangent drawn at a, i.e., $d[bta]$, is equal to the moment of the area under the curvature diagram between x_a and x_b taken about x_b.

A similar analysis shows that $d[atb]$ is equal to the moment of the area under the curvature diagram between x_a and x_b taken about x_a.

The formulations

$$\theta_b = \theta_a + \int_{x_a}^{x_b} \kappa(x)\, dx, \quad v_b = v_a + \theta_a(x_b - x_a) + \int_{x_a}^{x_b} [x_b - x]\kappa(x)\, dx$$

show clearly and explicitly that the net deflection at any point x is the superposition of a displacement due to:

1. a rigid body translation v_a,
2. a rigid body rotation θ_a, and
3. a pure deformation defined by $\kappa(x)$.

6.4 NOTES FOR APPLICATION OF THE MOMENT-AREA THEOREMS

1. Area is treated in an algebraic sense and the moment of a positive element of area about either x_a or x_b is positive.
2. The theorems apply to two points a and b only when $v(x)$ and $v'(x) = \theta(x)$ are continuous in the interval $x_a \leq x \leq x_b$.
3. To calculate the absolute deflection at a point x in the interval $x_a \leq x \leq x_b$ one must know in advance:

   a. the absolute deflection of some point in the interval, and
   b. either the rotation of some point in the interval or the absolute deflection of a second point in the interval.

4. The conditions cited in (3) above are given or can always be produced in the case of a fully restrained structure.
5. The deflection $v(x)$ is always measured parallel to and positive in the direction of the y-axis of the local coordinates associated with the member.

The formulation of the moment-area theorems shows clearly that the source of the deformation is of no consequence; however, in most practical situations the deformations are due to transverse loading and the subsequent bending of a member that is considered to be made of a material that obeys Hooke's law. In this case, as shown in texts on the mechanics of deformable bodies (see, for example, Popov), the curvature $\kappa(x)$ and the bending moment $M(x)$, taken positive according to the sign convention adopted in Chapter 3, are related by the rule $\kappa = M/EI$ in which E is the modulus of elasticity and I is the second moment of the area of the cross section or moment of inertia of the area about the centroidal axis of bending.

6.5 PROCEDURES FOR APPLICATION OF THE M-A THEOREMS TO DEFLECTION ANALYSIS

1. For each member of the beam under consideration identify the end points of each segment in which $v(x)$ and $\theta(x)$ are continuous.
2. Formulate the boundary conditions at supports, and the conditions of compatibility of displacements at the end points of the segments in which the continuity conditions cited in (1) above are satisfied.
3. Use the boundary and compatibility conditions together with the moment-area theorems and geometry to establish a reference tangent for each member of the structure.
4. Use the moment-area theorems to calculate deflections at desired points within each segment in which $v(x)$ and $\theta(x)$ are continuous.

Example 6.1

The cantilever beam shown in Figure 6.6a is bent so that the curvature varies linearly over the length of the beam as shown in Figure 6.6b. Calculate the deflection at points b and c and calculate the rotation of the tangent at point b.

$v(x)$ and $v'(x) = \theta(x)$ are continuous for $0 < x < L$. The boundary conditions are $v(0) = v_a = 0$, $v'(0) = \theta_a = 0$. Therefore, the tangent at a coincides with the x-axis, and the points b' and c' on the tangent drawn from a are located as shown in Figure 6.6c. By the second theorem $d[bta]$ is equal to the moment of the area under the curvature diagram between a and b, taken about b. Thus,

$$v_b - v_{b'} = d[bta] = \left(\frac{2L}{3}\right)\left(\frac{1}{2}\right)\left(\frac{6A}{L^2}\right)L = 2A$$

By the second theorem, the deflection of c from the tangent drawn at a is equal to the moment of the area under the curvature diagram between a and c about c. The area is treated as a composite of two triangles as shown in Figure 6.6b. Therefore,

$$v_c - v_{c'} = d[cta] = \left(\frac{2}{3}\frac{L}{2}\right)\left(\frac{1}{2}\right)\left(\frac{6A}{L^2}\right)\left(\frac{L}{2}\right) + \left(\frac{1}{3}\frac{L}{2}\right)\left(\frac{1}{2}\right)\left(\frac{3A}{L^2}\right)\left(\frac{L}{2}\right) = \frac{5A}{8}$$

By the first theorem, the rotation of the tangent drawn at b with respect to the rotation of the tangent drawn at a is equal to the area under the curvature diagram between a and b. Therefore,

$$\theta_{ba} = \theta_b - \theta_a = \left(\frac{1}{2}\right)\left(\frac{6A}{L^2}\right)L = \frac{3A}{L}$$

Using either good sense (the preferred way) or the definitions of $d[bta]$, $d[cta]$, and θ_{ba}, or the formulas that have been derived, one can plot the points b and c, and the tangent at b as shown in Figure 6.6c. Finally, the deflection curve can be sketched in as shown in Figure 6.6d. Note that

136 Chapter 6 Displacement-Deformation Analysis of Structures

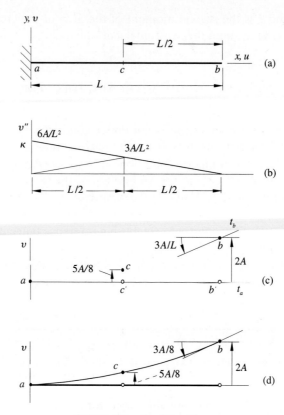

FIGURE 6.6 Cantilever beam.

the curve is sketched after the points b and c have been located and the tangent at b drawn. The cantilever beam has the same curvature function $\kappa(x)$ as was calculated for the same cantilever beam treated in the introduction to beams.

Example 6.2

The simply supported beam shown in Figure 6.7a is bent so that the curvature varies linearly over the length of the beam as shown in Figure 6.7b. Calculate the deflection at c and the rotation of the tangents at a and b.

$v(x)$ and $v'(x) = \theta(x)$ are continuous for $0 < x < L$. The boundary conditions are $v(0) = v_a = v(L) = v_b = 0$. The points a and b are located on the x-axis as shown in Figure 6.7c. The tangent at a (or the tangent at b) can be established as a reference tangent with the aid of the second theorem and the definition of $d[bta]$ (or $d[atb]$). By the second theorem, $d[bta]$ is equal to the moment of the area under the curvature diagram between a and b taken about b. Thus,

$$v_b - v_{b'} = d[bta] = \left(\frac{2L}{3}\right)\left(\frac{1}{2}\right)\left(\frac{6A}{L^2}\right)L = 2A$$

Therefore, the point b' on the tangent drawn from a is at distance $2A$ below the point b as shown in Figure 6.7c. This location of the point b' coincides with the location determined by making use of the fact that $v_b = 0$ and substituting into the above expression for $d[bta]$. The point c' on t_a

Section 6.5 Procedures for Application of the M-A Theorems to Deflection Analysis

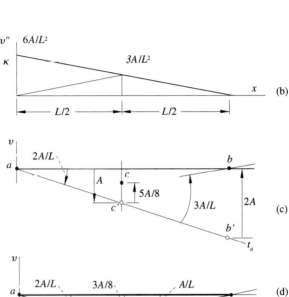

FIGURE 6.7 Deflection of a simple beam.

and the rotation of the tangent drawn at a are easily determined from small-angle geometry or, alternatively, by means of previously derived formulas. In either case it is found that $\theta_a = -2A/L$ and $v_{c'} = -A$. By the second theorem $d[cta]$ is equal to the moment of the area under the curvature diagram between a and c taken about c. Thus,

$$v_c - v_{c'} = d[cta] = \left(\frac{2}{3}\frac{L}{2}\right)\left(\frac{1}{2}\right)\left(\frac{6A}{L^2}\right)\left(\frac{L}{2}\right) + \left(\frac{1}{3}\frac{L}{2}\right)\left(\frac{1}{2}\right)\left(\frac{3A}{L^2}\right)\left(\frac{L}{2}\right) = \frac{5A}{8}$$

and point c on the deflection curve is located above the point c' on t_a as shown in Figure 6.7c. By the first theorem, the rotation of the tangent drawn at b with respect to the tangent drawn at a is equal to the area under the curvature diagram between a and b. Therefore,

$$\theta_{ba} = \theta_b - \theta_a = \left(\frac{1}{2}\right)\left(\frac{6A}{L^2}\right)L = \frac{3A}{L}$$

Again, using good sense (the preferred way) and the definition of θ_{ba} or the formulas that have been derived, the tangent at b is oriented as shown in Figure 6.7c. From the geometry shown in Figure 6.7c, it is easily determined that $v_c = -3A/8$ and $\theta_b = A/L$. Finally, although it is not required and clearly not necessary, the deflection curve is sketched as shown in Figure 6.7d. It is to be noted that, as expected, the deflection of point c coincides with that obtained in the introduction by direct integration of the differential equation of the deflection curve.

Example 6.3

Calculate the deflection at b and at d and the rotation of the tangent at c of the beam shown in Figure 6.8a. Because the bending is induced by loading, an essential preliminary step is to construct the curvature diagram. From beam bending theory the moment-curvature relation is $EI\kappa(x) = M(x)$. The bending moment diagram consistent with the sign convention adopted for bending moment and the coordinates used for deflection calculation is shown in Figure 6.8b, and the curvature diagram is shown in Figure 6.8c. The discontinuity in the curvature diagram is due to the discontinuity in I.

$v(x)$ and $v'(x) = \theta(x)$ are continuous for $0 < x < 24'$. The boundary conditions are $v(0) = v_a = v(18) = v_c = 0$. The points a and c are located on the x-axis as shown in Figure 6.8d. The tangent at c is fixed with the aid of the second theorem. By the second theorem $d[atc]$ is equal to the moment of the area under the curvature diagram between a and c taken about a. Thus,

$$EI_0 d[atc] = \left(\frac{2}{3} \times 6\right)\left(\frac{78 \times 6}{2}\right) + \left(6 + \frac{1}{3} \times 12\right)\left(\frac{39 \times 12}{2}\right)$$
$$- \left(6 + \frac{2}{3} \times 12\right)\left(\frac{27 \times 12}{2}\right) = 1008 \text{ k-ft}^3$$

On the $EI_0 v$ diagram point a' lies 1008 k-ft³ below point a as shown in Figure 6.8d. Geometry is used to locate the points b' and d' on the tangent drawn at c as shown in Figure 6.8d. The deflection of b from the tangent drawn at c is, by the second theorem

$$EI_0 d[btc] = \left(\frac{1}{3} \times 12\right)\left(\frac{39 \times 12}{2}\right) - \left(\frac{2}{3} \times 12\right) \times \left(\frac{27 \times 12}{2}\right) = -360 \text{ k-ft}^3$$

and the deflection of d from the tangent drawn at c is given by

$$EI_0 d[dtc] = -\left(\frac{2}{3} \times 6\right)\left(\frac{27 \times 6}{2}\right) = -324 \text{ k-ft}^3$$

On the basis of the definition of $d[btc]$ point b is determined to lie 360 k-ft³ below b'. Similarly, point d lies 324 k-ft³ below point d'. From small-angle geometry the rotation of the tangent at c is given by $EI_0 \theta_c = 56$ k-ft². The required results are shown in Figure 6.8e in which the rough deflection curve was sketched in after the points were plotted.

Example 6.4

Calculate the deflection at point c of the structure shown in Figure 6.9a, on page 140. Before proceeding to the solution process, it is pertinent to note that the structure and, therefore, its analysis can be approached from two slightly different points of view. First, the structure may be viewed as a single beam in which special conditions exist as a consequence of the pin at b. Second, the structure may be viewed as consisting of two distinct beams that are joined together by the pin at b.

When the first view is taken, the structure is understood to have a single deflection curve with a discontinuity in the tangent at b as a consequence of the pin. The moment-area theorems may be applied to the deflection analysis; however, the analysis must be arranged so that the point at which the pin is located is not between the end points of any segment to which the theo-

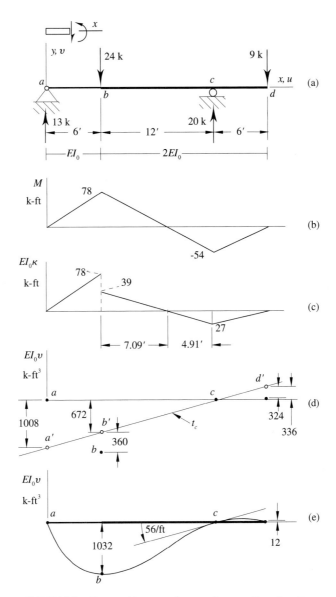

FIGURE 6.8 Beam with an overhang and non-uniform bending stiffness.

rems are applied. The reason is that the theorems were derived subject to the condition that $v(x)$ and $v'(x)$ are continuous in the interval and, in general, these conditions can not be satisfied if the interval includes a pin.

When the second view is taken, the deflection curve of the structure is understood to be composed of two distinct curves joined together at the location of the pin. Because each curve satisfies the conditions used in the derivation of the moment-area theorems, the theorems can be applied between any two points associated with one of the curves. The condition of compatibility

140 Chapter 6 Displacement-Deformation Analysis of Structures

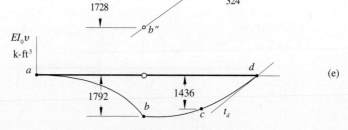

FIGURE 6.9 Interior pin.

of displacement at the pin which is employed to join the curves gives a composite curve that has a discontinuity in slope at the location of the pin. The example problem is solved from both points of view.

First View The bending moment and curvature diagrams are shown in Figure 6.9b and Figure 6.9c, respectively. Observe that there is a discontinuity in the curvature diagram and the discontinuity is due to the discontinuity in I.

$v(x)$ and $v'(x)$ are continuous for $0 < x < 12'$ and for $12 < x < 24'$. The deflection v is continuous at $x = 12'$, i.e., $v(12^-) = v(12^+)$. Because of the pin the slope at $x = 12'$ may be (and

Section 6.5   Procedures for Application of the M-A Theorems to Deflection Analysis   141

generally is) discontinuous. At a, since the support is fixed, $v_a = 0$ and $\theta_a = 0$. Because of the roller at d, $v_d = 0$. By inspection the points a and d lie on the x-axis and the tangent at a is horizontal as shown in Figure 6.9d.

As was discussed above, the moment-area theorems cannot be applied between two points, one of which is in segment ab, and one of which is in segment bcd. This eliminates the possibility of using the second theorem for a direct calculation of the deflection at c on the basis of the boundary conditions for ab. Because only one boundary condition in the form of a prescribed quantity is given for bcd, but two known conditions must be available to fix the displacements in the segment, there is no direct way of obtaining the deflection at c on the basis of the theorems and what is known about the deflections of bcd. The problem is resolved by observing that sufficient information is given about ab to calculate the deflection of b. From the condition of continuity of deflection at b, i.e., $v(12^-) = v(12^+)$, the calculated value of $v_b = v(12^-)$ becomes the known value of $v(12^+)$. With the deflection at a and the deflection of the b-end of bcd now known, the moment-area theorems and geometry can be used to calculate the deflection. By the second theorem,

$$EI_0 d[bta] = EI_0 v_b - EI_0 v_{b'} = -(4+4)(24 \times 8) - \left(\frac{2}{3} \times 4\right)\left(\frac{48 \times 4}{2}\right) = -1792 \text{ k-ft}^3$$

Since $v_{b'} = 0$, point b is located as shown in Figure 6.9d. With the deflection of b and of d known, the second theorem can be applied to bcd to establish the tangent at d. The moment of the area under the curvature diagram can be calculated from formulas, if available, or directly with the aid of definitions and the calculus. The direct calculation is employed in this example. Let $v_{b''}$ be the deflection of the point b'' on the tangent drawn at d. Then,

$$EI_0 d[btd] = EI_0 v_b - EI_0 v_b'' = \int_0^{12} x_1(12x_1 - x_1^2)\, dx_1 = 1728 \text{ k-ft}^3$$

With this result and geometry, points b'' and c' are established as shown in Figure 6.9d. By the second theorem $d[ctd]$ is given by

$$EI_0 d[ctd] = \int_6^{12} (x_1 - 6)(12x_1 - x_1^2)\, dx_1 = 324 \text{ k-ft}^3$$

Thus, the point c lies 324 k-ft³ above the point c' as shown in Figure 6.9d. Deflections are shown in Figure 6.9e along with a rough sketch of the deflection curve.

Second View   The members ab and bcd are treated as distinct beams that have been joined to form the structure. Each member is assigned its own local coordinate system as indicated in Figure 6.10. Each of the two members satisfy the continuity conditions for application of the moment-area theorems. The boundary conditions for member ab are $v_a = 0$ and $\theta_a = 0$ and, therefore, the two essential conditions for fixing the deflections in ab absolutely are available. Member bcd has only one prescribed boundary condition; namely, $v_d = 0$. Accordingly, there is no way to proceed directly with the calculation of the deflection of point c. Either the displacement or the rotation of some point in span bcd must be established before the deflection of point c can be calculated. The compatibility of displacement at the pin provides the desired information. The compatibility condition is that the deflection of b, considered as the b-end of ab and the deflection of b, considered as the b-end of bcd, must be equal. With the aid of the second theorem and the boundary conditions for ab, the deflection of b, viewed as the b-end of ab is determined to be $v_b = -1792$ k-ft³/EI_0 as indicated in Figure 6.10d. Therefore, as indicated in

142 Chapter 6 Displacement-Deformation Analysis of Structures

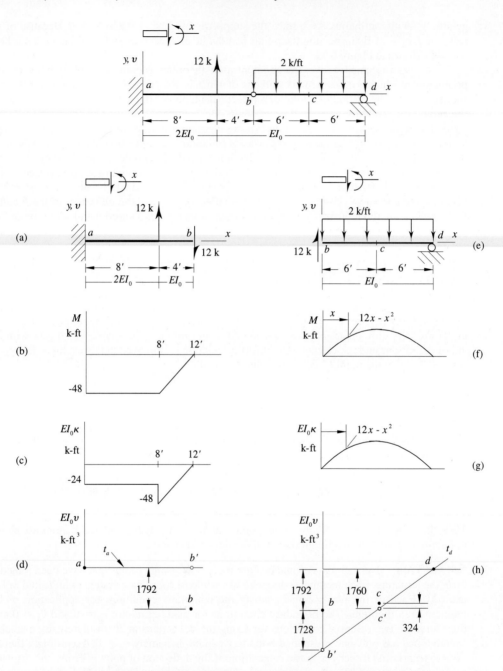

FIGURE 6.10 Beam with an interior pin viewed as a two-member structure.

Figure 6.10h, the deflection of the b-end of bcd is now known to be $v_b = -1792$ k-ft$^3/EI_0$. Having established the second boundary condition for bcd, one application of the second theorem is used to locate point b' on the tangent drawn at c. Geometry and a second application of the second theorem are used to establish the position of point c as illustrated in Figure 6.10h. Observe that there is no need to sketch a deflection curve to obtain the desired results and for this reason no curve has been drawn.

6.6 FRAMES

The moment-area theorems are directly applicable only to the calculation of the transverse component of deflection $v(x)$ and rotations $\theta(x)$ due to bending, measured in the local x-y coordinate axes associated with the member under consideration. In the case of a frame, the members may have any orientation in space and undergo axial as well as transverse displacements. In practice one is generally required to calculate components of displacement referenced to a system of master or global coordinates. Clearly, the calculation of frame displacements due to bending offers more of a challenge than the calculation of beam displacements. The challenge, however, is met by taking advantage of the fact that displacements are vector quantities, by taking advantage of the principle of superposition, and with the aid of small-angle geometry.

Basic ideas regarding local and global coordinates are illustrated in Figure 6.11. Local coordinates and rectangular displacement components referred to the local coordinates are identified by lower-case letters. Upper-case letters identify global coordinates and rectangular components of displacement referred to the global coordinates. In Figure 6.11, p identifies a point in a body in the undisplaced state, p' identifies the same point in the deformed, displaced state, and **s** is the displacement of the point. u and v are the rectangular components of the displacement referred to the local coordinates and U and V are the rectangular components of the displacement referred to the global coordinates. Rotation in the global coordinates is measured positive in the same sense as in the local coordinates. Typically, there is a need to express U and V in terms of u and v and/or vice versa.

The desired relations follow immediately from the vector diagram shown in Figure 6.11.

$$U = u\cos\alpha - v\sin\alpha, \quad V = u\sin\alpha + v\cos\alpha$$
$$u = U\cos\alpha + V\sin\alpha, \quad v = -U\sin\alpha + V\cos\alpha$$

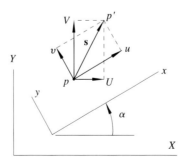

FIGURE 6.11 Local and global coordinates.

Frequently, to complete the solution of a problem it is necessary to subject a deformed body to a rigid-body rotation about a point. A basic limitation imposed on all structures considered in this text is that *all* displacements and rotations are infinitesimally small. Under these conditions the displacements due to the rigid rotation may be calculated using small-angle geometry and the geometry of the undisplaced body. Displacements due to the rotation may then be combined (vectorially) with preexisting displacements to obtain total displacements. The procedure is readily justified with the aid of the displacement diagram shown in Figure 6.12 in which displacements and rotations have been grossly exaggerated for clarity.

In Figure 6.12, p is the position of a point in the undisplaced body. Vector **pp′**, the magnitude of which is infinitesimally small compared to r, is the displacement of p due to a deformation of the body. Angle ψ is the small rotation of the radius Op due to the deformation. Angle θ defines the small rigid-body rotation about pivot point O. Vector **p′p″** is the additional displacement of p due to the rotation, and **pp″** is the total displacement of p. Vector **pp*** is the displacement that p, in the undeformed body, would have experienced due solely to the rotation θ. Because the difference between ρ and r is small and because ψ is also the angle between **p′p″** and **pp***, **p′p″** $\approx$ **pp***. The equality **p′p″** = **pp*** is used in practice.

6.7 PROCEDURES

A distinct member in a frame is understood to be identified by the condition that in the local coordinates, $v(x)$, $v'(x) = \theta(x)$, and $u(x)$ are continuous over the entire length of the member. Because only bending is considered, i.e., the deformation is

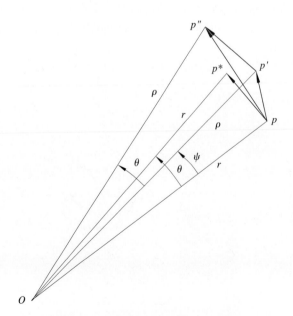

FIGURE 6.12 Displacements due to rotation.

inextensional, $u(x) =$ a constant for any one member. The procedures for the displacement analysis of a frame with the aid of the moment-area theorems are essentially those which have been stated previously and which are repeated below. The bending deformations are assumed to be known.

1. For each member of the beam or frame under consideration identify the end points of each segment in which $v(x)$ and $\theta(x)$ are continuous.
2. Formulate the boundary conditions at supports, and the conditions of compatibility of displacements at the end points of the segments in which the continuity conditions cited in (1) above are satisfied. In the case of a frame, axial as well as transverse components of displacement must be considered.
3. Use the boundary and compatibility conditions together with the moment-area theorems and geometry to fix a tangent for each member of the structure.
4. Use the moment-area theorems to calculate deflections at desired points within each segment in which $v(x)$ and $\theta(x)$ are continuous.

As will be seen from examples, the displacement analysis of a frame is more demanding than in the case of a beam, primarily because the local coordinates for the individual members are not parallel to the global coordinates.

Example 6.5

Determine the horizontal and vertical components of displacement and the rotation of the tangent at c of the rigid frame shown in Figure 6.13a. The frame is composed of the two members: ab and bc. The local coordinates are shown in Figure 6.13a and the curvature diagrams are shown in Figure 6.13b and Figure 6.13c, respectively. The global coordinates are chosen as shown in Figure 6.13a. The location of the global coordinates is inconsequential in the current application because they are used only for the purpose of coordinating directions. A component of displacement at the end of a member is denoted by an appropriate letter with a subscript that identifies the end of the member and, in brackets, a pair of letters that identify the member. Boundary and compatibility conditions are formulated in terms of global coordinates, and the relationships between the local and global coordinates are used to express the conditions in terms of quantities associated with the local coordinates. Rotations in the local and global coordinates are measured positive in the sense determined by application of the right-hand rule about a z-axis in the direction of $x \otimes y$ (which is, of course, the positive sense for rotation as set out in the development of the moment-area theorems).

The boundary and compatibility conditions in terms of global quantities are:

$$U_{a[ab]} = U_a = 0, \quad V_{a[ab]} = V_a = 0, \quad \theta_{a[ab]} = \theta_a = 0$$

$$U_{b[ab]} = U_{b[bc]} = U_b, \quad V_{b[ab]} = V_{b[bc]} = V_b, \quad \theta_{b[ab]} = \theta_{b[bc]} = \theta_b$$

in which U_b, V_b, and θ_b, which denote the common values for the displacements at the ends of the members, are understood to be the components of displacement of the joint at b. From the relationships between the global coordinates and the local coordinates for ab the boundary conditions at a become

$$U_a = -v_{a[ab]} = 0, \quad V_a = u_{a[ab]} = 0, \quad \theta_a = \theta_{a[ab]} = 0$$

146 Chapter 6 Displacement-Deformation Analysis of Structures

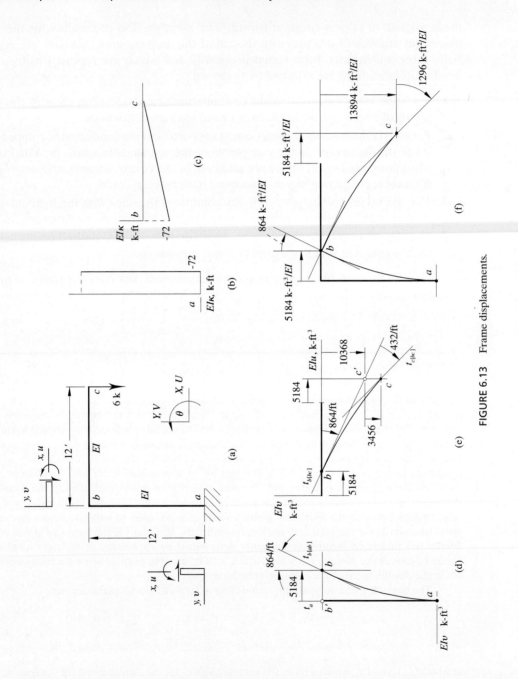

FIGURE 6.13 Frame displacements.

From the relationships between the global coordinates, the local coordinates for member *ab* and the local coordinates for member *bc*, the compatibility conditions at joint *b* become

$$-v_{b[ab]} = u_{b[bc]} = U_b, \quad u_{b[ab]} = v_{b[bc]} = V_b, \quad \theta_{b[ab]} = \theta_{b[bc]} = \theta_b$$

To satisfy the condition of inextensional deformation, the axial component of the displacements of the ends of a member must be equal, i.e.,

$$u_{b[ab]} = u_{a[ab]}, \quad u_{c[bc]} = u_{b[bc]}$$

Axial as well as transverse displacements are involved. Accordingly, to calculate absolute displacements a boundary condition on axial displacement as well as two boundary conditions on transverse displacement must be known. None of the required quantities is prescribed for member bc. The required conditions are known for member ab and with the aid of the moment-area theorems, the condition of inextensional deformation, and the compatibility conditions for joint b, the required conditions can be established for bc. Because the displacement and rotation of joint a are zero, the tangent at a must be as shown in Figure 6.13d. With the aid of the second theorem it is determined that $EIv_{b[ab]} = -5148$ k-ft^3. The first theorem gives $EI\theta_{b[ab]} = -864$ k-ft^2. Since the deformation is inextensional, $u_{b[ab]} = 0$. The compatibility equations then give $EIu_{b[bc]} = 5148$ k-ft^3, $EIv_{b[bc]} = 0$, and $EI\theta_{b[bc]} = -864$ k-ft^2. In this way, b and $t_{b[bc]}$ are established as shown in Figure 6.13e. Geometry, the moment-area theorems, and the condition of inextensional deformation then yield the remaining results displayed in Figure 6.13e. The results of the calculations are displayed in Figure 6.13f. The values of the required quantities are:

$$EIU_c = 5184 \text{ k-ft}^3, \quad EIV_c = -13824 \text{ k-ft}^3, \quad EI\theta_c = -1296 \text{ k-ft}^2$$

Example 6.6

The support conditions and loading for the frame of the previous example are changed to be as shown in Figure 6.14a. Determine the displacement at a and the rotation of the tangent at a and at c using global coordinates as shown in Figure 6.14a.

The local coordinates and the curvature diagrams for each of the two members are shown in Figure 6.14b and Figure 6.14c, respectively. It is to be noted that the curvature diagrams for the members of the frame in Figure 6.13 and the frame in this example are the same. Because the conditions of compatibility at joint b are also the same, the deformed geometry of the two frames coincide. It follows that the displacements of the two frames differ only by a rigid-body displacement. Clearly, solution of the problem can be obtained in two ways: (1) by direct application of the basic ideas involved in the moment-area method as in the previous example; or (2) by superposition of a rigid-body displacement on the displacements of the frame when point a was built-in and point c was free. Both approaches are employed herein. The boundary conditions and the conditions of compatibility at joint b are independent of the techniques of solution and, in terms of global quantities, are:

$$V_{a[ab]} = V_a = 0, \quad U_{c[bc]} = U_c = 0, \quad V_{c[bc]} = V_c = 0$$

$$U_{b[ab]} = U_{b[bc]} = U_b, \quad V_{b[ab]} = V_{b[bc]} = V_b, \quad \theta_{b[ab]} = \theta_{b[bc]} = \theta_b$$

Direct Solution From the relationships between the global coordinates and the local coordinates for the members, the boundary and compatibility conditions become

$$V_a = u_{a[ab]} = 0, \quad U_c = u_{c[bc]} = 0, \quad V_c = v_{c[bc]} = 0$$

$$-v_{b[ab]} = U_b = u_{b[bc]}, \quad u_{b[ab]} = V_b = v_{b[bc]}, \quad \theta_{b[ab]} = \theta_b = \theta_{b[bc]}$$

The condition of inextensional deformation for each of the members is

$$u_{b[ab]} = u_{a[ab]}, \quad u_{c[bc]} = u_{b[bc]}$$

148 Chapter 6 Displacement-Deformation Analysis of Structures

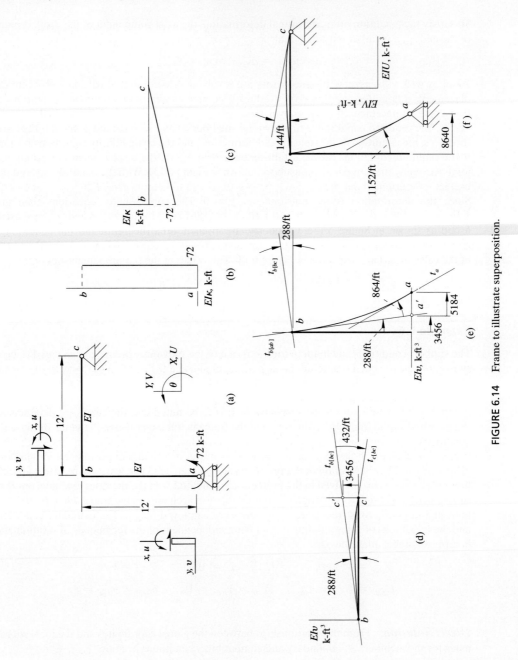

FIGURE 6.14 Frame to illustrate superposition.

The boundary conditions together with the conditions of inextensional deformation require that

$$u_{b[bc]} = 0, \qquad v_{b[bc]} = 0$$

Consequently, neither joint c nor joint b displaces. With the aid of the second theorem the tangent at the b end of bc, $t_{b[bc]}$, is established to be as shown in Figure 6.14d. The tangent at the b-end of ab, $t_{b[ab]}$ is then established with the aid of the condition of compatibility of rotation at joint b.

Geometry and the second theorem are used to locate the displaced position of point a as shown in Figure 6.14e. A summary of the required results is displayed in Figure 6.14e. The values of the required quantities are

$$EIU_a = 8640 \text{ k-ft}^3, \quad EI\theta_a = 1152 \text{ k-ft}^2, \quad EI\theta_c = -144 \text{ k-ft}^2$$

Superposition The displacements of the frame when the a-end of member ab is built in and the c-end of member bc is free, displayed in Figure 6.13, are reproduced in Figure 6.15a. In the analysis it is important to keep in mind that the figures show the displacements to a grossly exaggerated scale. Analytically, the displacements are treated as infinitesimally small. Denote with a prime the displacements shown in Figure 6.15a. Then,

$$EIU'_a = 0, \quad EIV'_a = 0, \quad EI\theta'_a = 0$$
$$EIU'_b = 5184 \text{ k-ft}^3, \quad EIV'_b = 0, \quad EI\theta'_b = -864 \text{ k-ft}^2$$
$$EIU'_c = 5184 \text{ k-ft}^3, \quad EIV'_c = -13824 \text{ k-ft}^3, \quad EI\theta'_c = -1296 \text{ k-ft}^2$$

Now select p as any convenient reference point that moves rigidly with the structure as shown in Figure 6.15b. Let U_p and V_p define the translation of the structure and let θ_p define the rotation about p. Let q by any other point in the structure. Then, by the discussion of rigid-body displacements in Chapter 5 and the principle of superposition,

$$U_q = U_p - (Y_q - Y_p)\theta_p + U'_q$$
$$V_q = V_p + (X_q - X_p)\theta_p + V'_q$$
$$\theta_q = \theta_p + \theta'_q$$

where θ_q is understood to represent the rotation of a line of the nature of a tangent drawn through q. With a taken as the point p and with q taken as $a, b,$ and c in succession, the foregoing relations give

$$EIU_a = EIU_p, \qquad\qquad EIU_b = EIU_p - 12EI\theta_p + 5184 \text{ k-ft}^3$$
$$EIV_a = EIV_p, \qquad\qquad EIV_b = EIV_p$$
$$EI\theta_a = EI\theta_p, \qquad\qquad EI\theta_b = EI\theta_p - 864 \text{ k-ft}^2$$
$$EIU_c = EIU_p - 12EI\theta_p + 5184 \text{ k-ft}^3$$
$$EIV_c = EIV_p + 12EI\theta_p - 13824 \text{ k-ft}^3$$
$$EI\theta_c = EI\theta_p - 1296 \text{ k-ft}^2$$

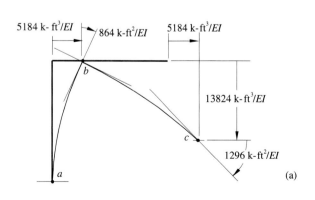
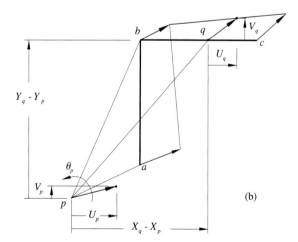

FIGURE 6.15 (a) Known displacements, (b) rigid body displacements.

Substitution of the expressions for V_a, U_c, and V_c into the boundary conditions $V_a = 0$, $U_c = 0$, $V_c = 0$ provides three equations for the determination of U_p, V_p, and θ_p. The results are: $EIU_p = 8640$ k-ft³, $EIV_p = 0$, $EI\theta_p = 1152$ k-ft². Back substitution into the equations gives

$$EIU_a = 8640 \text{ k-ft}^3, \quad EI\theta_a = 1152 \text{ k-ft}^2, \quad EI\theta_c = -144 \text{ k-ft}^2$$

which coincides with the results obtained in the direct solution.

The selection of a as the reference point was arbitrary; any other point could just as well have been taken and the results would be the same. For example, with b. as the reference point, the superposition equations become

$$EIU_a = EIU_p + 12EI\theta_p,$$
$$EIV_a = EIV_p,$$
$$EI\theta_a = EI\theta_p,$$
$$EIU_c = EIU_p + 5184 \text{ k-ft}^3$$
$$EIV_c = EIV_p + 12EI\theta_p - 13824 \text{ k-ft}^3$$
$$EI\theta_c = EI\theta_p - 1296 \text{ k-ft}^2$$

$$EIU_b = EIU_p + 5184 \text{ k-ft}^3$$
$$EIV_b = EIV_p$$
$$EI\theta_b = EI\theta_p - 864 \text{ k-ft}^2$$

With the aid of the boundary conditions it is found that: $EIU_p = -5184$ k-ft³, $EIV_p = 0$, $EI\theta_p = 1152$ k-ft². Substitution into the superposition equations yields the results as obtained previously.

Example 6.7

Calculate the horizontal and vertical components of displacement of joint c and the vertical component of displacement of point b of the rigid frame shown in Figure 6.16.

The relationships between displacements in the local coordinates and in the global coordinates for members ac and cd are

$$ac: U = 0.6u - 0.8v, \quad V = 0.8u + 0.6v, \quad \text{and } cd: U = u, \quad V = v$$

respectively. The boundary conditions in terms of displacements referred to local coordinates are

$$u_{a[ac]} = 0, \quad v_{a[ac]} = 0, \quad v_{d[cd]} = 0$$

and the conditions of compatibility of displacements at joint c are

$$0.6u_{c[ac]} - 0.8v_{c[ac]} = U_c = u_{c[cd]}$$
$$0.8u_{c[ac]} + 0.6v_{c[ac]} = V_c = v_{c[cd]}$$
$$\theta_{c[ac]} = \theta_c = \theta_{c[cd]}$$

The boundary conditions are such that a direct solution (in the sense of producing a sequence of numerical results that represent answers) is not possible. Accordingly, superposition of a rigid-body displacement on any known displaced state that is consistent with the deformations and compatibility conditions is employed to solve the problem. A displaced state that satisfies the

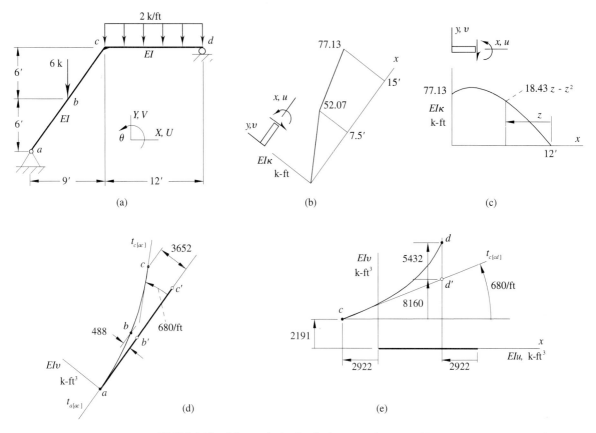

FIGURE 6.16 Direct solution for displacements is not possible.

requirements and is easy to obtain is that which satisfies the displacement boundary conditions at a and, in addition, is such that the rotation of the tangent at the a-end of member ac is zero. Displacements that satisfy the foregoing conditions and the condition of inextensional deformation are displayed in Figures. 6.16d and 6.16e. Transverse components of displacements and rotations were generated by direct application of the moment-area theorems. The figures reflect the fact that for inextensional deformation all points on a given member undergo exactly the same axial component of displacement. A summary of the essential results is given in Table 6.1 in which a prime is

TABLE 6.1 Base displacements

Point	EIU' k-ft^3	EIV' k-ft^3	$EI\theta'$ k-ft^3
a	0	0	0
b	–	293	–
c	−2922	2191	–
d	−2922	15783	–

used to identify the displacements for the assumed boundary conditions at a. With point a as the reference point the superposition equations reduce to

$$EIU_a = EIU_p, \qquad EIV_b = EIV_p + 4.5EI\theta_p + 293 \text{ k-ft}^3$$

$$EIV_a = EIV_p$$

$$EI\theta_a = EI\theta_p$$

$$EIU_c = EIU_p - 12EI\theta_p - 2922 \text{ k-ft}^3, \qquad EIV_d = EIV_p + 21EI\theta_p + 15783 \text{ k-ft}^3$$

$$EIV_c = EIV_p + 9EI\theta_p + 2191 \text{ k-ft}^3$$

With the aid of the boundary conditions expressed in the form $U_a = 0, V_a = 0, V_d = 0$, it is found that $U_p = 0, V_p = 0, EI\theta_p = -752$ k-ft^2. Substitution into the superposition equations yields

$$EIV_b = -3090 \text{ k-ft}^3, \quad EIU_c = 6100 \text{ k-ft}^3, \quad EIV_c = -4570 \text{ k-ft}^3$$

Example 6.8

Calculate the horizontal and vertical components of displacement at points b, c, d, and f for the frame shown in Figure 6.17.

The purpose of the example is to focus attention on application of superposition of rigid-body displacements on known displacements, which are consistent with the known deformations but do not satisfy all boundary conditions and all conditions of compatibility, to obtain displacements that do satisfy all boundary conditions and all conditions of compatibility. Accordingly, details of the application of the moment-area theorems and geometry to obtain basic results used below are omitted; the details are left as exercises for the reader.

The boundary conditions at supports a and e are

$$U_a = 0, \quad V_a = 0, \quad U_e = 0, \quad V_e = 0$$

The compatibility conditions at joints b, d, and c are

$$U_{b[ab]} = U_{b[bc]}, \qquad V_{b[ab]} = V_{b[bc]}, \qquad \theta_{b[ab]} = \theta_{b[bc]}$$

$$U_{d[cd]} = U_{d[de]}, \qquad V_{d[cd]} = V_{d[de]}, \qquad \theta_{d[cd]} = \theta_{d[de]}$$

$$U_{c[bc]} = U_{c[cd]}, \qquad V_{c[bc]} = V_{c[cd]}$$

The boundary conditions at a together with the compatibility conditions at b and the condition of inextensional deformation are sufficient to define the displacements of all points in members ab and bc to within an arbitrary rigid rotation of abc about point a. Similarly, the boundary conditions at e together with the compatibility conditions at d and the condition of inextensional deformation are sufficient to define the displacements of all points in members cd and de to within an arbitrary rigid rotation of cde about point e. The compatibility conditions of displacement at c are satisfied only when the rigid rotations of abc about a and of cde about e take on the particular values that solve the displacement problem. Herein, these rotations are obtained by superposition.

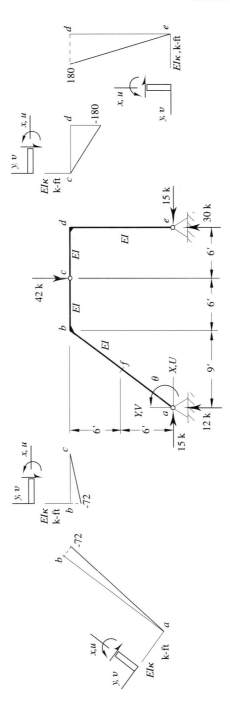

FIGURE 6.17 A more involved displacement analysis problem.

154 Chapter 6 Displacement-Deformation Analysis of Structures

TABLE 6.2 Base displacements for *abc*

Point	EIU' k-ft^3	EIV' k-ft^3	$EI\theta$ k-ft^2
a	0	0	0
f	270	−202.5	−
b	2160	−1620	−540
c	2160	−5724	−

A prime is used to identify base displacements of *abc* that satisfy the boundary conditions $U_a = 0$, $V_a = 0$, $\theta_a = 0$. These displacements, which are readily obtained with the aid of the moment-area theorems, are displayed in Table 6.2. Displayed in Table 6.3 are the corresponding base displacements of *cde* that satisfy the conditions $U_e = 0, V_e = 0, \theta_e = 0$.

Key base displacements for the two parts of the frame are shown in Figure 6.18 in which the point *c* considered as part of *abc* is identified as $c(abc)$, and the point *c* considered as part of *cde* is identified as $c(cde)$.

The superposition equations for *abc*, in general form, are

$$U_q = U_p - (Y_q - Y_p)\theta_p + U'_q$$
$$V_q = V_p + (X_q - X_p)\theta_p + V'_q$$
$$\theta_q = \theta_p + \theta'_q$$

in which the subscript *p* identifies the reference point for *abc* and the subscript *q* identifies any other point in *abc*. U_p, V_p, and θ_p are the components of the rigid-body displacement. With *a* as the reference point *p*, on the basis of the dimensions of the structure and the base data in Table 6.2, the specific equations for the displacements of *a*, *f*, *b*, and *c* in *abc* become

$$EIU_a = EIU_p,$$
$$EIV_a = EIV_p,$$
$$EI\theta_a = EI\theta_p$$
$$EIU_f = EIU_p - 6EI\theta_p + 270 \text{ k-ft}^3$$
$$EIV_f = EIV_p + 4.5EI\theta_p - 202.5 \text{ k-ft}^3$$
$$EIU_b = EIU_p - 12EI\theta_p + 2160 \text{ k-ft}^3, \quad EIU_c = EIU_p - 12EI\theta_p + 2160 \text{ k-ft}^3$$
$$EIV_b = EIV_p + 9EI\theta_p - 1620 \text{ k-ft}^3, \quad EIV_c = EIV_p + 15\,EI\theta_p - 5724 \text{ k-ft}^3$$

Except for point *a*, the equations for rotation have been omitted because rotations are not required. The superposition equations for *cde* are, in general form,

$$U_q = U_{p*} - (Y_q - Y_{p*})\theta_{p*} + U'_q$$
$$V_q = V_{p*} + (X_q - X_{p*})\theta_{p*} + V'_q$$
$$\theta_q = \theta_{p*} + \theta'_q$$

TABLE 6.3 Base displacements for *cde*

Point	EIU' k-ft^3	EIV' k-ft^3	$EI\theta'$ k-ft^2
c	−4320	−8640	−
d	−4320	0	1080
e	0	0	0

in which the subscript $p*$ now identifies the reference point for cde, the subscript q identifies any other point in cde, and U_{p*}, V_{p*}, and θ_{p*} are the components of the rigid-body displacement of cde. With e as the reference point $p*$, on the basis of the dimensions of the structure and the base data in Table 6.3, the specific equations for the displacements of c, d, and e in cde become

$$EIU_c = EIU_{p*} - 12EI\theta_{p*} - 4320 \text{ k-ft}^3, \qquad EIU_d = EIU_{p*} - 12EI\theta_{p*} - 4320 \text{ k-ft}^3$$

$$EIV_c = EIV_{p*} - 6EI\theta_{p*} - 8640 \text{ k-ft}^3, \qquad EIV_d = EIV_{p*}$$

$$EIU_e = EIU_{p*}$$

$$EIV_e = EIV_{p*}$$

$$EI\theta_e = EI\theta_{p*}$$

From the boundary conditions at a and at e, it follows that $U_p = V_p = U_{p*} = V_{p*} = 0$. Compatibility of the U and V components of displacement of abc and of cde at c requires that

$$EIU_c = -12EI\theta_p + 2160 \text{ k-ft}^3 = -12EI\theta_{p*} - 4320 \text{ k-ft}^3$$

$$EIV_c = 15EI\theta_p - 5724 \text{ k-ft}^3 = -6EI\theta_{p*} - 8640 \text{ k-ft}^3$$

The solution is

$$EI\theta_p = \frac{108}{7} \text{ k-ft}^2, \qquad EI\theta_{p*} = -\frac{3672}{7} \text{ k-ft}^2$$

Back substitution into the superposition equations gives

$$EIU_f = \frac{1242}{7} \text{ k-ft}^3, \qquad EIV_f = -\frac{1863}{14} \text{ k-ft}^3$$

$$EIU_b = \frac{13824}{7} \text{ k-ft}^3, \qquad EIV_b = -\frac{10368}{7} \text{ k-ft}^3$$

$$EIU_c = \frac{13824}{7} \text{ k-ft}^3, \qquad EIV_c = -\frac{38448}{7} \text{ k-ft}^3$$

$$EIU_d = \frac{13824}{7} \text{ k-ft}^3, \qquad EIV_c = 0$$

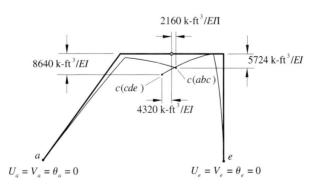

FIGURE 6.18 Base displacements of point c.

156 Chapter 6 Displacement-Deformation Analysis of Structures

6.8 COMBINATION APPROACH

Often deflection problems such as in Example 6.8 are most efficiently solved by employing a combination of the virtual-work method of displacement calculation, which is taken up in Chapter 7, and the moment-area method. As shown in Chapter 7, with the aid of the principle of virtual work it can be established that $7EI\theta_a = 108$ k-ft^2. With knowledge of θ_a, the boundary conditions, the compatibility conditions, the condition of inextensional deformation, the moment-area theorems, and geometry, calculation of the desired results is, in fact, a relatively easy, uncomplicated process.

The results obtained by application of the idea described in the preceding paragraph to the frame in Example 6.4 (Figure 6.17) are displayed in Figure 6.19. With $u_a = v_a = 0$ and $7EI\theta_a = 108$ k-ft^2, two applications of the moment-area theorems to member ab, together with the condition of inextensional deformation and geometry, yield

$$EIU_b = \frac{13824}{7}\text{ k-ft}^3, \quad EIV_b = -\frac{10368}{7}\text{ k-ft}^3, \quad EI\theta_b = -\frac{3672}{7}\text{ k-ft}^2$$

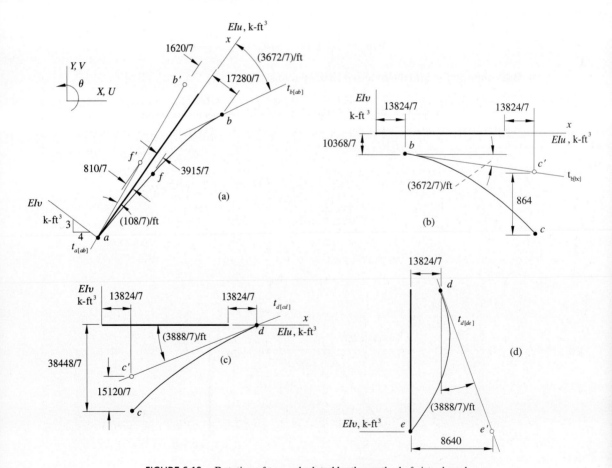

FIGURE 6.19 Rotation of $t_{a[ab]}$ calculated by the method of virtual work.

The displacement of the pin at c is obtained from the conditions of compatibility at joint b together with the condition of inextensional deformation, geometry, and the moment-area theorems.

From the information shown in Figure 6.19b, the displacements at c are

$$EIU_c = \frac{13824}{7} \text{ k-ft}^3, \quad EIV_c = -\frac{38448}{7} \text{ k-ft}^3$$

From compatibility of displacement at pin c, the condition of inextensional deformation of member cd, the boundary conditions at e, and the condition of inextensional deformation of member de, it is determined that $U_d = U_c$ and $V_d = 0$. With these results, the rotation of the tangent at the d-end of member cd is found from geometry and one application of the second moment-area theorem. Again, compatibility of displacements at joint d, the condition of inextensional deformation, and geometry, can be applied to member de to calculate U_e. As indicated in Figure 6.19d, it is determined (by calculation) that $U_e = 0$. The result is exactly what was expected.

KEY POINTS

1. At the most fundamental level, determination of displacements when deformations are given is a geometry problem the solution of which requires integration of differential deformation-displacement equations subject to boundary conditions and compatibility conditions.

2. Displacements in a framed structure composed of straight members can be obtained as the superposition of displacements due to bending deformation and displacements due to axial deformation, when the deformations and rotations are small.

3. For the great majority of rigid frames, the contribution of axial deformation to displacement is negligibly small compared to displacements due to bending deformation.

4. The moment-area theorems are a consequence of a particular approach to integrating the differential deformation-displacement equations when deformations and rotations are small. The theorems apply between any two points in an interval only when the deflection and rotation functions are continuous in the interval.

5. Boundary and compatibility conditions that are necessary for a structure to be statically determinate are always sufficient in number and type to permit calculation of displacements by a direct geometric method such as the moment-area method.

6. In the case of small deformations and rotations, the superposition of a small rigid-body displacement, based on the undeformed geometry of a structure, on displacements that are consistent with the deformations always generates displacements that are consistent with the same deformations.

PROBLEMS

6.1 a) Identify the end points of the segments in which the deflection v and the rotation $v' = \theta$ are continuous. b) Formulate the boundary conditions. c) Use the moment-area theorems and geometry to determine the deflection and the rotation at B and at C. Show directions for the deflections and rotations.

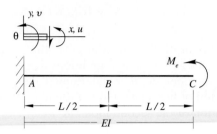

PROBLEM 6.1

6.2 a) Identify the end points of the segments in which the deflection v and the rotation $v' = \theta$ are continuous. b) Formulate the boundary conditions. c) Use the moment-area theorems and geometry to determine the deflection and the rotation at A and at B. Show directions for the deflections and rotations.

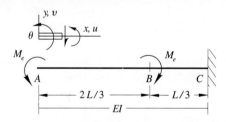

PROBLEM 6.2

6.3 a) Identify the end points of the segments in which the deflection v and the rotation $v' = \theta$ are continuous. b) Formulate the boundary conditions. c) Use the moment-area theorems and geometry to determine the deflection and the rotation at A and at B. Show directions for the deflections and rotations.

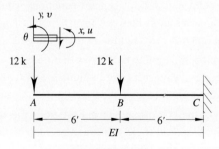

PROBLEM 6.3

6.4 a) Identify the end points of the segments in which the deflection v and the rotation $v' = \theta$ are continuous. b) Formulate the boundary conditions. c) Use the moment-area theorems and geometry to determine the deflection at C and the rotation at B. Show directions for the deflections and rotations.

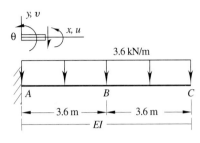

PROBLEM 6.4

6.5 a) Identify the end points of the segments in which the deflection v and the rotation $v' = \theta$ are continuous. b) Formulate the boundary conditions. c) Use the moment-area theorems and geometry to determine the deflection at B and at C and the rotation at A. Show directions for the deflections and rotation.

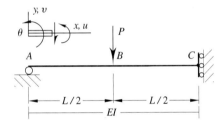

PROBLEM 6.5

6.6 a) Identify the end points of the segments in which the deflection v and the rotation $v' = \theta$ are continuous. b) Formulate the boundary conditions. c) Use the moment-area theorems and geometry to determine the rotation at A and at B and the deflection at B. Show directions for the deflection and rotations.

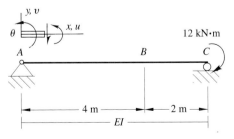

PROBLEM 6.6

6.7 a) Identify the end points of the segments in which the deflection v and the rotation $v' = \theta$ are continuous. b) Formulate the boundary conditions. c) Use the moment-area theorems and geometry to determine the deflection at B and the rotation at C. Show directions for the deflection and rotation.

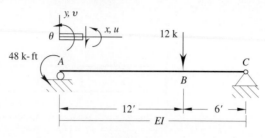

PROBLEM 6.7

6.8 a) Identify the end points of the segments in which the deflection v and the rotation $v' = \theta$ are continuous. b) Formulate the boundary conditions. c) Use the moment-area theorems and geometry to determine the rotation at B and the deflection at A and C. Show directions for the deflections and rotation.

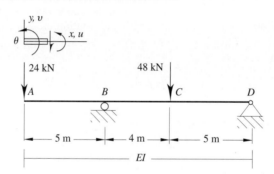

PROBLEM 6.8

6.9 a) Identify the end points of the segments in which the deflection v and the rotation $v' = \theta$ are continuous. b) Formulate the boundary conditions. c) Use the moment-area theorems and geometry to determine the deflection at $A, C,$ and E. Show directions for the deflections.

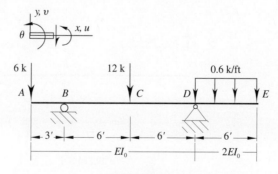

PROBLEM 6.9

6.10 a) Identify the end points of the segments in which the deflection v and the rotation $v' = \theta$ are continuous. b) Formulate the boundary conditions. c) Use the moment-area theorems and geometry to determine the deflection at B and C and the rotation at A. Show directions for the deflections and rotation.

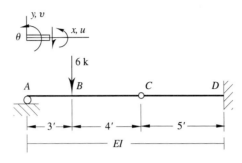

PROBLEM 6.10

6.11 a) Identify the end points of the segments in which the deflection v and the rotation $v' = \theta$ are continuous. b) Formulate the boundary conditions. c) Use the moment-area theorems and geometry to determine the deflection at B and D and the rotation at C. Show directions for the deflections and rotation.

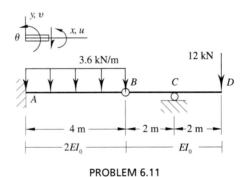

PROBLEM 6.11

6.12 a) Identify the end points of the segments in which the deflection v and the rotation $v' = \theta$ are continuous. b) Formulate the boundary conditions. c) Use the moment-area theorems and geometry to determine the deflection at $B, C,$ and D. Show directions for the deflections.

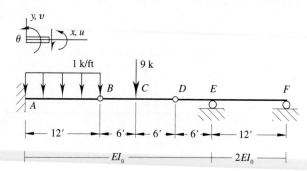

PROBLEM 6.12

6.13 The connection at B enforces the conditions that the axial component of displacement and the rotation at the B-end of member AB and at the B-end of member BD, respectively, be equal; the transverse components of displacement need not be continuous. Use the moment-area theorems and geometry to determine: the deflection of the B-end of member AB, the rotation at B, the deflection at the B-end of member BD, and the deflection at C.

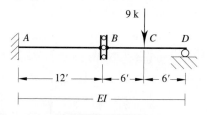

PROBLEM 6.13

6.14 The connection at D enforces the conditions that the axial component of displacement and the rotation at the D-end of member ACD and at the D-end of member DF, respectively, be equal; the transverse components of displacement need not be continuous. Use the moment-area theorems and geometry to determine: the deflection of the D-end of member ACD, the rotation at D, the deflection at the D-end of member DF, and the deflection at E.

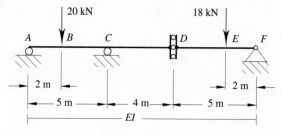

PROBLEM 6.14

6.15 Use the moment-area theorems and geometry to determine the deflection at B and at E.

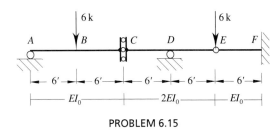

PROBLEM 6.15

6.16 Use the moment-area theorems, conditions of compatibility, boundary conditions, and geometry to determine the deflection of B and D.

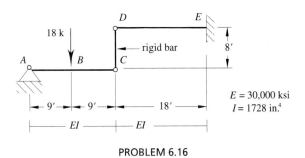

PROBLEM 6.16

6.17 Use the moment-area theorems, conditions of compatibility, boundary conditions, and geometry to determine the deflection of B and D.

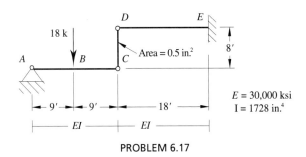

PROBLEM 6.17

6.18 Use the moment-area theorems, the boundary conditions, geometry, the condition of inextensional deformation and the relationship between the local and global coordinates to determine U and V at B. Evaluate U_B and V_B for the given values of E and I. Start the solution by showing that point A does not displace.

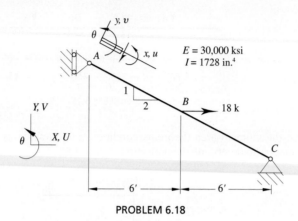

PROBLEM 6.18

6.19 Use the moment-area theorems, the boundary conditions, geometry, the condition of inextensional deformation, and the relationship between the local and global coordinates to determine U and V at B and at D. EI is constant.

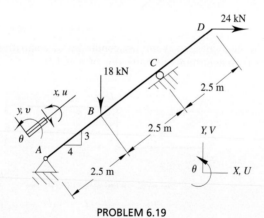

PROBLEM 6.19

6.20 Define a set of local coordinates. Then use the moment-area theorems, the boundary conditions, geometry, and the condition of inextensional deformation to determine the horizontal and vertical components of displacement at D. The beam is uniform, weighs 520 lb/ft, and has bending stiffness EI.

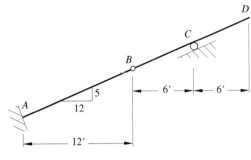

PROBLEM 6.20

6.21 Define a set of local coordinates. Then use the moment-area theorems, the boundary conditions, geometry, and the condition of inextensional deformation to determine the horizontal and vertical components of displacement at B. EI is constant.

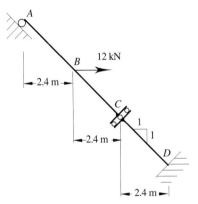

PROBLEM 6.21

6.22–6.30 General instructions: Define local coordinates for each member. Then use the moment-area theorems, the condition of inextensional deformation, the compatibility conditions, the boundary conditions, and geometry to determine the required quantities.

6.22 Determine the rotation of joint B and the horizontal and vertical components of displacement of C. Follow the general instructions for problems 6.22–6.30.

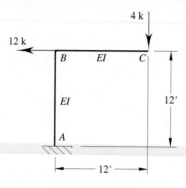

PROBLEM 6.22

6.23 Determine the rotation of joint B and the horizontal and vertical components of displacement of A. Follow the general instructions for problems 6.22–6.30.

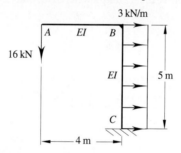

PROBLEM 6.23

6.24 Determine the rotation of joint B and the horizontal and vertical components of displacement of C. Follow the general instructions for problems 6.22–6.30.

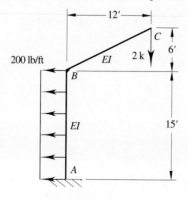

PROBLEM 6.24

6.25 Determine the rotation of joint B, the horizontal component of displacement of B, and the vertical component of displacement of C. Follow the general instructions for problems 6.22–6.30.

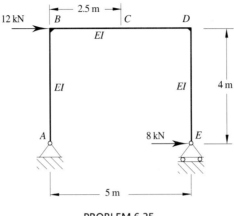

PROBLEM 6.25

6.26 Determine the horizontal and vertical components of displacement of A. Follow the general instructions for problems 6.22–6.30.

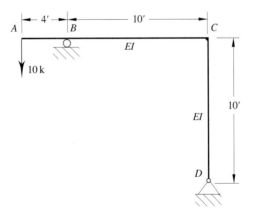

PROBLEM 6.26

6.27 Determine the horizontal component of displacement of *A* and the vertical component of displacement of *C*. Follow the general instructions for problems 6.22–6.30.

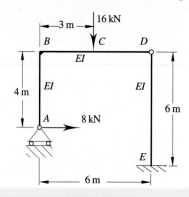

PROBLEM 6.27

6.28 Determine the vertical component of displacement at *B* and the horizontal component of displacement at *D*. Follow the general instructions for problems 6.22–6.30.

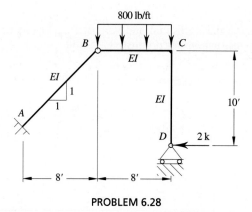

PROBLEM 6.28

6.29 Determine the vertical component of displacement at *A* and at *D*. Follow the general instructions for problems 6.22–6.30.

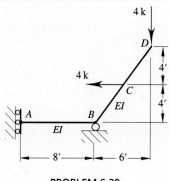

PROBLEM 6.29

6.30 Determine the vertical component of displacement at A and at C. Follow the general instructions for problems 6.22–6.30.

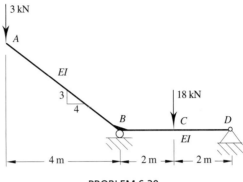

PROBLEM 6.30

6.31–6.39 General instructions: Superposition is required to solve these problems. Frame examples 6 and 7 and the associated text material should be reviewed before attempting solution. In each problem use the assigned point as reference point and use the conditions $U_p = V_p = \theta_p = 0$ to construct the associated known (reference) displacements.

6.31 Determine the horizontal component of displacement at B and at C, and the vertical component of displacement at D. Take A as the reference point.

6.32 Determine the horizontal component of displacement at B and at C, and the vertical component of displacement at D. Take C as the reference point.

6.33 Determine the horizontal component of displacement at B and at C, and the vertical component of displacement at D. Take E as the reference point.

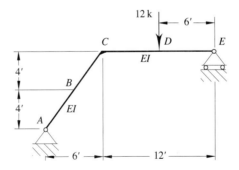

PROBLEMS 6.31, 6.32 and 6.33

6.34 Determine the vertical component of displacement at B and at C, and the horizontal component of displacement at D. Take A as the reference point.

6.35 Determine the vertical component of displacement at B and at D, and the horizontal component of displacement at D. Take C as the reference point.

6.36 Determine the vertical component of displacement at B and at E, and the horizontal component of displacement at D. Take E as the reference point.

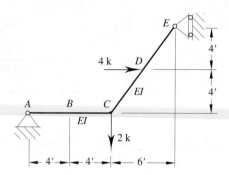

PROBLEMS 6.34, 6.35 and 6.36

6.37 Determine the horizontal component of displacement at B and at F, and the vertical component of displacement at D. Take A as the reference point.

6.38 Determine the horizontal component of displacement at B and at F, and the vertical component of displacement at D. Take C as the reference point.

6.39 Determine the horizontal component of displacement at B and at F, and the vertical component of displacement at D. Take E as the reference point.

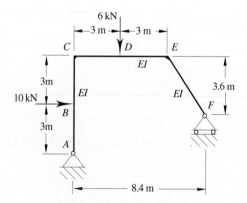

PROBLEMS 6.37, 6.38 and 6.39

6.40 The loads induce a vertical displacement at E of 1260 k-ft^3/EI downward. Use this result, the moment-area theorems, the condition of inextensional deformation, the boundary and compatibility conditions, and geometry to determine the vertical component of displacement at B and at C, and the horizontal component of displacement at D. The bending stiffness EI is constant.

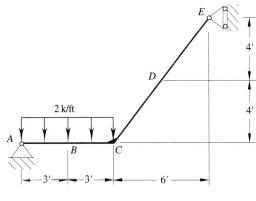

PROBLEM 6.40

6.41 The horizontal component of displacement at E is 2182 kN · m^3/EI to the right. Use this result, the moment-area theorems, the condition of inextensional deformation, the boundary and compatibility conditions, and geometry to determine the vertical component of displacement at D and the horizontal component of displacement at B and at F. The bending stiffness EI is constant.

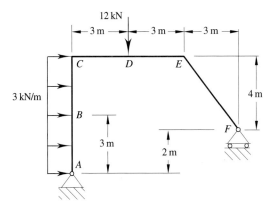

PROBLEM 6.41

6.42 The horizontal component of displacement at D is 3495 k-ft^3/EI to the right. Use this result, the moment-area theorems, the condition of inextensional deformation, the boundary and compatibility conditions, and geometry to determine the vertical component of displacement at C and at D, and the horizontal component of displacement at E. The bending stiffness EI is constant.

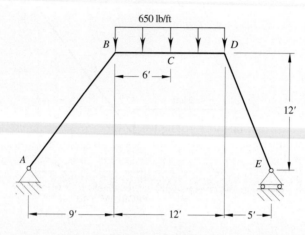

PROBLEM 6.42

CHAPTER 7

Energy Methods (Part 2)

7.1 INTRODUCTION

The initial focus of attention here is application of energy methods to the deformation analysis of bodies that undergo infinitesimal displacements. The principle of virtual work and basic applications to the calculation of displacements of plane-framed structures is the starting point. (Appendix A contains supplemental notes on the principle.) Subsequently, the concepts of strain energy and complementary strain energy are introduced. The principle of stationary potential energy, which is useful in the analysis of statically indeterminate structures, is developed. Betti's theorem, Maxwell's reciprocal theorem, and Castigliano's theorems, all of which have practical applications, are developed and applied to problems of interest.

7.2 DISPLACEMENTS BY VIRTUAL WORK

The basic ideas are developed with the aid of a sketch. Consider a body in equilibrium under the action of internal forces, denoted generically by the Greek letter σ, and external forces $Q_1, Q_2, \ldots, Q_n$ as shown in Figure 7.1a.

As implied by the sketch, the supports are such that the body cannot undergo a rigid displacement and there are no displacements at the supports. Such deformations as may occur are assumed to be so small that they need not be taken into account in writing the equations of equilibrium. In other words, the body is treated as being in equilibrium in the undeformed state. The assumption is fully justified and

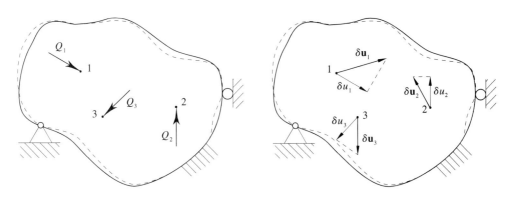

FIGURE 7.1 Virtual Work.

employed in the analysis of the great body of practical problems of interest. In general, the equations of equilibrium do not determine a unique set of internal forces such that equilibrium will prevail. Accordingly, it is to be understood that σ denotes *any* set of internal forces such that the equations of equilibrium are satisfied. In passing it is noted that it is only in the case of a statically determinate system that the equations of equilibrium define a unique σ. It is assumed that a set of internal forces, σ, has been constructed so that the body is in equilibrium under the assigned forces $Q_1, Q_2, \ldots, Q_n$. Let the body be subjected to an arbitrary virtual displacement field for which the deformations are denoted generically by $\delta\varepsilon$ and for which the corresponding displacements at the points where the forces $Q_1, Q_2, \ldots, Q_n$ act are $\delta\mathbf{u}_1, \delta\mathbf{u}_2, \ldots, \delta\mathbf{u}_n$ with components $\delta u_1, \delta u_2, \ldots, \delta u_n$ in the directions of $Q_1, Q_2, \ldots, Q_n$, respectively. For clarity the displacements are shown in highly exaggerated scale in Figure 7.1b. It is assumed that $\delta\varepsilon$ is given. The internal and external forces and the deformations and displacements satisfy the conditions for application of the principle of virtual work, which is formulated as $\delta W_e = \delta W_\sigma$. Since there are no displacements at the supports, the virtual work of the external forces is

$$\delta W_e = Q_1\,\delta u_1 + Q_2\,\delta u_2 + \cdots + Q_n\,\delta u_n = \Sigma Q_k\,\delta u_k$$

Therefore, the virtual work equation can be written in the more explicit form

$$\Sigma Q_k\,\delta u_k = \delta W_\sigma$$

It will be shown that with σ and $\delta\varepsilon$ known, δW_σ can be evaluated. Formulas for the evaluation of δW_σ will be derived. Accordingly, δW_σ may be treated as a known quantity. Let the forces be assigned the values $Q_1 = Q = 1$, and $Q_k = 0$ for all $k \neq 1$. Then

$$\delta W_e = \Sigma Q_k\,\delta u_k = \delta u_1 = \delta u = \delta W_\sigma$$

Observe that the last expression simply states that the component of the virtual displacement at point 1 in the direction of $Q_1 = 1$ is numerically equal to the virtual work δW_σ done by any distribution of internal forces σ for which the body is in equilibrium under $Q = $ a unit load at point 1 in the direction of Q_1 on the given virtual deformation defined by $\delta\varepsilon$. This observation is the basis for the virtual-work method of calculating a component of displacement u at a point in a body that undergoes infinitesimal deformation ε when the deformation is given. Before going on to the method, consider another special case of loading.

Let the loads now consist of a pair of equal and opposite forces as shown in Figure 7.2. Then, the virtual work reduces to

$$\delta W_e = Q\,\delta u_1 + Q\,\delta u_2 = QL\left(\frac{\delta u_1 + \delta u_2}{L}\right) = M\,\delta\theta = \delta W_\sigma$$

where M is the moment of the couple formed by the forces and $\delta\theta$ is the rotation of the chord of that deformed line connecting the points 1 and 2 which, prior to the virtual displacement, was straight and in the direction 1, 2. If point 2 is now replaced by a new point 2 at, say, the midpoint of L, and if the forces Q are now doubled, the virtual work will reduce to

$$M\,\delta\theta = \delta W_\sigma$$

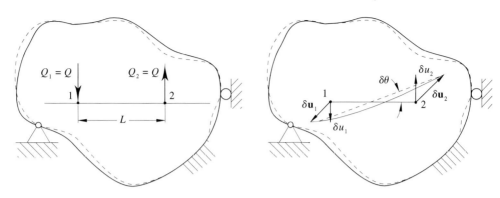

FIGURE 7.2 Rotation of a chord.

where the moment of the new couple is the same as previously given, i.e., $M = QL$, and $\delta\theta$ is now understood to be the rotation of the chord of that deformed line connecting the points 1 and 2 which, prior to the virtual displacement, was straight and in the direction 1, 2. The process of taking point 2 closer and closer to point 1 while choosing Q so that M has a constant value may be repeated indefinitely. In the limit, as point 2 approaches point 1, $\delta\theta$ becomes the rotation of the tangent of a deformed line which prior to the virtual displacement was in the direction 1, 2. By taking Q so that $QL = 1$, i.e., by taking Q so that the pair of forces form a unit couple, the virtual work expression for the rotation $\delta\theta$ takes the same form as the expression for the displacement δu.

7.3 PROCEDURE

Within the framework of the basic assumptions employed in structural analysis, deformations and related displacements due to any specific cause always qualify as virtual deformations and displacements. Thus, a simple procedure to calculate a component of displacement u in a given direction at any point from known deformation ε due to any cause, is:

1. Place a unit load $Q = 1$ at the point in question in the direction in question.
2. Construct any set of internal forces σ so that the body is in equilibrium in the undeformed state under the unit load.
3. Use the known deformation defined by ε as the virtual deformation and obtain the component of displacement u from

$$\delta W_e = Qu = 1u = u = \delta W_\sigma$$

7.4 EVALUATION OF δW_σ

Specific expressions for the evaluation of δW_σ are needed to implement the method. A formulation suitable for the evaluation of δW_σ for straight beam elements that undergo axial and bending deformation is derived below. The formulation is adequate for the overwhelming majority of practical problems in the displacement analysis of framed structures. The desired formulation is obtained with the aid of the principle of virtual work itself. With no loss in generality, the principle is applied to a beam segment that is in equilibrium under the action of distributed axial and transverse loads and end forces. Again, with no loss in generality, the reference axis of the beam is assumed to be continuous in the sense that axial displacements, transverse displacements, and slopes are continuous over the length of the segment. Forces and virtual displacements that satisfy the conditions are displayed in Figure 7.3.

Because the segment is in equilibrium, the differential equations of equilibrium and the force boundary conditions must be satisfied. As indicated in Figure 7.3a, the representations (sign convention) of the internal forces and external forces coincide with those employed in Chapter 3, in which the differential equations of equilibrium were derived. Thus, from previous developments, the differential equations of equilibrium and force boundary conditions are

$$\frac{dN}{dx} = N' = -p(x), \qquad \frac{dM}{dx} = M' = V(x), \qquad \frac{dV}{dx} = V' = -w(x)$$

$$N_a = N(0^+), \qquad M_a = M(0^+), \qquad V_a = V(0^+)$$

$$N_b = N(L^-), \qquad M_b = N(L^-), \qquad V_b = V(L^-)$$

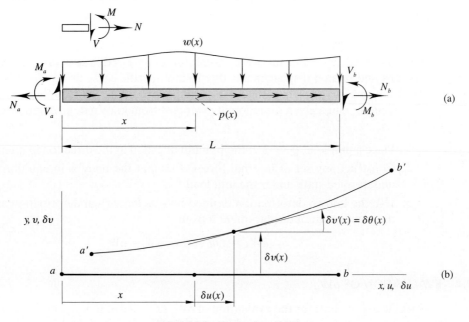

FIGURE 7.3 Evaluation of δW_σ.

From the displacement-deformation analysis in Chapter 6, the virtual displacement-virtual deformation relations are

$$\frac{d\,\delta u}{dx} = \delta u' = \delta\varepsilon(x)$$

$$\delta\theta = \frac{d\,\delta v}{dx} = \delta v'(x), \quad \delta\kappa = \frac{d^2\,\delta v}{dx^2} = \frac{d\,\delta\theta}{dx} = \delta v''(x) = \frac{d\,\delta v'}{dx}$$

The virtual work of the external forces is

$$\delta W_e = -\int_0^L w\,\delta v\,dx - [V_b\,\delta v(L) - V_a\,\delta v(0)] + [M_b\,\delta v'(L) - M_a\,\delta v'(0)]$$
$$+ \int_0^L p\,\delta u\,dx + [N_b\,\delta u(L) - N_a\,\delta u(0)]$$

The work expression is simplified and put into a more useful form in steps with the aid of the differential equations of equilibrium and integration by parts. First replace w by its equivalent, $-V'$, to get

$$-\int_0^L w\,\delta v\,dx = \int_0^L V'\,\delta v\,dx = V\,\delta v\Big]_0^L - \int_0^L V\,\delta v'\,dx$$

Next, replace V by its equivalent, M', and integrate by parts to obtain

$$-\int_0^L w\,\delta v\,dx = V\,\delta v\Big]_0^L - \int_0^L M'\,\delta v'\,dx = V\,\delta v\Big]_0^L - M\,\delta v'\Big]_0^L + \int_0^L M\,\delta v''\,dx$$

With the aid of the force boundary conditions and the relationship $\delta v'' = \delta\kappa$, the last equation can be rewritten to read

$$-\int_0^L w\,\delta v\,dx - [V_b\,\delta v(L) - V_a\,\delta v(0)] + [M_b\,\delta v'(L) - M_a\,\delta v'(0)]$$
$$= \int_0^L M\,\delta\kappa\,dx$$

Now replace p by its equivalent, $-N'$, and again integrate by parts to get

$$\int_0^L p\,\delta u\,dx = -\int_0^L N'\,\delta u\,dx = -N\,\delta u\Big]_0^L + \int_0^L N\,\delta u'\,dx$$

Again, with the aid of the force boundary conditions and the relationship $\delta u' = \delta\varepsilon$, rearrangement of the foregoing equation yields

$$\int_0^L p\,\delta u\,dx + [N_b\,\delta u(L) - N_a\,\delta u(0)] = \int_0^L N\,\delta\varepsilon\,dx$$

Insertion of the foregoing expressions for the virtual work of the distributed loads into the original expression for δW_e, which is equal to δW_σ, gives

$$\delta W_e = \delta W_\sigma = \int_0^L M\, \delta\kappa\, dx + \int_0^L N\, \delta\varepsilon\, dx = -\delta W_i$$

From the formulation it is clear that δW_σ is the sum (or superposition) of two parts, one of which is due to bending as measured by $\delta\kappa$ and the other of which is due to axial deformation as measured by $\delta\varepsilon$. For a complete structure δW_σ is obtained by summing up the values over all members. Therefore, for a complete structure

$$\delta W_e = \delta W_\sigma = \sum_m \int_0^L M\, \delta\kappa\, dx + \sum_m \int_0^L N\, \delta\varepsilon\, dx$$

M and N in the expression for δW_σ are components of the resultant over a cross section of the stresses or the internal forces in the beam. For this reason many writers refer to δW_σ as the internal virtual work. So as to not confuse δW_σ with δW_i, which in this work is called the internal virtual work, from this point on δW_σ will be referred to as the virtual work of the stresses.

To adapt the expression for the virtual work of the stresses to the problem of calculating displacements when the deformations κ and ε due to any given cause are known, first recall that by the assumptions of the small deformation theory of structures, the displacements and corresponding deformations due to the given cause qualify as virtual displacements and deformations. Thus, the foregoing relations are valid when δu, δv, $\delta\kappa$, and $\delta\varepsilon$ are replaced by u, v, κ, and ε, respectively. With the substitutions, the work relationship becomes

$$\delta W_e = \delta W_\sigma = \sum_m \int_0^L M\kappa\, dx + \sum_m \int_0^L N\varepsilon\, dx$$

where δW_e is now understood to be the virtual work of the external forces on the displacements corresponding to the deformations κ and ε.

To avoid a future situation in which one symbol could represent two fundamentally different quantities, denote by m and n any bending moment and axial force distributions such that the structure is in equilibrium under the unit load applied at the point of interest in the direction of interest. Then, for the prescribed conditions,

$$\delta W_e = 1 \cdot u = \sum_m \int_0^L m\kappa\, dx + \sum_m \int_0^L n\varepsilon\, dx$$

where u is the component of displacement at the point of application of the unit load in the direction of the unit load.

7.5 BASIC APPLICATIONS

In most practical situations, the deformations κ and ε are due to loads applied to a structure for which the material obeys Hooke's law. In such cases the work equation

7.6 BEAM

Assume that due to differential heating the cantilever beam shown in Figure 7.4 is bent so that the curvature is defined by the relation

$$v'' = \kappa = \frac{6A}{L^3}(L - x)$$

and that it is required to calculate the deflection and rotation of the tip of the beam.

Either direct integration of the differential equation or application of the moment-area theorems yields the results

$$v(L) = 2A, \quad v'(L) = \theta(L) = \frac{3A}{L}$$

To obtain the deflection at the tip by the method of virtual work, apply a unit load as shown in Figure 7.4b. The axial tension due to the loading is $n(x) = 0$. The sign convention for $m(x)$ must be consistent with the sign convention used in the derivation of the expression for the virtual work of the stresses. (See Figure 7.3.) The required sign convention is illustrated in Figure 7.4. The bending moment due to the unit load is given by

$$m(x) = -(L - x)$$

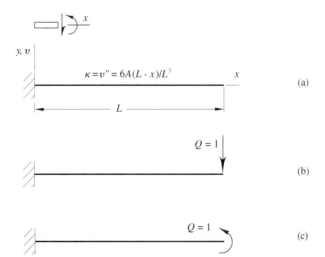

FIGURE 7.4 Deformation due to differential heating/cooling.

Because $n(x) = 0$, and there is only one member, by the principle of virtual work the deflection at the tip of the beam in the direction of the unit load is

$$u = \int_0^L m\kappa \, dx = \int_0^L -(L-x)\left[\frac{6A}{L^3}(L-x)\right] dx = -2A$$

The negative sign indicates that the deflection is in the direction opposite to that of the unit load. In other words, the deflection is 2A in the direction of the v-axis. Thus, the principle of virtual work gives the same result as that obtained by the moment-area method. (This, of course, must be the case.)

To obtain the rotation at the tip, apply a unit couple as shown in Figure 7.4c. The bending moment due to the unit couple is $m(x) = 1$ and by the principle of virtual work, the rotation θ of the tangent at the tip in the direction of the unit couple is

$$\theta = \int_0^L m\kappa \, dx = \int_0^L \frac{6A}{L^3}(L-x) \, dx = \frac{3A}{L}$$

which coincides with the value obtained previously using the moment-area method.

7.7 TRUSS

Frequently, long bridge trusses are cambered upward so that the net deflection of the bottom chord due to heavy dead load and the camber is essentially zero. Camber may be produced by fabricating the members of the bottom chord so that they are shorter than their theoretical lengths by the amount that they will deform under dead load. The amount by which a member is fabricated shorter than its theoretical length can be treated as an axial deformation of the member. A practical problem of interest is that of calculating displacements of assigned joints of a truss as a consequence of the axial deformations associated with the cambering process.

The truss shown in Figure 7.5 is to be cambered by cutting each member of the bottom chord to a length that is slightly less than the theoretical length L. All other members are to be cut to their correct respective lengths. The amount by which a member is to be cut short is expressed as a fraction p of the theoretical length L. Let the amounts be pL for members ab and fg, $2pL$ for members bc and ef, and $3pL$ for members cd and de. The problem of interest is to determine p so that the displacement of joint d due to the camber is a prescribed fraction p^* of L; in other words, p is to be determined so that the displacement of joint d is p^*L where p^* is prescribed.

To obtain the displacement of joint d, apply a unit load as shown in Figure 7.5b. The member tensions due to the unit load are as shown in Figure 7.5b. Because the tension n is constant over the length of a member and the bending moment due to the unit load is zero for all members, the method of virtual work takes the special formulation

$$u = \sum_m \int_0^L n\varepsilon \, dx = \sum_m n \int_0^L \varepsilon \, dx = \sum_m n \, \Delta L$$

where ΔL is the elongation of the member. Displacement calculations for trusses are arranged in tabular form. Because only the members of the bottom chord undergo deformation, an abbreviated table is sufficient for the example. The calculations are displayed in Table 7.1. It is to be noted that the entries in the column headed by ΔL are

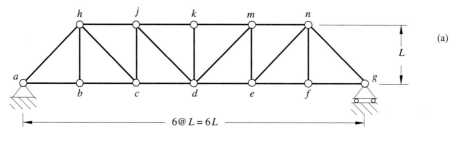

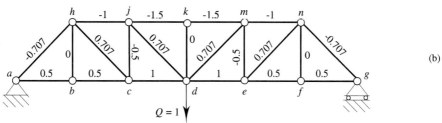

FIGURE 7.5 Cambered truss.

TABLE 7.1 Cambered truss calculation

Bar	n	ΔL	$n\,\Delta L$
ab	0.5	$-pL$	$-0.5pL$
bc	0.5	$-2pL$	$-1.0pL$
cd	1.0	$-3pL$	$-3.0pL$
de	1.0	$-3pL$	$-3.0pL$
ef	0.5	$-2pL$	$-1.0pL$
fg	0.5	$-pL$	$-0.5pL$
			$-9.0pL$

negative. The reason is that ΔL is positive for elongation, but the members are cut short by the indicated amounts. The last entry in the last column of the table is the sum of terms in the column, and by the principle of virtual work this is equal to the displacement of joint d due to the member shortenings displayed in the column headed ΔL. The minus sign signifies that the displacement is in the direction opposite to that of the unit load which is as it must be. From the conditions in the problem it follows that $p^* = 9p$.

7.8 DISPLACEMENTS DUE TO APPLIED LOADS

If a structure is elastic in the sense that Hooke's law is satisfied, the deformations due to applied loads are given by

$$\kappa = \frac{M}{EI}, \quad \varepsilon = \frac{N}{EA}$$

where M and N are the bending moment and axial force due to the applied loads. The virtual-work method for calculating displacements may now be formulated as

$$u = \sum_m \int_0^L \frac{mM}{EI} dx + \sum_m \int_0^L \frac{nN}{EA} dx$$

It is to be noted that the foregoing formulation is independent of the sign convention used in the representations of the internal forces. The only requirements are that the same coordinates and same sign conventions be employed in the evaluation of m and M, and that the same coordinates and same sign conventions be employed in the evaluation of n and N.

7.9 BEAMS AND FRAMES

Ordinarily, beams and frames are loaded and proportioned so that the contribution of any axial deformation, if present, to displacement is negligibly small. Accordingly, the practice is to take into account only the bending deformation in calculating displacements. In any event, the structure of the work equation shows that the contributions of bending deformation and of axial deformation may be evaluated separately and superimposed to obtain total displacement.

Example 7.1

Calculate the deflection at the tip and at the midpoint of the cantilever beam shown in Figure 7.6.

The way in which the beam is supported and loaded, and the nature of the calculations (integrations) suggest choosing the coordinates and sign convention for bending moment as shown in Figure 7.6a. With this choice, the bending moment due to the distributed load is

$$M = \frac{wx^2}{2} \qquad 0 \le x \le L$$

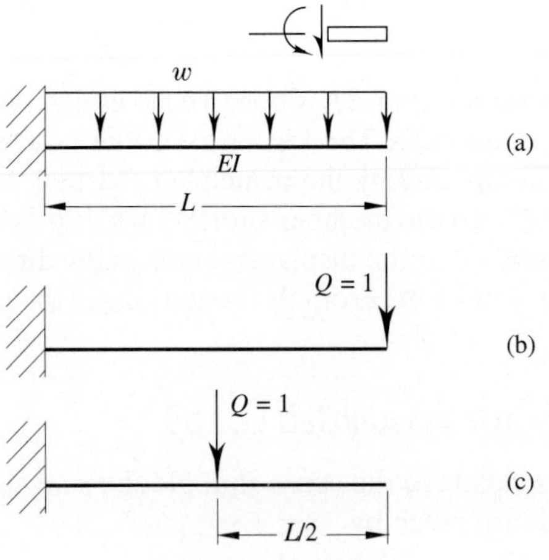

FIGURE 7.6 Elastic beam.

To obtain the displacement at the tip of the beam, place the unit load as shown in Figure 7.6b. Then,

$$m = x \qquad 0 \leq x \leq L$$

and the deflection of the tip in the direction of the unit load is

$$u = \int_0^L \frac{mM}{EI} dx = \int_0^L \frac{wx^3}{2EI} dx = \frac{wL^4}{8EI}$$

To calculate the deflection at the midpoint, the unit load is placed as shown in Figure 7.6c. The bending moment due to the unit load is given by

$$m = 0 \qquad 0 \leq x \leq \frac{L}{2}, \qquad m = x - \frac{L}{2} \qquad \frac{L}{2} \leq x \leq L$$

and the displacement is

$$u = \int_0^L \frac{mM}{EI} dx = \int_{L/2}^L \left(x - \frac{L}{2}\right)\left(\frac{wx^2}{2EI}\right) dx = \frac{17wL^4}{384EI}$$

Example 7.2

Calculate the deflection at point b of the simply supported beam shown in Figure 7.7. For the purpose of the displacement calculation, it is convenient to view the beam as made of two members, ab and bc, that are joined rigidly at b. For the given loading,

$$\text{member } ab: M = (1800 \text{ lb})x - (150 \text{ lb/ft})x^2 \qquad 0 \leq x \leq 4 \text{ ft}$$
$$\text{member } bc: M = (1800 \text{ lb})x - (150 \text{ lb/ft})x^2 \qquad 0 \leq x \leq 8 \text{ ft}$$

With the unit load placed at b as shown in Figure 7.7b,

$$\text{member } ab: m = 2x/3 \qquad 0 \leq x \leq 4 \text{ ft}$$
$$\text{member } bc: m = x/3 \qquad 0 \leq x \leq 8 \text{ ft}$$

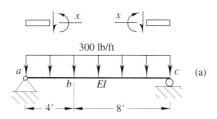

FIGURE 7.7 Simple beam.

The deflection at b is

$$u = \int_0^{4\,\text{ft}} \frac{1}{EI}\left(\frac{2x}{3}\right)(1800x - 150x^2)\,dx + \int_0^{8\,\text{ft}} \frac{1}{EI}\left(\frac{x}{3}\right)(1800x - 150x^2)\,dx = \frac{70{,}400}{EI}\,\text{lb-ft}^3$$

In this example, dimensions of various terms have been retained throughout the calculation. In the following examples, dimensions are carried along in intermediate calculations only to the extent necessary. Dimensions will always be supplied for final results.

Example 7.3

Calculate the deflection at point c of the structure shown in Figure 7.8. The structure is treated as being composed of three beams; ab, bc, and cd. The coordinates and sign conventions used to determine the bending moments are as shown in Figure 7.8a. The expressions for the bending moments are:

$$\text{beam } ab: \quad M = -12x \qquad 0 \le x \le 4, \qquad M = -48 \qquad 4 \le x \le 12$$
$$\text{beam } bc: \quad M = 12x - x^2 \qquad 0 \le x \le 6$$
$$\text{beam } cd: \quad M = 12x - x^2 \qquad 0 \le x \le 6$$

With the unit load placed at c as shown in Figure 7.8b, the bending moments are:

$$\text{beam } ab: \quad m = -x/2 \qquad 0 \le x \le 12$$
$$\text{beam } bc: \quad m = x/2 \qquad 0 \le x \le 6$$
$$\text{beam } cd: \quad m = x/2 \qquad 0 \le x \le 6$$

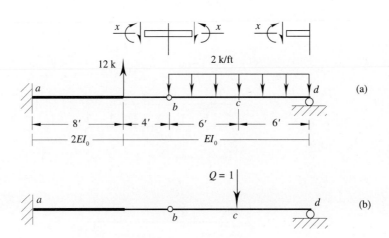

FIGURE 7.8 Compound beam.

The integrals needed to evaluate the displacement are:

$$\text{beam } ab: \int_0^L \frac{mM}{EI} dx = \int_0^4 \frac{1}{EI_0}\left(-\frac{x}{2}\right)(-12x) \, dx + \int_4^{12} \frac{1}{2EI_0}\left(-\frac{x}{2}\right)(-48) \, dx$$

$$= \frac{128}{EI_0} + \frac{768}{EI_0} = \frac{896 \text{ k-ft}^3}{EI_0}$$

$$\text{beam } bc: \int_0^L \frac{mM}{EI} dx = \int_0^6 \frac{1}{EI_0}\left(\frac{x}{2}\right)(12x - x^2) \, dx = \frac{270 \text{ k-ft}^3}{EI_0}$$

$$\text{beam } cd: \int_0^L \frac{mM}{EI} dx = \int_0^6 \frac{1}{EI_0}\left(\frac{x}{2}\right)(12x - x^2) \, dx = \frac{270 \text{ k-ft}^3}{EI_0}$$

Thus, the deflection at c is

$$u = \frac{896 \text{ k-ft}^3}{EI_0} + \frac{270 \text{ k-ft}^3}{EI_0} + \frac{270 \text{ k-ft}^3}{EI_0} = \frac{1436 \text{ k-ft}^3}{EI_0}$$

The result agrees with that obtained previously by the moment-area method.

Example 7.4

Calculate the horizontal component of displacement of joint c of the frame shown in Figure 7.9. Disregard axial deformations in the calculation.

The frame is treated as being composed of two members: ac and cd. The coordinates and sign conventions used to determine bending moments are as shown in Figure 7.9a. The expressions for the bending moments due to the applied loads are:

$$\text{member } ac: M = \frac{243x}{35} \quad 0 \le x \le 7.5, \quad M = \frac{945}{35} + \frac{117x}{35} \quad 7.5 \le x \le 15$$

$$\text{member } cd: M = \frac{129x}{7} - x^2 \quad 0 \le x \le 12$$

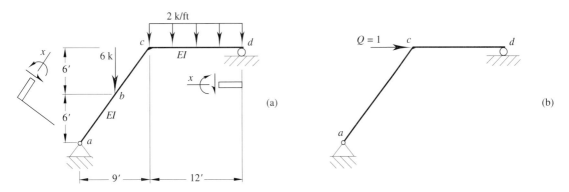

FIGURE 7.9 Frame displacement.

With the unit load placed as shown in Figure 7.9b, the bending moments are:

$$\text{member } ac: m = \frac{16x}{35} \quad 0 \le x \le 15, \quad \text{member } cd: m = \frac{4x}{7} \quad 0 \le x \le 12$$

For member ac,

$$\int_0^L \frac{mM}{EI} dx = \int_0^{7.5} \frac{1}{EI}\left(\frac{16x}{35}\right)\left(\frac{243x}{35}\right) dx + \int_{7.5}^{15} \frac{1}{EI}\left(\frac{16x}{35}\right)\left(\frac{945}{35} + \frac{117x}{35}\right) dx$$

$$= \frac{446.3}{EI} + \frac{2545.7}{EI} = \frac{2992.0}{EI}$$

For member cd,

$$\int_0^L \frac{mM}{EI} dx = \int_0^{12} \frac{1}{EI}\left(\frac{4x}{7}\right)\left(\frac{129x}{7} - x^2\right) dx = \frac{3103.4}{EI}$$

The horizontal component of displacement at c in the direction of the unit load is

$$u = \sum_m \int_0^L \frac{mM}{EI} dx = \frac{2992.0}{EI} + \frac{3103.4}{EI} = \frac{6095.4}{EI} \text{ k-ft}^3$$

The result is in agreement with that obtained for the displacement by the moment-area method. The slight difference is due to round-off error in the numerical work. It can be shown that in principle the results are identical.

Observation If the problem were expanded to include calculation of the displacement at b (or any other point) as well as at c, the simplest approach would be to use the now-known result for the horizontal component of displacement at c, geometry, the condition of inextensional deformation, and the moment-area theorems. It is readily perceived that, in general, the most efficient way of calculating displacements of frames is by means of a combination of the method of virtual work and the moment-area method.

Example 7.5

The frame of the last example treated in Chapter 6 (Figure 6.17) is reproduced herein as Figure 7.10. It was shown in Chapter 6 that all of the results that were required could be obtained rather easily if the rotation of the tangent at the a-end of member ab is known. The problem here

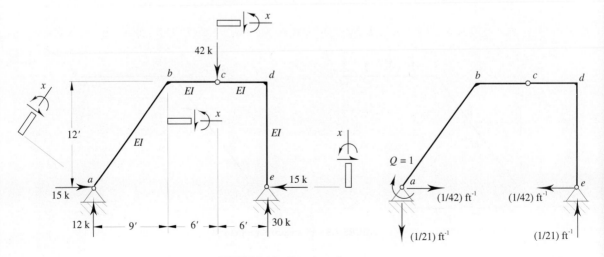

FIGURE 7.10 Rotation of a tangent.

is to calculate the rotation by the method of virtual work. From Figure 7.10, the equations of the bending moments due to the applied loads are:

member ab: $M = -72x/15 \quad 0 \le x \le 15,$ member bc: $M = -72 + 12x \quad 0 \le x \le 6$
cd: $M = -30x \quad 0 \le x \le 6,$ ed: $M = 15x \quad 0 \le x \le 12$

The corresponding bending moments due to a unit clockwise couple applied at a as shown in Figure 7.10 are:

member ab: $m = 1 - x/21 \quad 0 \le x \le 15,$ member bc: $m = 2/7 - x/21 \quad 0 \le x \le 6$
member bc: $m = -x/21 \quad 0 \le x \le 6,$ member ed: $m = x/42 \quad 0 \le x \le 12$

The required integrals are

member ab: $\int_0^L \frac{mM}{EI} dx = \int_0^{15} \frac{1}{EI}\left(1 - \frac{1}{21}x\right)\left(\frac{-72}{15}x\right) dx = -\frac{1980}{7EI}$

bc: $\int_0^L \frac{mM}{EI} dx = \int_0^6 \frac{1}{EI}\left(\frac{2}{7} - \frac{1}{21}x\right)(-72 + 12x) dx = -\frac{288}{7EI}$

cd: $\int_0^L \frac{mM}{EI} dx = \int_0^6 \frac{1}{EI}\left(-\frac{1}{21}x\right)(-30x) dx = \frac{720}{7EI}$

ed: $\int_0^L \frac{mM}{EI} dx = \int_0^{12} \frac{1}{EI}\left(\frac{1}{42}x\right)(15x) dx = \frac{1440}{7EI}$

Therefore, the rotation of the tangent at a in the sense of the unit couple is

$$\theta = \sum_m \int_0^L \frac{mM}{EI} dx = \frac{1}{EI}\left(-\frac{1980}{7} - \frac{288}{7} + \frac{720}{7} + \frac{1440}{7}\right) = -\frac{108}{7EI} \text{ k-ft}^2$$

The fact that θ is negative means that the sense of the rotation is opposite to the sense of the unit couple, i.e., the rotation is in the counterclockwise direction.

7.10 TRUSSES

For simplicity, only joint displacements are considered. This is the problem most commonly of interest. In this case the unit load is placed at a joint, the bending moments m are zero for all members, and from the formulation

$$u = \sum_m \int_0^L \frac{mM}{EI} dx + \sum_m \int_0^L \frac{nN}{EA} dx \rightarrow \sum_m \int_0^L \frac{nN}{EA} dx$$

it is clear that bending of the members of a truss contributes nothing to the displacements. Ordinarily the members of a truss have constant cross-sectional areas and the loads that induce the deformations are applied at the joints. For this common situation, the bar tensions N are constant over the lengths of the members and the formulation is further simplified to

$$u = \sum_m \int_0^L \frac{nN}{EA} dx = \sum_m \frac{nNL}{EA}$$

Thus, the problem of calculating joint displacement reduces to that of calculating the

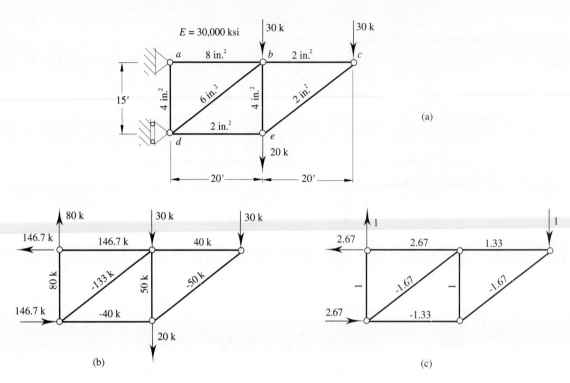

FIGURE 7.11 Vertical displacement of joint c.

bar forces N due to the applied loads, the bar forces n due to the unit load, and the formation of a sum of products. In practice, the calculations are carried out in tabular form. The tabular form is convenient and useful in checking calculations.

Example 7.6

Calculate the vertical component of displacement of joint c of the truss shown in Figure 7.11.

The bar tensions N due to the applied loads are displayed on the line diagram shown in Figure 7.11b. The unit load is placed at joint c in the vertical direction as shown in Figure 7.11c. The bar tensions n due to the unit load are displayed on the line diagram shown in Figure 7.11c. Because E is a constant, the displacement calculations may be arranged as shown in Table 7.2.

From Table 7.2,

$$Eu = \sum_m \frac{nNL}{A} = 53994 \text{ k/in.}$$

With $E = 30{,}000$ ksi, the displacement is $u = 1.80$ inches.

Example 7.7

Calculate the horizontal component of displacement of joint h of the truss shown in Figure 7.12. The problem is not much different than the previous problem. The main purpose here is to emphasize that the bar tensions N and n are signed quantities and that, in general, their signs (in

Section 7.10 Trusses

TABLE 7.2 Vertical displacement of joint c

Bar	L ft	A in²	N kip	NL/A kip/in	n —	nNL/A kip/in
ab	20	8	146.7	4400	2.67	11733
ad	15	4	80	3600	1	3600
bc	20	2	40	4800	1.33	6400
bd	25	6	−133.3	−6667	−1.67	11111
be	15	4	50	2250	1	2250
ce	25	2	−50	−7500	−1.67	12500
de	20	2	−40	−4800	−1.33	6400
						53994

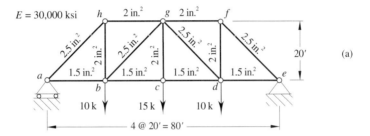

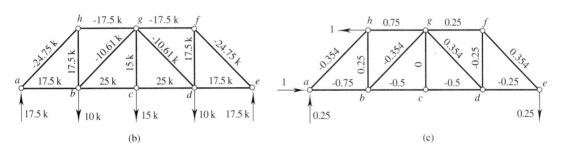

FIGURE 7.12 Horizontal displacement of joint h.

contrast to those in the preceding problem) are not the same. The bar forces corresponding to the applied loads are shown in Figure 7.12b. To obtain the horizontal component of displacement at h, the unit load is placed at h in the horizontal direction as shown in Figure 7.12c, in which the corresponding bar forces are also displayed.

The solution is arranged in tabular form as shown in Table 7.3.
From Table 7.3,

$$Eu = \sum_m \frac{nNL}{A} = -8900 \text{ k/in.}$$

With $E = 30{,}000$ ksi, the displacement is $u = -0.307$ inches. The negative sign means that the displacement is in the sense opposite to that of the unit load. In other words, the displacement of joint h has a horizontal component of 0.307 inches to right.

TABLE 7.3 Horizontal displacement of joint h

Bar	L ft	A in^2	N k	NL/A k/in	n —	nNL/A k/in
ab	20	1.5	17.5	2800	−0.75	−2100
ah	28.28	2.5	−24.75	−3360	−0.354	1189
bc	20	1.5	25	4000	−0.5	−2000
bg	28.28	2.5	−10.61	−1440	−0.354	510
bh	20	2	17.5	2100	0.25	525
cd	20	1.5	25	4000	−0.5	−2000
cg	20	2	15	1800	0	0
de	20	1.5	17.5	2800	−0.25	−700
df	20	2	17.5	2100	−0.25	−525
dg	28.28	2.5	−10.61	−1440	0.354	−510
ef	28.28	2.5	−24.75	−3360	0.354	−1189
fg	20	2	−17.5	−2100	0.25	−525
gh	20	2	−17.5	−2100	0.75	−1575
						−8900

Example 7.8

Determine the increase in the distance between joints a and f of the truss shown in Figure 7.12 under the loading shown. Problems of this type are of practical importance in the analysis of trusses that are internally statically indeterminate.

The required quantity is the sum of the component of the displacement of f in the direction from a to f, and the component of the displacement of a in the direction from f toward a. The two components of displacement may be obtained separately by the method of virtual work by placing a unit load at f and a unit load at a, respectively, as shown in Figure 7.13a and Figure 7.13b, and then adding. It is much easier, however, to obtain the sum of the two components directly in one operation by taking advantage of the principle of superposition and using the combined loading as shown in Figure 7.13c. Because the two unit loads form a self-equilibrating system of forces, all of the reactions due to these loads are zero and the corresponding bar ten-

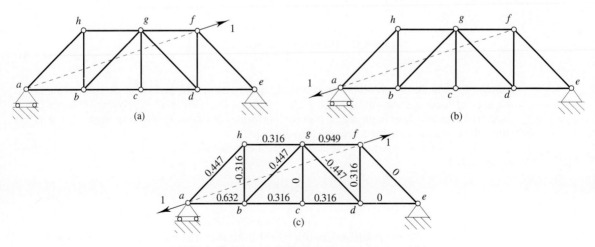

FIGURE 7.13 Increase in distance between joints a and f.

TABLE 7.4 Increase in distance between joints a and f

Bar	L ft	A in^2	N k	NL/A k/in	n —	nNL/A k/in
ab	20	1.5	17.5	2800	0.632	1770
ah	28.28	2.5	−24.75	−3360	0.447	−1502
bc	20	1.5	25	4000	0.316	1264
bg	28.28	2.5	−10.61	−1440	0.447	−644
bh	20	2	17.5	2100	−0.316	−664
cd	20	1.5	25	4000	0.316	1264
cg	20	2	15	1800	0	0
de	20	1.5	17.5	2800	0	0
df	20	2	17.5	2100	0.316	664
dg	28.28	2.5	−10.61	−1440	−0.447	644
ef	28.28	2.5	−24.75	−3360	0	0
fg	20	2	−17.5	−2100	0.949	−1993
gh	20	2	−17.5	−2100	0.316	−664
						139

sions are as shown in Figure 7.13c. The analysis is again arranged in tabular form with the results as displayed in Table 7.4. From Table 7.4,

$$E \, \Delta s_{af} = \sum_m \frac{nNL}{A} = 139 \text{ k/in.}$$

where Δs_{af} is the increase in the distance between joints a and f. Using $E = 30{,}000$ ksi, $\Delta s_{af} = 0.00463$ inches.

7.11 COMBINED LOADS

Consider a generalized plane frame structure in equilibrium under two different sets of loads as shown in Figure 7.14a and 7.14b.

Let M and N identify the internal forces corresponding to the loads shown in Figure 7.14a, and let M^* and N^* identify the internal forces corresponding to the loads

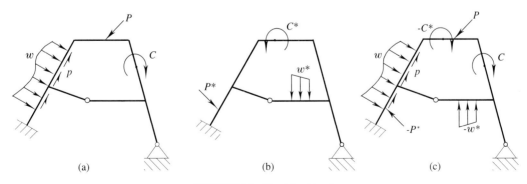

FIGURE 7.14 Combined loads.

shown in Figure 7.14b. Because the structure is in equilibrium the virtual work of the external forces shown in Figure 7.14a on a given virtual displacement is

$$\delta W_e = \sum_m \int_0^L (-w\,\delta v + p\,\delta u)\,dx + \sum_P (P\,\delta u_P) + \sum_C (C\,\delta\theta_C)$$

$$= \sum_m \int_0^L (M\,\delta\kappa + N\,\delta\varepsilon)\,dx = \delta W_\sigma$$

Let δW_e^* denote the virtual work of the external forces shown in Figure 7.14b on the same virtual displacement. Then,

$$\delta W_e^* = \sum_m \int_0^L (-w^*\,\delta v + p^*\,\delta u)\,dx + \sum_{P^*} (P^*\,\delta u_{P^*}) + \sum_{C^*} (C^*\,\delta\theta_{C^*})$$

$$= \sum_m \int_0^L (M^*\,\delta\kappa + N^*\,\delta\varepsilon)\,dx = \delta W_\sigma^*$$

The difference in the external virtual works is

$$\delta W_e - \delta W_e^* = \sum_m \int_0^L [(M - M^*)\,\delta\kappa + (N - N^*)\,\delta\varepsilon]\,dx = \delta W_\sigma - \delta W_\sigma^*$$

in which $M - M^*$ and $N - N^*$ are internal force distributions such that the structure is in equilibrium under the combined loads shown in Figure 7.14a, and the reverse of the loads shown in Figure 7.14b. The combined loading is illustrated in Figure 7.14c. If the loads in Figure 7.14b are chosen to be the same as those shown in Figure 7.14a, then the net external loading is zero and

$$0 = \sum_m \int_0^L [(M - M^*)\,\delta\kappa + (N - N^*)\,\delta\varepsilon]\,dx$$

The result here is not trivial because it is only in the case of a statically determinate structure that there is a unique system of internal forces such that the structure will be in equilibrium for a given set of applied loads. Before going on to a specific example to illustrate the point, it is observed that the relationship still holds if the virtual displacements are replaced by displacements due to any specific cause. The reason is that within the framework of the theory, displacements due to any cause qualify as virtual displacements. In passing it is noted that the foregoing relation, which is independent of material properties, can be used to establish uniqueness of the solution for the internal forces, the member deformations, and the displacements of a structure due to any loading when the structure is supported so that it cannot undergo a rigid displacement and the material obeys Hooke's law.

To solidify these ideas, consider the beam shown in Figure 7.15. The conditions of support are such that the beam is statically indeterminate to the second degree. Because there are no axial loads, $N = 0$.

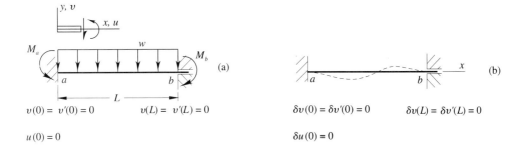

FIGURE 7.15 Statically-indeterminate beam.

A distribution of bending moments such that the beam is in equilibrium under the distributed load w is given by

$$M = -M_a\left(1 - \frac{x}{L}\right) - M_b\frac{x}{L} + \frac{wLx}{2} - \frac{wx^2}{2}$$

where M_a and M_b are arbitrary. With $w^* = w$ and $N^* = 0$, another bending moment distribution that satisfies the equilibrium requirement is given by

$$M^* = -M_a^*\left(1 - \frac{x}{L}\right) - M_a^*\frac{x}{L} + \frac{wLx}{2} - \frac{wx^2}{2}$$

where M_a^* and M_b^* are arbitrary. The difference in the bending moment distributions is

$$M - M^* = -(M_a - M_a^*)\left(1 - \frac{x}{L}\right) - (M_b - M_b^*)\frac{x}{L}$$

An arbitrary virtual displacement must be such that δv, $\delta v'$, and δu are continuous and satisfy the boundary conditions shown in Figure 7.15b. Because $\delta\kappa = \delta v''$ and $N = N^* = 0$,

$$\int_0^L [(M - M^*)\,\delta\kappa + (N - N^*)\,\delta\varepsilon]\,dx = \int_0^L (M - M^*)\,\delta v''\,dx$$

$$= -(M_a - M_a^*)\int_0^L \left(1 - \frac{x}{L}\right)\delta v''\,dx$$

$$- (M_b - M_b^*)\int_0^L \frac{x}{L}\delta v''\,dx$$

Now consider the second of the two integrals in the foregoing equation. Integration by parts gives

$$\int_0^L \frac{x}{L}\delta v''\,dx = \left[\frac{x}{L}\delta v'\right]_0^L - \frac{1}{L}\int_0^L \delta v'\,dx = \left[\frac{x}{L}\delta v'\right]_0^L - \frac{1}{L}\left[\delta v\right]_0^L = 0$$

The last result follows from the boundary conditions on the virtual displacement. The same type of calculation shows that the first integral also has value zero. Therefore,

$$\int_0^L [(M - M^*)\, \delta\kappa + (N - N^*)\, \delta\varepsilon]\, dx = 0$$

7.12 STRAIN ENERGY AND INTERNAL FORCE-DEFORMATION RELATIONS

When external work W_e is done on a body the body is deformed and acquires an ability to do mechanical work. The work that the deformed body can do is conceived as being stored within the body in the form of recoverable internal energy. By the laws of thermodynamics, the recoverable energy U cannot exceed the work W_e, i.e., $U \leq W_e$. It is understood that a body is hyperelastic[1] when $U = W_e$ for all possible modes of deformation. In this case U is a function of the configuration of the body. In mechanics U is called the strain energy of deformation; it is also called the potential energy of deformation. The total strain energy of a body is the sum of the strain energies of its parts. The distribution of the strain energy over the space occupied by the body is defined by a strain energy density function u. The stress-strain or internal force-deformation relations for an elastic body can be derived from the strain energy density function with the aid of the principle of virtual work.

The local deformation of a beam is measured by the curvature κ and the strain ε of the reference axis. The strain energy density is understood to be the strain energy per unit of length of the reference axis and is given the general form

$$\hat{u} = \hat{u}(\kappa, \varepsilon)$$

Let M and N be the bending moment and axial tension at a section of a beam that is in equilibrium, and let κ and ε be the curvature and the strain of the reference axis at the section. Because the beam is elastic, M and N are functions of κ and ε, i.e.,

$$M = M(\kappa, \varepsilon), \quad N = N(\kappa, \varepsilon)$$

These equations are understood to be the internal force-deformation relations.

Let the beam be subjected to an arbitrary virtual displacement (with transverse and axial components $\delta v = \delta v(x)$, $\delta u = \delta u(x)$, respectively) that induces virtual deformations $\delta\kappa$ and $\delta\varepsilon$. Then, by the principle of virtual work, the incremental work of the internal forces on the virtual deformations per unit of length of the beam, δw_σ, is

$$\delta w_\sigma = M\, \delta\kappa + N\, \delta\varepsilon$$

For a (hyper)elastic beam,

$$\delta w_\sigma = \delta\hat{u} = \frac{\partial \hat{u}}{\partial \kappa}\, \delta\kappa + \frac{\partial \hat{u}}{\partial \varepsilon}\, \delta\varepsilon$$

---

[1] The statement that a body is elastic means only that the internal forces are functions of the deformations. Without exception the elastic structures in structural mechanics are understood to be hyperelastic. The practice of using the words hyperelastic and elastic interchangeably is common.

Because κ and ε are independent variables and because the relation $\delta w_\sigma = \delta \hat{u}$ must hold for all $\delta \kappa$ and all $\delta \varepsilon$, it follows that

$$M = \frac{\partial \hat{u}}{\partial \kappa} = M(\kappa, \varepsilon), \quad N = \frac{\partial \hat{u}}{\partial \varepsilon} = N(\kappa, \varepsilon)$$

At this point, it is important to note that the statement that a body is elastic means more than just $\delta w_\sigma = \delta \hat{u}$; it also means, by virtue of the equality and knowledge of the formulation of $\hat{u}$ as a function of the deformations, that δw_σ can be evaluated with the aid of the calculus. In the special case of Hooke's law, $M = EI\kappa$ and $N = EA\varepsilon$, and

$$\delta w_\sigma = M\,\delta\kappa + N\,\delta\varepsilon = EI\kappa\,\delta\kappa + EA\varepsilon\,\delta\varepsilon$$

$$= \frac{\partial}{\partial \kappa}[(EI\kappa^2 + EA\varepsilon^2)/2]\,\delta\kappa + \frac{\partial}{\partial \varepsilon}[(EI\kappa^2 + EA\varepsilon^2)/2]\,\delta\varepsilon$$

From the relationship between δw_σ and $\delta \hat{u}$, and the above result, it follows that the beam is (hyper)elastic and

$$\hat{u} = (EI\kappa^2 + EA\varepsilon^2)/2$$

Starting with the expression for $\hat{u}$, the internal force-deformation relations are derived to be

$$M = \frac{\partial \hat{u}}{\partial \kappa} = EI\kappa, \quad N = \frac{\partial \hat{u}}{\partial \varepsilon} = EA\varepsilon$$

7.13 CONSERVATIVE APPLIED FORCES

The statement that the applied forces on a structure are prescribed is understood to mean that each force is constant in magnitude and in direction. It is demonstrated in statics that the work done by a constant force on a displacement from a point a to another point b is independent of the path taken from a to b (i.e., the work done depends only on the positions of the points a and b). A force for which the work done is independent of the path is said to be conservative. An important characteristic of a conservative force is that it possesses a potential energy function, Ω, which depends only on the position of the force. The potential energy function has the property that the increment in Ω on the displacement from a to b, $\Delta\Omega = \Omega_b - \Omega_a$, is equal to the negative of the work done by the force on the displacement. Only changes in the potential energy function of a force play a role in the application of energy methods to problems in mechanics. This means that with respect to applications the potential energy function need be known only to within an arbitrary additive constant. In applications to structural analysis, the constant is usually taken to be zero so that the potential energy function simply reduces to the negative of the work done by the force on its displacement from any convenient reference position, i.e., in applications the relationship between the potential energy function of a conservative force and the work done by the force is expressed as $\Omega = -W$. From the foregoing discussion it follows that the virtual work of a conservative force can be expressed in terms of the first order increment of its potential energy function in the form $\delta W = -\delta \Omega$. The relationship is not trivial. The basic expression for the evaluation of the virtual work is $\delta W = F\delta u$, where δu is the component of the virtual displacement in the direction of F, whereas $\delta \Omega$ is evaluated with the aid of the calculus, i.e.,

$$\delta \Omega = \left(\frac{\partial \Omega}{\partial s}\right)_u \delta u$$

where $(\partial \Omega/\partial s)_u$ is understood to be the derivative of Ω taken with respect to a coordinate s in the direction of δu.

7.14 PRINCIPLE OF STATIONARY POTENTIAL ENERGY

The principle applies only to a body which made of a hyperelastic material and which is subjected to conservative forces. For such a body a total potential energy function V is defined to be the sum of the potential energy function Ω of the external forces and the strain energy function U for the body, i.e.,

$$V \equiv \Omega + U$$

Since Ω and U are functions of the configuration or, alternatively, functions of the displacements of the body from a reference position, so is V. If the body is subjected to a virtual displacement, V, Ω, and U undergo virtual changes or variations δV, $\delta \Omega$ and δU, and the variations are related by

$$\delta V = \delta \Omega + \delta U$$

From their respective defining relations, $\delta \Omega = -\delta W_e$ and $\delta U = \delta W_\sigma$. Therefore,

$$\delta V = -\delta W_e + \delta W_\sigma$$

By the principle of virtual work the body is in an equilibrium position if and only if $\delta W_e = \delta W_\sigma$ for arbitrary virtual displacements; however, by the previous relationship it follows that the body is in an equilibrium position if and only if $\delta V = 0$ for arbitrary virtual displacements. The equilibrium condition $\delta V = 0$ for arbitrary virtual displacements is an analytical statement of the principle of stationary potential energy.

The principle of stationary potential energy, which is of fundamental importance in the matrix analysis of structures and in the Finite Element method of analysis, derives its name from the fact that a function of several variables $z_1, z_2, \ldots, z_n$ is understood to be stationary at the point $(z_1, z_2, \ldots, z_n)$ when the values of the variables at the point are such that the first-order change in the function is zero for arbitrary infinitesimal changes in the variables. An essential condition for a function of several variables to be a maximum or a minimum at a point is that the function be stationary at the point.

The principle of stationary potential energy and the related principle of minimum potential energy are of most practical use in the analysis of statically indeterminate structures, wherein they are employed to derive equations of equilibrium in terms of displacement variables. The principle of minimum potential energy for a plane frame structure for which the material obeys Hooke's law is taken up next.

7.15 PRINCIPLE OF MINIMUM POTENTIAL ENERGY

The principle of minimum potential energy asserts that a body is in a position of stable equilibrium if and only if the total potential energy V is a relative minimum in that position. The development that follows has two purposes. The first is to establish the stability of equilibrium of a plane-framed structure that (a) is made of a material that obeys Hooke's law, (b) is restrained so that it cannot undergo a rigid displacement, and (c) undergoes infinitesimally small displacements. (All structures considered in this work and in traditional introductory courses in structural mechanics satisfy this limitation.) The second purpose of the development is to provide explicit formulations for the potential energies.

Consider a generalized structure as shown in Figure 7.16. For each member, introduce local coordinates and sign conventions for loads, internal forces, and displacements as set out previously in Figure 7.3.

Let u_P be the component of displacement at the point of application of P in the direction of P, and let θ_C be the rotation of the tangent of the member at the point of application of C in the direction of C. The total work done by the external forces on their displacements is the negative of the potential energy, Ω. Because the external forces are prescribed

$$\Omega = -\sum_m \int_0^L (-wv + pu)\, dx - \sum_P P u_P - \sum_C C \theta_C$$

The material obeys Hooke's law. Accordingly, the total strain energy is given by

$$U = \sum_m \int_0^L \frac{1}{2}(EI\kappa^2 + EA\varepsilon^2)\, dx$$

and the total potential energy is $V = \Omega + U$. Let the structure be subjected to an arbitrary virtual displacement and let the subscript 1 denote the energies of the

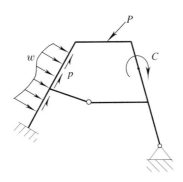

FIGURE 7.16 Stable equilibrium.

structure in the new configuration. Then,

$$\Omega_1 = -\sum_m \int_0^L \int [-w(v + \delta v) + p(u + \delta u)] \, dx$$

$$- \sum_P P(u_P + \delta u_P) - \sum_C C(\theta_C + \delta \theta_C)$$

$$U_1 = \sum_m \int_0^L \frac{1}{2}[EI(\kappa + \delta \kappa)^2 + EA(\varepsilon + \delta \varepsilon)^2] \, dx$$

and $V_1 = \Omega_1 + U_1$. Let $\Delta V = V_1 - V$ be the exact increment in the total potential energy as a consequence of the virtual displacement. Then, from the foregoing expressions for the energies,

$$\Delta V = -\left(\sum_m \int_0^L (-w \, \delta v + p \, \delta u) \, dx + \sum_P P \, \delta u_P + \sum_C C \delta \theta_C\right)$$

$$+ \sum_m \int_0^L (EI \kappa \, \delta \kappa + EA \varepsilon \, \delta \varepsilon) \, dx + \sum_m \int_0^L \frac{1}{2}(EI \, \delta \kappa^2 + EA \, \delta \varepsilon^2) \, dx$$

The first group of terms on the right side of the expression for ΔV is nothing more than $-\delta W_e$. The integrand in the second group of terms is $\delta \hat{u} = \delta w_\sigma$. Therefore, the second group of terms is equal to δW_σ. Thus, the increment in the potential energy reduces to

$$\Delta V = -\delta W_e + \delta W_\sigma + \sum_m \int_0^L \frac{1}{2}(EI \, \delta \kappa^2 + EA \, \delta \varepsilon^2) \, dx$$

If the original position is an equilibrium position then by the principle of virtual work, $-\delta W_e + \delta W = \delta V = 0$ for arbitrary virtual displacements and

$$\Delta V = \sum_m \int_0^L \frac{1}{2}(EI \, \delta \kappa^2 + EA \, \delta \varepsilon^2) \, dx > 0$$

unless $\delta \kappa = 0$ and $\delta \varepsilon = 0$ everywhere. Accordingly, within the framework of the assumed conditions, it follows that the total potential energy is an absolute minimum when the structure is in an equilibrium position, and by the principle of minimum potential energy the equilibrium is stable.

7.16 COMPLEMENTARY STRAIN ENERGY AND FORCE-DEFORMATION RELATIONS

Assume that the force-deformation relations

$$M = M(\kappa, \varepsilon), \quad N = N(\kappa, \varepsilon)$$

can be solved to obtain the deformations κ and ε as functions of the internal force resultants M and N, so that it is possible to express κ and ε in the form

$$\kappa = \kappa(M, N), \quad \varepsilon = \varepsilon(M, N)$$

Then one can introduce a new energy density function $\hat{u}^*$ defined by

$$\hat{u}^* = M\kappa + N\varepsilon - \hat{u} = \hat{u}^*(M, N)$$

$\hat{u}^*$ is called the complementary energy density. If either ε or κ is always zero, a simple graphical interpretation can be given to the strain energy density $\hat{u}$ as the area under an internal force-deformation diagram, and the complementary energy density $\hat{u}^*$ can be given a related interpretation in terms of area. The geometric interpretation of $\hat{u}^*$ is of little value and is omitted here. The complementary energy density function has the special property that

$$\frac{\partial \hat{u}^*}{\partial M} = \kappa, \quad \frac{\partial \hat{u}^*}{\partial N} = \varepsilon$$

The property is easily verified with the aid of the chain rule of the calculus and the properties of the strain energy density function. Consider $\partial \hat{u}^*/\partial M$:

$$\frac{\partial \hat{u}^*}{\partial M} = \kappa + M\frac{\partial \kappa}{\partial M} + N\frac{\partial \varepsilon}{\partial M} - \left(\frac{\partial \hat{u}}{\partial \kappa}\frac{\partial \kappa}{\partial M} + \frac{\partial \hat{u}}{\partial \varepsilon}\frac{\partial \varepsilon}{\partial M}\right)$$

$$= \kappa + M\frac{\partial \kappa}{\partial M} + N\frac{\partial \varepsilon}{\partial M} - \left(M\frac{\partial \kappa}{\partial M} + N\frac{\partial \varepsilon}{\partial M}\right) = \kappa$$

The same type of calculation is used to verify that $\varepsilon = \partial \hat{u}^*/\partial N$.

7.17 THE SPECIAL CASE OF HOOKE'S LAW

In the case of Hooke's law,

$$\hat{u} = \frac{EI\kappa^2}{2} + \frac{EA\varepsilon^2}{2}, \quad M = EI\kappa, \quad N = EA\varepsilon$$

$$\hat{u}^* = M\kappa + N\varepsilon - \hat{u} = \frac{M^2}{2EI} + \frac{N^2}{2EA} = \frac{EI\kappa^2}{2} + \frac{EI\varepsilon^2}{2} = \hat{u}$$

$$\frac{\partial \hat{u}^*}{\partial M} = \frac{M}{EI} = \kappa, \quad \frac{\partial \hat{u}^*}{\partial N} = \frac{N}{EA} = \varepsilon$$

Note that the special relationship $\hat{u}^* = \hat{u}$, valid only for Hooke's law, does not assert that the strain energy and complementary strain energy density functions are the same; they are not. The equation states that the value of $\hat{u}^*$ calculated from

$$\hat{u}^* = \frac{M^2}{2EI} + \frac{N^2}{2EA}$$

and the value of $\hat{u}$ calculated from

$$\hat{u} = \frac{EI\kappa^2}{2} + \frac{EA\varepsilon^2}{2}$$

are equal when κ, ε, M, and N satisfy the internal force-deformation relations $M = EI\kappa$ and $N = EA\varepsilon$.

7.18 BETTI'S THEOREM

Consider a structure so restrained that it cannot undergo a rigid-body displacement, and let the structure be in equilibrium under two different sets of loads identified by subscript labels p and q as shown in Figure 7.17. Let M_p and N_p be the internal force resultants due to the "p" loads, and let M_q and N_q be the internal force resultants due to the "q" loads.

The displacements due to the p loads and the displacements due to the q loads qualify as virtual displacements. Because all forces and displacements satisfy the conditions for which the principle of virtual work applies, the work of the external p loads on the displacements produced by the external q loads is

$$W_e^{pq} = \sum_m \int_0^L (-w_p v_q + p_p u_q)\, dx + \sum_P (P_p u_q) + \sum_C (C_p \theta_q)$$

$$= \sum_m \int_0^L (M_p \kappa_q + N_p \varepsilon_q)\, dx = W_\sigma^{pq}$$

and the work of the external q loads on the displacements produced by the external p loads is

$$W_e^{qp} = \sum_m \int_0^L (-w_q v_p + p_q u_p)\, dx + \sum_P (P_q u_p) + \sum_C (C_q \theta_p)$$

$$= \sum_m \int_0^L (M_q \kappa_p + N_q \varepsilon_p)\, dx = W_\sigma^{qp}$$

Now assume the material obeys Hooke's law. Then,

$$M_p = EI\kappa_p, \quad N_p = EA\varepsilon_p, \quad M_q = EI\kappa_q, \quad N_q = EA\varepsilon_q$$

$$W_e^{pq} = W_\sigma^{pq} = \sum_m \int_0^L (EI\kappa_p \kappa_q + EA\varepsilon_p \varepsilon_q)\, dx$$

$$= \sum_m \int_0^L (EI\kappa_q \kappa_p + EA\varepsilon_q \varepsilon_p)\, dx = W_\sigma^{qp} = W_e^{qp}$$

Therefore, in this very special case,

$$W_e^{pq} = W_e^{qp}$$

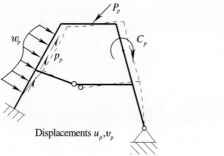
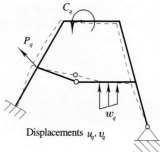

FIGURE 7.17 Betti's theorem.

The foregoing equation is an analytical statement of Betti's theorem. In words, the theorem asserts that:

> If a linear elastic structure is acted upon by two different sets of loads denoted by p and q, respectively, the work of the p system of loads on displacements induced by the q system of loads is equal to the work done by the q system of loads on displacements induced by the p system of loads.

Betti's theorem is the foundation for Maxwell's reciprocal theorem which guarantees the existence of certain symmetries in parameters that characterize a structure.

7.19 MAXWELL'S RECIPROCAL THEOREM

Let the p system of loads consist of just the unit load applied at a point 1 in the direction 1 as shown in Figure 7.18a, and let the q system of loads consist of just the unit load applied at a point 2 in the direction 2 as shown in Figure 7.18b. The displacement induced at point 2 by the unit load at point 1 is denoted $\boldsymbol{\mu}_{21}$ and the component of the displacement in the direction 2 is denoted μ_{21}. Similarly, the displacement induced at point 1 by the unit load at point 2 is denoted $\boldsymbol{\mu}_{12}$ and the component of the displacement in the direction 1 is denoted μ_{12}.

The work of the force in Figure 7.18a on the displacement in Figure 7.18b is $1 \times \mu_{12}$ and the work of the force in Figure 7.18b on the displacement in Figure 7.18a is $1 \times \mu_{21}$. Accordingly, by Betti's theorem,

$$\mu_{12} = \mu_{21}$$

This equation represents Maxwell's reciprocal theorem. In words, the theorem states that:

> The component of displacement at a point one in direction one due to a unit load at a point two in direction two is equal to the component of displacement at the point two in direction two due to a unit load at point one in direction one.

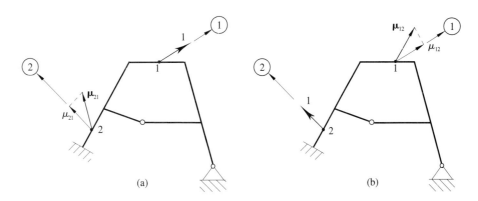

FIGURE 7.18 Maxwell's reciprocal theorem.

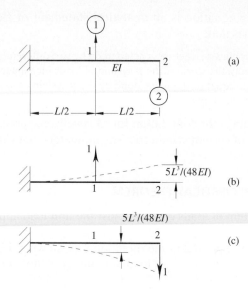

FIGURE 7.19 Maxwell's theorem; both unit loads are forces.

The theorem also applies if either or both of the unit loads is a unit couple. In the case of a couple the associated displacement is understood to be rotation at the point of application of the couple in the direction of the couple.

Example 7.9

Verify Maxwell's reciprocal theorem for the uniform cantilever beam with points 1 and 2 and directions 1 and 2 as shown in Figure 7.19a. The unit load at point 1 produces the displacement at point 2 as shown in Figure 7.19b (the result is easily verified with the aid of the second moment-area theorem). Because the displacement at point 2 is in the opposite sense to the direction 2, $\mu_{21} = -5L^3/(48EI)$. The unit load at point 2 induces displacements as shown in Figure 7.18c. Because the deflection at point 1 is in the opposite sense to the direction 1, $\mu_{12} = -5L^3/(48EI)$. Clearly, $\mu_{12} = \mu_{21}$, and the theorem is verified. It should be observed that if the direction of either one of the loads is changed, the coefficients μ_{12} and μ_{21} will both be positive and the theorem still holds.

Example 7.10

Verify Maxwell's reciprocal theorem for the uniform cantilever beam with points 1 and 2 and directions 1 and 2 as shown in Figure 7.20a. Because the direction 2 is that of a clockwise rotation, the unit load at 2 becomes a unit couple as shown in Figure 7.20c. The unit load at 1 induces a counterclockwise rotation of the tangent at 2 as shown in Figure 7.20b. Therefore, $\mu_{21} = -L^2/(8EI)$. The unit couple at 2 induces a displacement in the sense opposite to the direction 1 as shown in Figure 7.20c and $\mu_{12} = -L^2/(8EI)$. Because $\mu_{12} = \mu_{21}$, the theorem is verified.

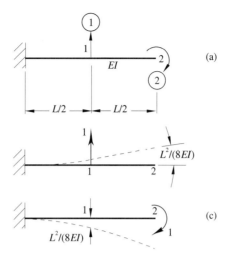

FIGURE 7.20 Maxwell's theorem for a unit force and a unit couple.

Note that Maxwell's theorem does not assert that a rotation is equal to a linear displacement. The theorem simply makes a statement about how the magnitude and direction of the rotation and of the linear translation are related. That the results are dimensionally correct is easily verified by keeping in mind that the unit load is dimensionless and the related bending moments have the dimension of length, and because the unit couple is dimensionless the related bending moments are also dimensionless.

7.20 CASTIGLIANO'S THEOREMS

The theorems are applicable only to a (hyper)elastic body that is restrained so that it cannot undergo a rigid-body displacement and for which the deformations and rotations are infinitesimally small. The theorems are established herein for plane-framed structures. To take full advantage of the theorem, the notion of loading a structure is expanded to include the prescription of non-zero displacements at isolated points. In most practical applications the prescribed non-zero displacements are associated with the support system for the structure, and the problems in which prescribed displacements come into play are referred to as problems in the analysis of structures with support settlement. Of course, if there is no settlement of a support the prescribed displacement is necessarily zero. In the development of the theorem, explicit use is made of the fact that at a given point in a given direction, either the component of displacement or force, but not both, can be prescribed. If the force is prescribed the displacement is included among the unknown quantities to be determined as part of the analysis. If the displacement is prescribed the associated force, which has its line of action parallel to the displacement, is an unknown in the sense of a reaction force to be determined as part of the analysis. The situation is clarified with the aid of the structure shown in Figure 7.21. In Figure 7.21 all loads with the exception of the reaction force R and the reactions at the supports are prescribed. The displacements at the

204 Chapter 7 Energy Methods (Part 2)

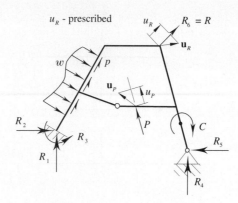

FIGURE 7.21 Forces in Castigliano's theorems.

supports have prescribed values of zero and u_R has a prescribed non-zero value. The reaction force R and the forces of reaction at the supports as well as the internal forces are among the unknown forces to be determined. Observe that even if there are no prescribed loads, the force R, the reactions at the supports, and the internal forces will be non-zero unless the structure, with R included as an unknown, is statically determinate.

The derivations involve the use of an energy function Φ which, for a plane-framed structure, is defined by the equation

$$\Phi \equiv \sum_m \int_0^L \hat{u}^* \, dx - \sum_r R u_R = \sum_m \int_0^L \hat{u}^*(M, N) \, dx - \sum_r R u_R$$
$$= \sum_m \int_0^L (M\kappa + N\varepsilon - \hat{u}) \, dx - \sum_r R u_R$$

in which M and N represent the true internal forces and R represents the true reaction forces due to the simultaneous action of the prescribed loads and prescribed displacements. The range index m indicates that the sum is over all members, and the range index r indicates that the sum is taken over all reaction forces. The forces are such that the equations of equilibrium, the force-deformation relations, the compatibility equations, and the prescribed displacement conditions are satisfied throughout the structure. The function Φ has no specific natural physical interpretation. It may be noted, however, that the first sum in the definition is the sum of the complementary energies of the parts of the structure. In the second sum, the term $R u_R$ can be interpreted as the work that R would do on the displacement u_R when R is thought of as an applied load.

7.21 THEOREM ON DEFLECTIONS

The theorem is developed on the basis of the assumption that M and N and all reaction forces can be expressed in terms of the concentrated force P. It follows that, by its defining equation, the energy function Φ can be expressed in the form $\Phi = \Phi(P)$. Let

$\delta\Phi$ be the infinitesimal change in Φ due to the infinitesimal change δP in P while all remaining prescribed loads remain constant. Then, by the calculus

$$\delta\Phi = \sum_m \int_0^L \delta\hat{u}^* \, dx - \sum_r \delta R \, u_R$$

$$= \sum_m \int_0^L [\delta M \, \kappa + \delta N \, \varepsilon + (M \, \delta\kappa + N \, \delta\varepsilon - \delta\hat{u})] \, dx - \sum_r \delta R \, u_R$$

which, because $\delta\hat{u} = M \, \delta\kappa + N \, \delta\varepsilon$, reduces to

$$\delta\Phi = \sum_m \int_0^L (\delta M \, \kappa + \delta N \, \varepsilon) \, dx - \sum_r \delta R \, u_R$$

The bending moments M, the tensions N, and the reaction forces R (which includes all support reactions) satisfy the equations of equilibrium written with respect to the undeformed geometry of the structure. This means that δM and δN and the forces δR (as well as the associated changes in the reactions at the supports) also satisfy the same equations of equilibrium when the only external loads acting on the structure are δP and the reaction forces δR as shown in Figure 7.22. The displacements and corresponding deformations due to the actual loads qualify as virtual displacements and deformations. Accordingly, it follows from the principle of virtual work that

$$\sum_m \int_0^L (\delta M \, \kappa + \delta N \, \varepsilon) \, dx = \delta P \, u_P + \sum_r \delta R \, u_R$$

where u_P is the component of displacement at the point of application of P in the direction of P.

Substitution into the expression for $\delta\Phi$, followed by a rearrangement of terms, yields the formula

$$u_P = \frac{\delta\Phi}{\delta P} \to \frac{\partial\Phi}{\partial P}$$

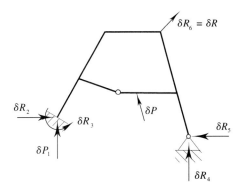

FIGURE 7.22 Incremental forces in Castigliano's theorem on deflections.

A similar analysis shows that the rotation of the tangent at the point of application of the couple C, in the direction of the couple, is given by

$$\theta_C = \frac{\partial \Phi}{\partial C}$$

The last two equations are analytical representations of Castigliano's theorem on deflections.

In application to the calculation of deflections of a statically determinate structure, there is no practical distinction between the use of Castigliano's theorem and the method of virtual work. The procedures for a statically indeterminate structure are significantly different, especially if the material does not obey Hooke's law. Due to the restrictions imposed in the derivation of the theorem, its application to statically indeterminate structures becomes increasingly difficult as the degree of indeterminacy increases. No more effort (and often less effort) is required to apply the method of virtual work to a statically indeterminate structure than to a statically determinate structure. The point of difficulty lies in the determination of the force increments δM and δN due to the load increment δP. In the case of a statically determinate structure, by the meaning of the term statically determinate, the internal forces and therefore their increments are determined by statics alone; they are independent of material properties. The calculations are the same as those used to determine the internal forces due to the unit load when the method of virtual work is applied to calculate deflections of a statically determinate structure. When Castigliano's theorem is used to calculate deflections of a statically indeterminate structure, the internal-force increments must be derived by application of the calculus to the internal-force functions that satisfy (a) the equations of equilibrium, (b) the force-deformation relations for the material, and (c) the conditions of compatibility for the structure. In the case of a statically indeterminate structure, the internal forces under the action of the unit load need only satisfy the equations of equilibrium when the method of virtual work is used to calculate deflections. The essence of the foregoing discussion may be summed up as follows: The difficulty of applying Castigliano's theorem to calculating a deflection in a statically indeterminate structure stems from the fact that with respect to procedures there is an infinite number of internal-force functions that can be used in the calculation, but the theorem requires the selection of one of the functions by a particular rule.

7.22 THEOREM ON DEFLECTIONS SPECIALIZED FOR HOOKE'S LAW

When the material obeys Hooke's law the energy function Φ reduces to

$$\Phi = \sum_m \int_0^L \left(\frac{M^2}{2EI} + \frac{N^2}{2EA} \right) dx - \sum_r R u_R$$

and the deflection is given by

$$u_P = \frac{\partial \Phi}{\partial P} = \sum_m \int_0^L \left(\frac{M}{EI} \frac{\partial M}{\partial P} + \frac{N}{EA} \frac{\partial N}{\partial P} \right) dx - \sum_r \frac{\partial R}{\partial P} u_R$$

By the principle of superposition,

$$M = M_1 + mP, \qquad N = N_1 + nP, \qquad R = R_1 + rP$$

where M_1 and N_1 are the internal forces and R_1 the reaction forces due to the prescribed displacements u_R and all loads except P; and, m and n are the internal forces and r the reaction forces when the assigned displacements u_R are zero and $P = 1$ is the only assigned load acting on the structure.[2] Clearly,

$$m = \frac{\partial M}{\partial P}, \qquad n = \frac{\partial N}{\partial P}, \qquad r = \frac{\partial R}{\partial P}$$

and the deflection is calculated from

$$u_P = \sum_m \int_0^L \left(\frac{mM}{EI} + \frac{nN}{EA} \right) dx - \sum_r r u_R$$

which is of exactly the same form as results when the principle of virtual work is applied to calculate u_P.

7.23 THE THEOREM OF LEAST WORK

Castigliano's theorem of least work, which is trivially true for a statically determinate structure, is of considerable practical value in the analysis of statically indeterminate structures. As in the development of the theorem on deflections, let M and N be the true internal forces induced in the members and let R represent the true reaction forces due to the simultaneous action of the prescribed loads and prescribed displacements. Recall that the true forces are such that the equations of equilibrium, the force-deformation relations, the compatibility equations, and the prescribed displacement conditions are satisfied throughout the structure. For the true forces,

$$\begin{aligned}\Phi &= \sum_m \int_0^L \hat{u}^* \, dx - \sum_r R u_R = \sum_m \int_0^L \hat{u}^*(M, N) \, dx - \sum_r R u_R \\ &= \sum_m \int_0^L (M\kappa + N\varepsilon - \hat{u}) \, dx - \sum_r R u_R\end{aligned}$$

Now let $\tilde{M} = M + \delta M$, $\tilde{N} = N + \delta N$, $\tilde{R} = R + \delta R$ be any other set of internal forces and reactions such that the equations of equilibrium (written, as is required for consistency with the small displacement theory of structures, with respect to the undeformed geometry of the structure) are satisfied for the same set of prescribed loads. If the structure is statically determinate there is only one set of internal forces and reactions that satisfy the equations of equilibrium for the prescribed loads. If the structure is statically indeterminate, there is an infinity of choices for $\tilde{M}$, $\tilde{N}$, and the reactions $\tilde{R}$ for each degree of indeterminacy. Under the conditions of the theory, it follows that

---

[2] These expressions are valid whether the structure is statically determinate or statically indeterminate. For the statically determinate structure, M, N, and R and m, n, and r are found by statics alone. To satisfy the conditions under which the theorem on deflections was derived, for the statically indeterminate structure all of these quantities must be constructed so as to satisfy Hooke's law and the compatibility conditions as well as the equations of equilibrium.

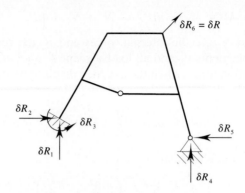

FIGURE 7.23 Incremental forces in Castigliano's theorem of least work.

the reaction increments and δR taken together with the incremental internal force resultants δM and δN comprise a system of forces that satisfy the equations of equilibrium written with respect to the undeformed geometry of the structure. The system of external incremental forces is shown in Figure 7.23.

Because the force increments are infinitesimally small the change $\delta\Phi$ in the function Φ can be evaluated by the rules of the calculus. Thus, as in the development of the theorem on deflections,

$$\delta\Phi = \sum_m \int_0^L \delta\hat{u}^* \, dx - \sum_r \delta R \, u_R$$
$$= \sum_m \int_0^L [\delta M \, \kappa + \delta N \, \varepsilon + (M \, \delta\kappa + N \, \delta\varepsilon - \delta\hat{u})] \, dx - \sum_r \delta R \, u_R$$

which, because $\delta\hat{u} = M \, \delta\kappa + N \, \delta\varepsilon$, reduces to

$$\delta\Phi = \sum_m \int_0^L (\delta M \, \kappa + \delta N \, \varepsilon) \, dx - \sum_r \delta R \, u_R$$

From the equilibrium conditions satisfied by δM, δN, δR, and the associated reaction increments, and from the principle of virtual work,

$$\sum_m \int_0^L (\delta M \, \kappa + \delta N \, \varepsilon) \, dx = \sum_r \delta R \, u_R$$

Therefore, $\delta\Phi = 0$. The condition $\delta\Phi = 0$ is an analytical statement of Castigliano's theorem of least work. The theorem states that

> Of all the internal force and reaction distributions that satisfy the equations of equilibrium, the true distribution is such that the energy function Φ is a minimum.

The derivation of the theorem as presented is valid as long as the material is hyperelastic. In other words, the theorem is valid for nonlinearly elastic materials. Hooke's law need not be satisfied. $\delta\Phi = 0$ is only a necessary condition for Φ to be a minimum. Satisfaction of the condition does not guarantee that the function is a minimum. However, it is easy to show that if the material obeys Hooke's law the true internal forces and reactions make the energy function Φ an absolute minimum.

When all of the prescribed displacements have value zero and the material obeys Hooke's law,

$$\Phi \to \sum_m \int_0^L \left(\frac{M^2}{2EI} + \frac{N^2}{2EA} \right) dx = \sum_m \int_0^L \left(\frac{EI\kappa^2}{2} + \frac{EA\varepsilon^2}{2} \right) dx = U = W_e$$

The equality of the energy function Φ and the work W_e when $u_R = 0$ and the material obeys Hooke's law is the motivation for the name of the theorem of least work.

7.24 APPLICATION OF CASTIGLIANO'S THEOREMS

Practical application of the theorem of least work is limited to the analysis of statically indeterminate structures for which the degree of indeterminacy is small and direct formulation of the compatibility conditions in terms of redundant forces is difficult. What makes the theorem so valuable in such situations is that it automatically takes into account the particular compatibility conditions to be satisfied for whatever choice is made for the redundant forces. Because full appreciation of the utility of the theorem comes with experience in the analysis of statically indeterminate structures by force methods, applications of the theorem are deferred to Chapter 9 in which force methods are taken up.

The analytical complexities and strong limitations associated with the theorem on deflections, and the inferiority of the theorem, compared to the method of virtual work, as a basis for computation, limits its practical utility to displacement calculation of statically determinate structures that satisfy Hooke's law. As will be observed in the example that follows, for a statically determinate structure for which Hooke's law applies, calculations based on the theorem are essentially the same as the calculations based on virtual work.

Example 7.11

Use Castigliano's theorem on deflections to calculate the vertically downward deflection at point b of the simple beam shown in Figure 7.24a. The beam and the problem are the same as those considered previously to illustrate application of the method of virtual work. To derive maximum benefit from this example, all aspects of the solution process by means of Castigliano's theorem on deflections should be compared with those in Example 7.2 in the section in this chapter dealing with the application of the method of virtual work to the calculation of beam deflections.

The absence of a vertical concentrated force at b does not pose a real problem with respect to application of the theorem on deflections. All that need be done to remedy the situation is to supply a force P at b as shown in Figure 7.24b, and evaluate the deflection for $P = 0$,

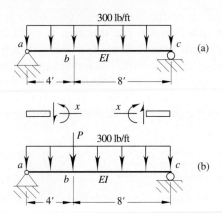

FIGURE 7.24 Theorem on deflections applied to a simple beam.

i.e., evaluate $(\partial \Phi/\partial P)_{P=0} = u_b$. To facilitate the calculations, the beam is treated as a structure composed of two members, *ab* and *bc*. The tension N in both members is zero. The coordinates and sign convention for evaluation of the bending moments are as shown in Figure 7.24b. The bending moments and their derivatives with respect to P are:

member *ab*: $M = 1800x - 150x^2 + \dfrac{2Px}{3}$, $\quad \dfrac{\partial M}{\partial P} = \dfrac{2x}{3} \quad 0 \leq x \leq 4$

member *bc*: $M = 1800x - 150x^2 + \dfrac{Px}{3}$, $\quad \dfrac{\partial M}{\partial P} = \dfrac{x}{3} \quad 0 \leq x \leq 8$

By the theorem on deflections, the deflection at *b* for $P = 0$ is

$$\left(\dfrac{\partial \Phi}{\partial P}\right)_{P=0} = u_b = \sum_m \int_0^L \left[\dfrac{M(\partial M/\partial P)}{EI}\right]_{P=0} dx$$

The bending moments and their derivatives evaluated for $P = 0$ are

member *ab*: $M = 1800x - 150x^2$, $\quad \dfrac{\partial M}{\partial P} = \dfrac{2x}{3} \quad 0 \leq x \leq 4$

member *bc*: $M = 1800x - 150x^2$, $\quad \dfrac{\partial M}{\partial P} = \dfrac{x}{3} \quad 0 \leq x \leq 8$

Substitution into the expression for the deflection at *b* gives

$$u_b = \int_0^4 \dfrac{1}{EI}(1800x - 150x^2)\left(\dfrac{2x}{3}\right) dx + \int_0^8 \dfrac{1}{EI}(1800x - 150x^2)\left(\dfrac{x}{3}\right) dx = \dfrac{70{,}400}{EI} \text{ lb-ft}^3$$

Observe that the final equation for the evaluation of u_b and the final result are exactly the same as those in the solution of the problem by the method of virtual work.

KEY POINTS

1. Within the framework of the small displacement theory of structures, displacements due to any specific cause always qualify as virtual displacements.

2. In applications of the method of virtual work to deflection calculations, the only requirements imposed on the reaction forces and the internal forces is that they satisfy the equations of equilibrium. For a statically determinate structure there is only one such set of forces. For a statically indeterminate structure, there is an infinity of sets of reaction forces and internal forces that satisfy all equilibrium requirements, and every set will yield the same value for the desired component of displacement.

3. Within the framework of the small displacement theory, the equilibrium of any structure that is supported so that a rigid-body displacement is not possible, and for which the material obeys Hooke's law, is always stable.

4. Included in the necessary conditions for application of Castigliano's theorems is the requirement that the material be hyperelastic; however, Hooke's law need not be satisfied.

PROBLEMS

Note: Problems 7.1–7.21 are the same as problems 6.1–6.21.

7.1 Use the method of virtual work to determine the vertical component of displacement and the rotation at B and at C.

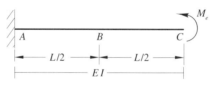

PROBLEM 7.1

7.2 Use the method of virtual work to determine the vertical component of displacement and the rotation at A and at B.

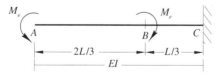

PROBLEM 7.2

7.3 Use the method of virtual work to determine the vertical component of displacement and the rotation at A and at B.

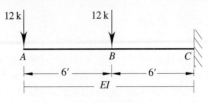

PROBLEM 7.3

7.4 Use the method of virtual work to determine the vertical component of displacement at C and the rotation at B.

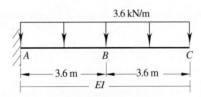

PROBLEM 7.4

7.5 Use the method of virtual work to determine the vertical component of displacement at B and at C and the rotation at A.

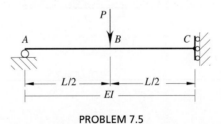

PROBLEM 7.5

7.6 Use the method of virtual work to determine the rotation at A and at B and the vertical component of displacement at B.

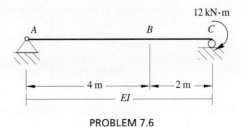

PROBLEM 7.6

7.7 Use the method of virtual work to determine the vertical component of displacement at B and the rotation at C.

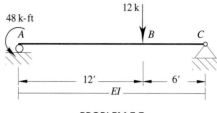

PROBLEM 7.7

7.8 Use the method of virtual work to determine the rotation at B and the vertical component of displacement at A and C.

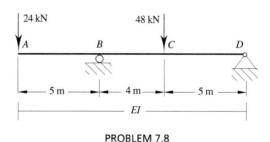

PROBLEM 7.8

7.9 Use the method of virtual work to determine the vertical component of displacement at A, C, and E.

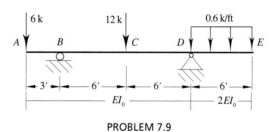

PROBLEM 7.9

7.10 Use the method of virtual work to determine the vertical component of displacement at B and C, and the rotation at A.

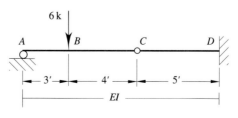

PROBLEM 7.10

7.11 Use the method of virtual work to determine the vertical component of displacement at B and D and the rotation at C.

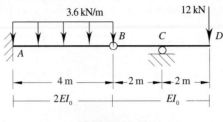

PROBLEM 7.11

7.12 Use the method of virtual work to determine the vertical component of displacement at B and D. Use virtual work or the moment-area method to determine the vertical component of displacement at C.

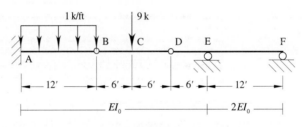

PROBLEM 7.12

7.13 Use the method of virtual work to determine the rotation at B. Use the method of virtual work or the moment-area method to determine the vertical component of displacement at the B-end of member AB and the B-end of member BD.

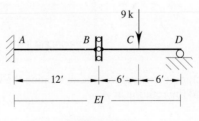

PROBLEM 7.13

7.14 Use the method of virtual work to determine the rotation at D. Use the method of virtual work or the moment-area method to determine the vertical component of displacement at the D-end of member ACD, at the D-end of member DF, and at E.

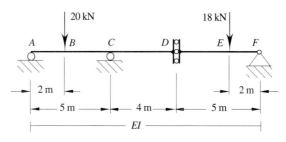

PROBLEM 7.14

7.15 Use the method of virtual work to determine the vertical component of displacement at B. Use the method of virtual work or the moment-area method to determine the vertical component of displacement at E.

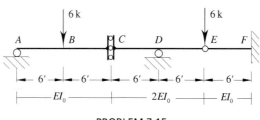

PROBLEM 7.15

7.16 Use the method of virtual work to determine the vertical component of displacement at D. Use the method of virtual work or the moment-area method to determine the vertical component of displacement at B.

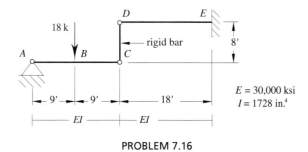

PROBLEM 7.16

7.17 Use the method of virtual work to determine the vertical component of displacement at D. Use the method of virtual work or the moment-area method to determine the vertical component of displacement at B.

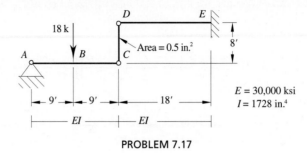

PROBLEM 7.17

7.18 Use the method of virtual work to determine the horizontal and vertical components of displacement of B. The deformation is inextensional.

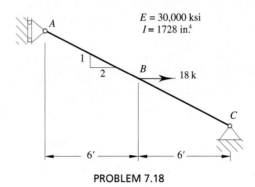

PROBLEM 7.18

7.19 Use the method of virtual work to determine the horizontal component of displacement at B. Use any combination of the method of virtual work and/or the moment-area method to determine the horizontal component of displacement at D and the vertical component of displacement at B and at D. The deformation is inextensional. EI is constant.

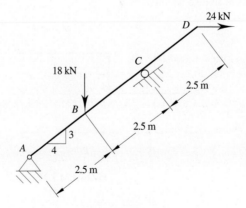

PROBLEM 7.19

7.20 Use the method of virtual work to determine the horizontal component of displacement at D. Use any combination of the method of virtual work and the moment-area method to determine the vertical component of displacement at D. The deformation is inextensional. The uniform beam has bending stiffness EI and weighs 520 lb/ft.

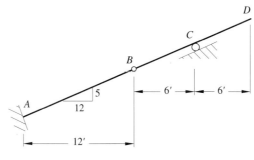

PROBLEM 7.20

7.21 Use the method of virtual work to determine the horizontal component of displacement at B. Use any combination of the method of virtual work and the moment-area method to determine the vertical component of displacement at B. The deformation is inextensional. EI is constant.

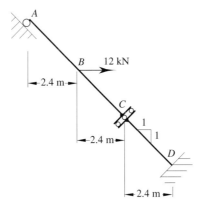

PROBLEM 7.21

7.22–7.30 General instructions: Use the method of virtual work to calculate at least one of the required quantities. Use any combination of the method of virtual work and the moment-area method to determine the remaining quantities. In all cases the deformation is inextensional. Problems 7.22–7.30 are the same as problems 6.22–6.30.

7.22 Determine the rotation of joint *B* and the horizontal and vertical components of displacement of *C*. Follow the general instructions for problems 7.22–7.30.

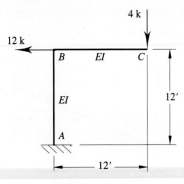

PROBLEM 7.22

7.23 Determine the rotation of joint *B* and the horizontal and vertical components of displacement of *A*. Follow the general instructions for problems 7.22–7.30.

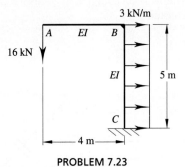

PROBLEM 7.23

7.24 Determine the rotation of joint *B* and the horizontal and vertical components of displacement of *C*. Follow the general instructions for problems 7.22–7.30.

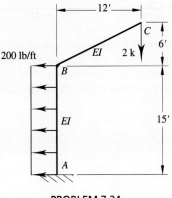

PROBLEM 7.24

7.25 Determine the rotation of joint B, the horizontal component of displacement of joint B, and the vertical component of displacement of C. Follow the general instructions for problems 7.22–7.30.

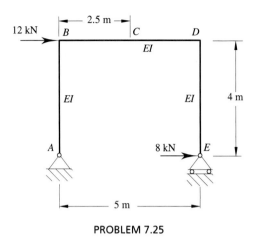

PROBLEM 7.25

7.26 Determine the horizontal and vertical components of displacement of A. Follow the general instructions for problems 7.22–7.30.

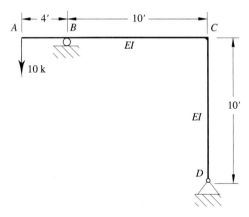

PROBLEM 7.26

7.27 Determine the horizontal component of displacement at *A* and the vertical component of displacement at *C*. Follow the general instructions for problems 7.22–7.30.

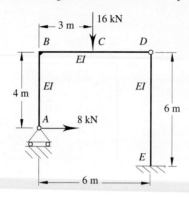

PROBLEM 7.27

7.28 Determine the vertical component of displacement at *B* and the horizontal component of displacement at *D*. Follow the general instructions for problems 7.22–7.30.

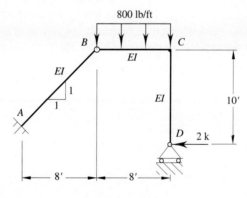

PROBLEM 7.28

7.29 Determine the vertical component of displacement at *A* and at *D*. Follow the general instructions for problems 7.22–7.30.

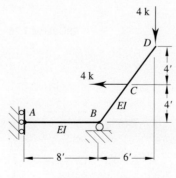

PROBLEM 7.29

7.30 Determine the vertical component of displacement at *A* and at *C*. Follow the general instructions for problems 7.22–7.30.

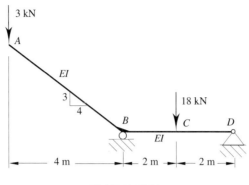

PROBLEM 7.30

7.31–7.36 General instructions: Use any combination of the method of virtual work and/or the moment-area method to obtain the required quantities.

7.31 Determine the horizontal component of displacement at *B* and at *C*, and the vertical component of displacement at *D*. Follow the general instructions for problems 7.31–7.36.

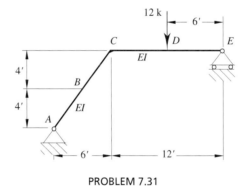

PROBLEM 7.31

7.32 Determine the vertical component of displacement at *B* and at *C*, and the horizontal component of displacement at *D*. Follow the general instructions for problems 7.31–7.36.

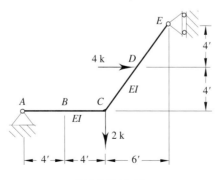

PROBLEM 7.32

7.33 Determine the horizontal component of displacement at B and at F, and the vertical component of displacement at D. Follow the general instructions for problems 7.31–7.36.

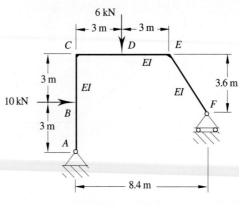

PROBLEM 7.33

7.34 Determine the vertical component of displacement at B and E and the horizontal component of displacement at D. EI is constant. Follow the general instructions for problems 7.31–7.36.

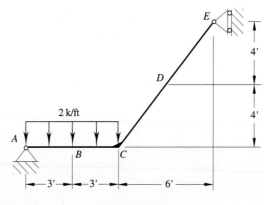

PROBLEM 7.34

7.35 Determine the vertical component of displacement at D and the horizontal component of displacement at B and at F. EI is constant. Follow the general instructions for problems 7.31–7.36.

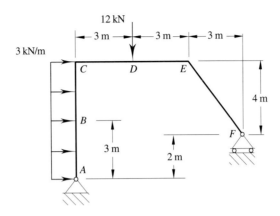

PROBLEM 7.35

7.36 Determine the vertical component of displacement at C and at D, and the horizontal component of displacement at E. EI is constant. Follow the general instructions for problems 7.31–7.36.

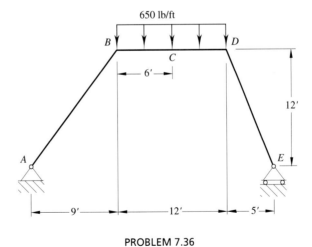

PROBLEM 7.36

7.37 Use the method of virtual work to determine the vertical component of displacement of joints g and h due to fabrication errors of $\Delta L_{hd} = -0.24$ inches and $\Delta L_{bc} = 0.36$ inches.

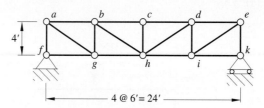

PROBLEM 7.37

7.38–7.43 General instructions: Use the method of virtual work to determine the required displacements. For each case of loading display the bar tensions on a line diagram of the truss. Arrange the calculations in tabular form as in the example problems.

7.38 Use the method of virtual work to determine the vertical component of displacement of joint g and the horizontal component of displacement of joint c. Follow the general instructions for problems 7.38–7.43.

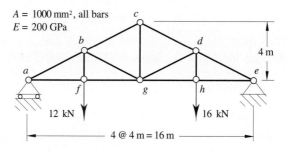

PROBLEM 7.38

7.39 Use the method of virtual work to calculate the horizontal and vertical components of displacement of joint f. Follow the general instructions for problems 7.38–7.43.

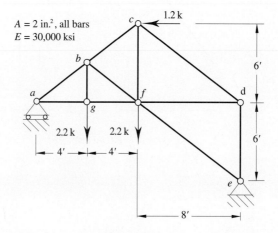

PROBLEM 7.39

7.40 Use the method of virtual work to calculate the vertical component of displacement of joints b and h. $E = 200$ GPa. Follow the general instructions for problems 7.38–7.43.

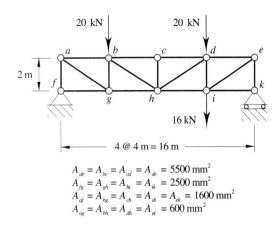

$A_{ab} = A_{bc} = A_{cd} = A_{de} = 5500$ mm^2
$A_{fg} = A_{gh} = A_{hi} = A_{ik} = 2500$ mm^2
$A_{af} = A_{bg} = A_{ch} = A_{di} = A_{ek} = 1600$ mm^2
$A_{ag} = A_{bh} = A_{dh} = A_{ei} = 600$ mm^2

PROBLEM 7.40

7.41 Use the method of virtual work to calculate the vertical component of displacement of joints f and g. $E = 200$ GPa. Follow the general instructions for problems 7.38–7.43.

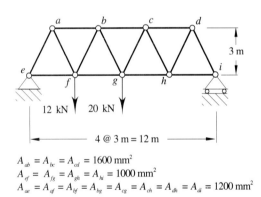

$A_{ab} = A_{bc} = A_{cd} = 1600$ mm^2
$A_{ef} = A_{fg} = A_{gh} = A_{hi} = 1000$ mm^2
$A_{ae} = A_{af} = A_{bf} = A_{bg} = A_{cg} = A_{ch} = A_{dh} = A_{di} = 1200$ mm^2

PROBLEM 7.41

7.42 Use the method of virtual work to calculate the vertical component of displacement of joint b and the horizontal component of displacement of joint f. $E = 30{,}000$ ksi. Follow the general instructions for problems 7.38–7.43.

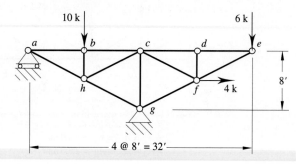

$A_{ab} = A_{bc} = A_{cd} = A_{de} = 2 \text{ in}^2$
$A_{ah} = A_{hg} = A_{gf} = A_{fe} = 3 \text{ in}^2$
$A_{bh} = A_{ch} = A_{cg} = A_{cf} = A_{df} = 1.5 \text{ in}^2$

PROBLEM 7.42

7.43 Use the method of virtual work to calculate the vertical component of displacement of joints e and f. $E = 200$ GPa. Follow the general instructions for problems 7.38–7.43.

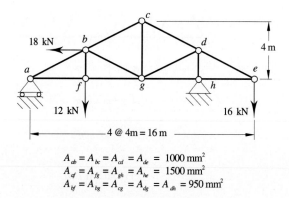

$A_{ab} = A_{bc} = A_{cd} = A_{de} = 1000 \text{ mm}^2$
$A_{af} = A_{fg} = A_{gh} = A_{he} = 1500 \text{ mm}^2$
$A_{bf} = A_{bg} = A_{cg} = A_{dg} = A_{dh} = 950 \text{ mm}^2$

PROBLEM 7.43

7.44 a) Derive the equation for the deflection function $v(x)$ for the cantilever beam under the action of the distributed load. b) Derive the equation for the deflection function $v(x)$ for the cantilever beam under the action of the load at the tip. c) Calculate the work of the distributed load on the deflection due to the concentrated load and compare it to the work done by the concentrated load on the deflection due to the distributed load. d) Is Betti's theorem satisfied?

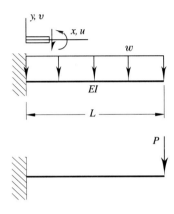

PROBLEM 7.44

7.45 Verify Betti's theorem for the two cases of loading shown.

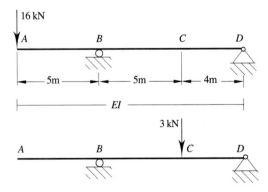

PROBLEM 7.45

7.46 Verify Betti's theorem for the two cases of loading shown.

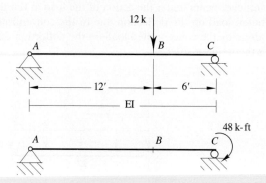

PROBLEM 7.46

7.47 Verify Betti's theorem for the two cases of loading shown.

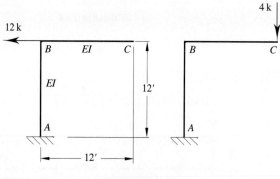

PROBLEM 7.47

7.48 Verify Betti's theorem for the two cases of loading shown. $A = 900$ mm^2 for all members. $E = 200$ GPa.

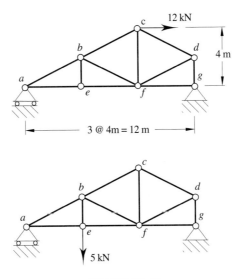

PROBLEM 7.48

7.49 Verify the reciprocal theorem.

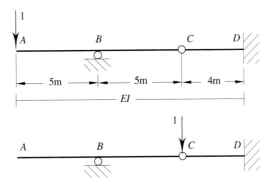

PROBLEM 7.49

7.50 Verify the reciprocal theorem.

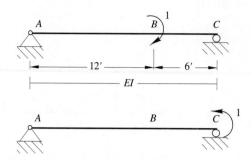

PROBLEM 7.50

7.51 Verify the reciprocal theorem.

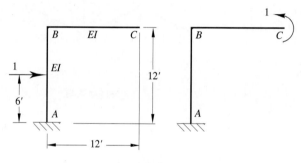

PROBLEM 7.51

7.52 Verify the reciprocal theorem. $A=2$ in.^{2} for all members. $E=30,000$ ksi.

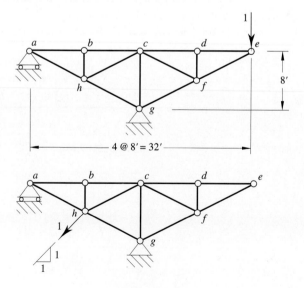

PROBLEM 7.52

CHAPTER 8

Statically Indeterminate Structures

8.1 INTRODUCTION

A review of basic ideas introduced in Chapter 1 is essential to any discussion of statically indeterminate structures. In the overwhelming majority of practical situations, a structure is designed so that deformations and displacements due to the applied loads are so small that the equations of equilibrium may be formulated on the basis of the undeformed geometry. Accordingly, a complete set of conditions for equilibrium is formed as a set of ordinary linear simultaneous equations that can be expressed in the form

$$a_{11}R_1 + a_{12}R_2 + \cdots + a_{1m}R_m = b_1$$
$$a_{21}R_1 + a_{22}R_2 + \cdots + a_{2m}R_m = b_2$$
$$\cdots$$
$$a_{n1}R_1 + a_{n2}R_2 + \cdots + a_{nm}R_m = b_n$$

in which: $R_1, R_2, \ldots, R_m$ are the unknown independent components of the forces of reaction and interaction; the coefficients $a_{11}, a_{12}, \ldots, a_{nm}$ of the unknown forces are constants; and the coefficients $b_1, b_2, \ldots, b_n$ are linear in the applied loads. For arbitrary loading, b can be made to take on any value. In the discussion to follow no distinction is made between a reaction force or a force of interaction; any one of these is simply referred to as a reaction component or an unknown force.

A fully restrained structure has been defined to be such that each of its members, considered as a rigid body, remains at rest for arbitrary applied loads. Alternatively, a deformable structure is fully restrained only if it cannot undergo rigid-body displacements. It follows from the definition that for such a structure there always exists at least one set of forces of reaction and interaction such that the equations of equilibrium are satisfied for arbitrary applied loads. Useful structures are designed to be fully restrained and these are the only kinds of structures under consideration here. In this case m is always equal to or greater than n and (if necessary, by suitable renaming of the unknown forces) the equations of equilibrium can always be rewritten in the form

$$a_{11}R_1 + a_{12}R_2 + \cdots + a_{1n}R_n = b_1 - a_{1,n+1}R_{n+1} - a_{1,n+2}R_{n+2} - \cdots - a_{1,m}R_m$$
$$a_{21}R_1 + a_{22}R_2 + \cdots + a_{2n}R_n = b_2 - a_{2,n+1}R_{n+1} - a_{2,n+2}R_{n+2} - \cdots - a_{2,m}R_m$$
$$\cdots$$
$$a_{n1}R_1 + a_{n2}R_2 + \cdots + a_{nn}R_n = b_n - a_{n,n+1}R_{n+1} - a_{n,n+2}R_{n+2} - \cdots - a_{n,m}R_m$$

in which det (A), the determinant of the coefficients of the unknown reaction components $R_1, R_2, \ldots, R_n$, is non-zero, i.e.,

$$\det(A) = \begin{vmatrix} a_{11}, a_{12}, \ldots, a_{1n} \\ a_{21}, a_{22}, \ldots, a_{2n} \\ \ldots \\ a_{n1}, a_{n2}, \ldots, a_{nn} \end{vmatrix} \neq 0$$

If $m = n$, the equations can be solved uniquely for $R_1, R_2, \ldots, R_n$ in terms of the applied loads. In this case the structure is said to be statically determinate because the unknown forces are determined uniquely from statics alone. For $m > n$, the fact that det $(A) \neq 0$ guarantees that one can always find values for $R_1, R_2, \ldots, R_n$ such that equilibrium will be satisfied for arbitrary loading and this will be so for all possible values for $R_{n+1}, R_{n+2}, \ldots, R_m$, including the values $R_{n+1} = R_{n+2} = \ldots, = R_m = 0$. Because the best that can be done solely on the basis of the equations of equilibrium is to solve for $R_1, R_2, \ldots, R_n$ in terms of $R_{n+1}, R_{n+2}, \ldots, R_m$, the reaction components and, therefore, the structure are said to be statically indeterminate, and $d = m - n$ is the degree of indeterminacy. (It may be, and often is, possible to solve for some of $R_1, R_2, \ldots, R_n$ in terms of the applied loads alone, but this is not relevant to the discussion at hand.) Because $R_{n+1}, R_{n+2}, \ldots, R_m$ are not essential for equilibrium, these reaction components and the mechanical devices that generate these reactions are understood to be redundant with respect to satisfaction of the conditions of equilibrium. It is observed that removal of the constraints or mechanical devices corresponding to the redundant reaction components produces a modified, statically determinate structure. The modified structure is referred to as the *primary* or *cut-back structure* corresponding to the redundant forces $R_{n+1}, R_{n+2}, \ldots, R_m$. In general there are several possible distinct choices for the redundant reaction components (in reality the number of choices is infinite) and each choice results in a different primary or cut-back structure. An important feature of the cut-back structure is that, with the exception of the points at which the redundant reaction components act, the compatibility conditions for the cut-back structure and the given or parent structure are the same.

Clearly, the analysis of a statically indeterminate structure will be more demanding than that of a statically determinate structure, and it is reasonable to expect that the effort involved increases with the degree of indeterminacy. In view of this, it is reasonable to inquire about the benefits of statically indeterminate construction. The most obvious benefit is the backup available to maintain equilibrium in the event of failure of a restraint. Two additional benefits are (1) improved distribution of internal-force resultants, which leads to greater economy in use of materials; and (2) reduced deflections. It is not within the scope of this work to discuss the relative merits of statically determinate and statically indeterminate construction beyond this point.

The conceptual process of generating an associated statically determinate cut-back structure from a given statically indeterminate structure by removing redundant constraints can be applied in reverse to generate the given statically indeterminate structure from an associated statically determinate cut-back structure by introducing the constraints. The function of a constraint (a connection or support) is to enforce a condition of compatibility of displacement that is consistent with the character of the constraint. Because the members that make up a framed structure are represented by their respective reference lines, the associated constraints or connections are understood to be point connections.

8.2 GENERAL PROCEDURE OF ANALYSIS

The general procedure for a complete analysis of any structure consists of formulating and solving simultaneously these governing equations:

1. the equations of equilibrium
2. the displacement-deformation relations
3. the force-deformation relations
4. the compatibility conditions

As noted above, for structures that undergo small deformations and displacements, the equations of equilibrium are linear in the unknown forces of reaction and interaction. The displacement-deformation equations and the compatibility equations are also linear in displacements and the measures of deformation. The force-deformation relations characterize the material. If the material obeys Hooke's law, as is assumed to be the case herein, the force-deformation equations are linear. It follows from the linearity of the governing equations that the principle of superposition applies.

8.3 MEMBER RIGIDITY

Often, for the purpose of reducing the total computational effort required in a manual analysis, one or more members in a framed structure may be assumed to be partially or wholly rigid. For example, common practice in the manual analysis of a frame is to assume that the members are axially rigid. Ordinarily, the goal envisioned at the time the assumption was introduced is achieved; i.e., *all* of the unknown forces of reaction and interaction have been determined with reduced computational effort. It is to be noted, however, that introducing too many or inappropriate material rigidities may put a complete analysis beyond reach. In other words, the statically indeterminate structure remains so and is not amenable to a complete analysis as long as all of the assumed rigidities are retained. The special and general cases of partially or wholly rigid members are distinguished by the fact that it is only in the special cases that the forces throughout the structure are determined completely by the deformations of the members, member elasticities, and conditions of equilibrium. The goal of a practical analysis is a complete determination of the unknown forces of reaction and interaction. Accordingly, implicit in this work is the assumption that forces throughout a structure are completely defined once the deformed state of the structure has been determined.

8.4 SOLUTION STRATEGIES

Solution of the governing equations for the analysis of a structure has generally been pursued along two distinct lines of attack, both of which are motivated by the structure of the equations and the objectives of structural analysis. The procedures are readily seen to be particular arrangements of elimination methods of solving simultaneous linear algebraic equations.

8.5 FORCE METHOD

In the first line of attack, $m - n$ unknown redundant reaction components are taken as the fundamental mechanical variables. The equations of equilibrium, the force-deformation relations, and the deformation-displacement relations are then used to express the $m - n$ compatibility equations associated with the chosen redundant reaction components in terms of the redundant reaction components. Solution of the equations yields the redundant reaction components. All remaining quantities are obtained by back substitution into the governing equations. One need not formulate all of the basic relations to arrive at the appropriate compatibility equations in terms of redundant forces. Names have been given to particular procedures for arriving at the desired equations in the desired form. Examples are the method of consistent deformation, the three-moment equation method for beams, and the method of least work. Methods of analysis in which forces are taken as the fundamental mechanical variables, and the compatibility conditions are expressed in terms of forces, are referred to as force methods of analysis.

8.6 DISPLACEMENT METHOD AND KINEMATIC INDETERMINACY

In the second line of attack, displacements are taken as the fundamental mechanical variables. Additional concepts are essential to relating the purely mathematical aspects of solving equations to practical applications of principles of mechanics. Basic ideas and related terminology are set out here. Details will be taken up later.

It will be shown that the mechanical state of any member in an elastic framed structure is determined completely by the loads between the ends of the member, the displacements at the ends of the member, and the rotations of the tangents at the ends of the member. Accordingly, before being joined to form the structure, each member of a plane frame is essentially a six-degree-of-freedom mechanical system. The plane frame obtained by joining the members and providing supports is a mechanical system that has degrees of freedom equal to six times the number of members minus the number of compatibility conditions associated with the joints and supports. The position coordinates (see Chapter 5) for such a structure may be taken to be the independent components of displacements of the joints.

A framed structure for which the displacements of all joints are prescribed is said to be kinematically determinate. A simple example of a kinematically determinate structure is a beam with built-in ends. A structure for which the displacements of one or more joints is determined by the applied loads is understood to be kinematically indeterminate. The degree of indeterminacy is equal to the number of independent position coordinates or components of joint displacements and rotations that are not prescribed. In other words the degree of kinematic indeterminacy of the structure and the degrees of freedom of the structure as a mechanical system are the same. With the aid of the principle of virtual work it can be shown that there is one independent equation of equilibrium corresponding to each degree of freedom in a mechanical system (see Chapter 5). In view of this and the preceding discussion it follows that for a structure that has n degrees of kinematic indeterminacy there are n independent equations of equilibrium that can be expressed in terms of n independent components of joint displacements. Solution of the equations yields the displacements. All remaining unknowns are obtained by back substitution into displacement-deformation relations, force-deformation relations, and equations of equilibrium.

One need not formulate all of the basic relations and go through a formal elimination process to derive the appropriate equations of equilibrium in terms of displacements. In fact, such a process is neither necessary nor desirable. Names have been given to particular procedures for arriving at the desired equations in the desired form. Examples are the slope-deflection method and the method of stationary potential energy (sometimes called the method of minimum potential energy). Methods of analysis in which displacements are taken as the fundamental mechanical variables and equations of equilibrium are expressed in terms of displacements are referred to as displacement methods of analysis. On the surface, the Hardy Cross method of moment distribution, which will be discussed, appears to be a force method of analysis. At the foundation level, moment-distribution is a displacement method of analysis.

8.7 FLEXIBILITY AND STIFFNESS: FLEXIBILITY COEFFICIENTS

The concepts of flexibility and flexibility coefficients (or, more simply, flexibilities) are useful and play an important role in the analysis of statically indeterminate structures by force methods. The basic idea of flexibility is brought out with the aid of a simple linear elastic spring model. The statement that a spring is linearly elastic means that the relationship between the applied force and the deformation of the spring as measured by elongation is linear as shown in Figure 8.1.

Flexibility of the spring refers to the amount of deformation that is induced by an applied force. It is understood that the greater the flexibility, the greater the deformation for a given applied force. The measure of the flexibility of a spring (not necessarily linearly elastic) at any stage of deformation is the flexibility coefficient at that stage which is defined to be equal to the increment of deformation per unit of increment of applied load. On the basis of this definition, for the linear elastic spring the flexibility (flexibility

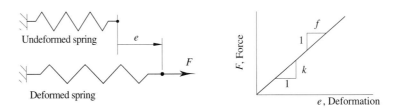

FIGURE 8.1 Flexibility and stiffness.

coefficient), denoted by f, is the same at all stages of deformation and is equal to the total deformation of the spring e divided by the force F acting on the spring.

The stiffness coefficient of a spring at any stage of deformation is the measure of the incremental resistance to incremental deformation at that stage. By definition the stiffness or stiffness coefficient k, at any stage of deformation, is the incremental force required per unit of incremental deformation. For the linear elastic spring the stiffness is a constant and is equal to the force acting on the spring divided by the elongation of the spring. Clearly, the stiffness and the flexibility are mutual reciprocals.

In the case of a spring that is restrained as shown in Figure 8.1, the flexibility of the spring relates the applied force to the displacement at the point of application of the force. An extended definition of flexibility, which embodies the basic idea involved in the case of a spring, is employed in structural analysis. The extended definition has its foundation in the character of the relationships that exist between forces and displacements of structures. In the case of a fully restrained structure for which the linear theory is applicable, the component of displacement at any point 1 measured in a given direction is always a linear function of the force applied in a given direction at any other point 2. The proportionality constant of the relationship measures the flexibility of the structure at point 1 in the selected direction at point 1 with respect to loading at point 2 in the selected direction at point 2. By definition the proportionality constant is the flexibility or flexibility coefficient of the structure at point 1 in the selected direction at point 1 with respect to loading at point 2 in the selected direction at point 2. This flexibility is also commonly referred to as the influence coefficient for displacement at point 1 in the specified direction at point 1 as a consequence of loading at point 2 in the specified direction at point 2. Because the relationships are linear, the flexibility coefficient for point 1 in the specified direction is equal to the component displacement induced at point 1 in the specified direction as a consequence of application of a unit load at point 2 in the selected direction at point 2. To clarify and reinforce the concept of flexibility, consider the structure shown in Figure 8.2.

In Figure 8.2b, the unit load is applied at point 2 in direction 2. The component of displacement at 2 in direction 2 due to the unit load is μ_{22} and the component of

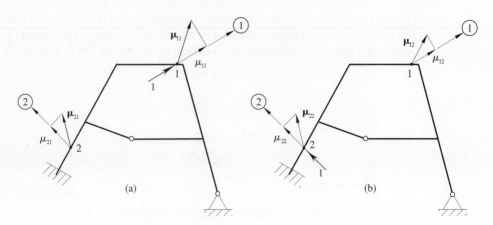

FIGURE 8.2 Flexibilities for a structure.

displacement at point 1 in direction 1 is μ_{12}. By the definition, the flexibility at point 2 in direction 2 due to loading at 2 in direction 2 is $f_{22} = \mu_{22}$ and the flexibility at 1 in direction 1 due to loading at 2 in direction 2 is $f_{12} = \mu_{12}$. The displacements at points 1 and 2 in the directions 1 and 2 when the unit load is placed at 1 in direction 1 are shown in Figure 8.2a. The associated flexibility coefficients are $f_{11} = \mu_{11}$ and $f_{21} = \mu_{21}$.

The loadings used in establishing flexibility coefficients satisfy the conditions under which Maxwell's reciprocal theorem is established. From the theorem, it follows that the *cross flexibilities* f_{21} and f_{12} must be equal. With the aid of Betti's theorem and the fact that the energy function Φ (see Chapter 7) is always positive, it is relatively easy to show that the *direct flexibilities* f_{11} and f_{22} are always positive and, in addition, $f_{11}f_{22} - (f_{12})^2$ is always positive.

8.8 STIFFNESS COEFFICIENTS

A generalization of the concept of stiffness and an extended definition of stiffness coefficient play an important role in the analysis of structures by displacement methods. As in the case of flexibility, the generalizations have their foundation in the character of the relationships that exist between forces and displacements in structures for which the linear theory is valid. Basic ideas relating to stiffness coefficients are brought out with the aid of sketches.

In Figure 8.3a, u_1 is a prescribed displacement at point 1 in direction 1. The prescribed displacement at point 2 is $u_2 = 0$ for all values of u_1. F_{11} and F_{21} are the forces required at points 1 and 2, respectively, to maintain the prescribed conditions. F_{11} is the rectangular component of F_{11} in direction 1, and F_{21} is the rectangular component of F_{21} in direction 2. The force components F_{11} and F_{21} are proportional to the displacement u_1. The proportionality constants k_{11} and k_{21} are understood to be stiffness coefficients corresponding to the directions 1 and 2 for the prescribed mode of displacement. The terminology is motivated by the fact that the relationship between each of the forces and the displacement u_1 is the same as the relationship between the spring force and the deformation of a linear elastic spring of stiffness k. The form of

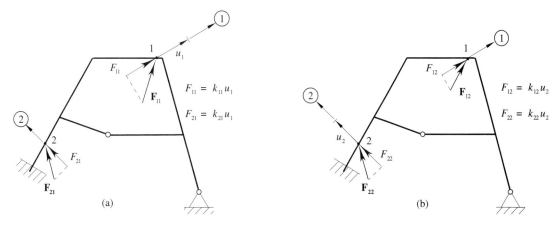

FIGURE 8.3 Stiffnesses for a structure.

the relationship between force and displacement suggests a procedure for calculating the stiffness coefficients. With $u_1 = 1$, $F_{11} = k_{11}$ and $F_{21} = k_{21}$. Clearly, the procedure is to assign $u_{11} = 1$, $u_2 = 0$ and then calculate the forces required at 1 and 2 to maintain the prescribed conditions. With prescribed displacements at points 1 and 2 as shown in Figure 8.3b, the corresponding stiffness coefficients for directions 1 and 2 are k_{12} and k_{22}, respectively.

The structure and force systems shown in Figure 8.3 satisfy the conditions for application of Betti's theorem. It follows from application of the theorem that $k_{12} = k_{21}$. It was shown in Chapter 7 that the strain energy of a structure for which the linear theory applies is always positive and equal to the external work done on the structure in taking it from the undeformed state to the deformed state. For the conditions shown in Figure 8.3a, the work done on the structure is $W_e = (1/2)k_{11}u_1^2 > 0$, from which it follows that $k_{11} > 0$. The same kind of calculation shows that $k_{22} > 0$ too. A similar but somewhat more involved calculation shows that $k_{11}k_{22} - (k_{12})^2 > 0$.

KEY POINTS

1. The governing equations applicable to the analysis of any structure consist of (a) the equations of equilibrium; (b) the displacement-deformation equations; (c) the force-deformation equations; and (d) the compatibility equations.
2. The principle of superposition applies to structures for which the linear theory is valid.
3. Two basic strategies are used in solving the governing equations. The strategies are identified according to the selection of the basic mechanical variables. A strategy based on forces as the basic mechanical variables is called a force method of analysis. A strategy based on displacements is called a displacement method.
4. The concept of flexibility is central to force methods of analysis, and the reciprocal concept of stiffness is central to displacement methods of analysis.

CHAPTER 9

Force Methods

9.1 METHOD OF CONSISTENT DEFORMATION

The observations regarding constraints and how a statically indeterminate structure may be generated from an associated cut-back structure form the basis for a fundamental scheme of structural analysis that is referred to as *the method of consistent deformation*. This is a force method of analysis because redundant forces are taken as the basic mechanical variables and conditions of compatibility are expressed in terms of forces.

To illustrate basic ideas, consider the beam shown in Figure 9.1a. It is obvious that the beam is fully restrained. Because the four unknown reaction forces are related by three independent equations of equilibrium, the beam is statically indeterminate to the first degree. Among the possible choices for the redundant reaction are A_1, A_2, and B_1. A_3 cannot be chosen as the redundant reaction because equilibrium cannot be sustained under arbitrary loads with $A_3 = 0$. That A_3 cannot be chosen as the redundant reaction is easily verified by simply noting that removal of that part of the support that prevents axial translation (the part that develops A_3) of the beam results in a structure that is incompletely constrained.

The cut-back structure corresponding to B_1 as the redundant reaction is the cantilever beam shown in Figure 9.1b. It is observed that, with the exception of the condition on displacement at the right end, the governing equations for both beams are the same. It follows that the mechanical state of the cantilever beam under the simultane-

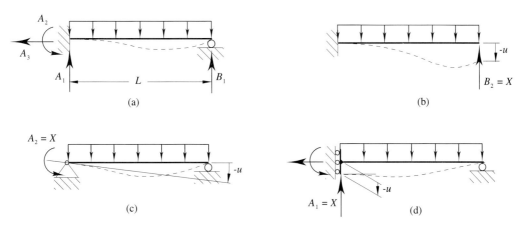

FIGURE 9.1 Cut-back structures.

ous action of B_1 and the applied load can be made to coincide with that of the given statically indeterminate beam by choosing B_1 so that the deflection of the tip of the cantilever beam is zero.

If A_2 is chosen as the redundant reaction, the cut-back structure is a simply supported beam as shown in Figure 9.1c. With the exception of the condition on rotation of the tangent at the left end, the governing equations for both beams are the same. Clearly, the mechanical state of the simply supported beam under the simultaneous action of the applied load and A_2 can be made to coincide with the mechanical state of the given statically indeterminate beam by choosing A_2 so that the net rotation of the tangent at the left end of the simply supported beam is zero.

The beam shown in Figure 9.1d is the cut-back structure corresponding to the selection of A_1 as the redundant reaction. The support shown at the left end of the beam is capable of developing an axial force and a couple as indicated in the figure. The cut-back structure is statically determinate. Again, it is observed that with the exception of the condition on displacement at the left end, the governing equations for the given beam and the cut-back structure are the same. By choosing A_1 so that the deflection at the left support is zero, the mechanical state of the beam in Figure 9.1d can be made to coincide with that of the given statically indeterminate beam.

The ideas and reasoning embodied in the beam example carry over to continuous beams, frames, and trusses with more than one degree of indeterminacy. The frame shown in Figure 9.2a is statically indeterminate to the second degree. The cut-back or primary statically determinate structure corresponding to the choice of the reaction couple of moment X_1 at the left support and the horizontal reactive force X_2 at the right support as the redundant reactive forces is shown in Figure 9.2b. The mechanical state of the given frame and the mechanical state of the cut-back structure will coincide if X_1 and X_2 are chosen so that under the combined action of the applied loads, X_1, and X_2, the net rotation u_1 of the tangent at the left support and the net horizontal displacement u_2 at the right support are zero.

The beam in Figure 9.1 and the frame in Figure 9.2 are examples of structures that are understood to be externally statically indeterminate because the reactions cannot be determined from statics alone. The reactions at the supports of the statically indeterminate structure shown in Figure 9.3 can be found from statics. Accordingly, the

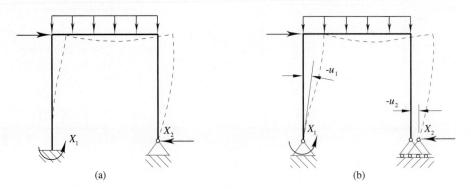

FIGURE 9.2 Statically indeterminate frame.

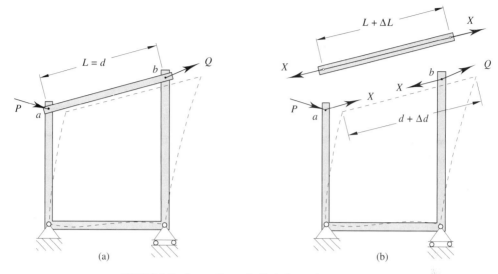

FIGURE 9.3 Internally statically-indeterminate structure.

structure is understood to be statically determinate externally and statically indeterminate internally. The cut-back structure, corresponding to selection of the tension in the elastic link as the redundant force, is the frame shown in Figure 9.3b. The mechanical states of the primary structure and the link (considered together as a mechanical system) will be the same as that of the given structure if the redundant force X is chosen so that ΔL, the increase in the length of the link, is equal to the increase in the distance Δd between the pins at points a and b of the frame.

9.2 USE OF SUPERPOSITION

Evaluation of the mechanical state of a cut-back structure under the simultaneous action of the applied loads and redundant forces is most conveniently handled by application of the principle of superposition. The effects produced on a cut-back structure by the applied loads and by each redundant force are evaluated separately. By the principle of superposition, the effect produced by redundant X is simply X times the effect when the redundant force is assigned a unit value. The separate effects are then added algebraically. The redundant forces are obtained by solving the simultaneous equations that result when the combined action of all forces on the cut-back structure is to be such that the compatibility conditions for the parent statically indeterminate structure are satisfied.

9.3 EXAMPLE PROBLEMS

The principles involved in the method of consistent deformation are the same for all types of structures. For easy reference, however, examples relating to beams, frames, and trusses are placed in separate groups.

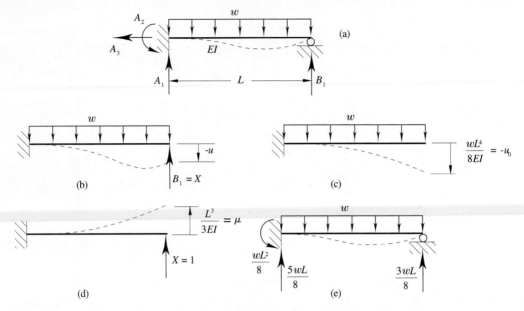

FIGURE 9.4 First solution.

Example 9.1

Determine the reactions for the propped cantilever beam shown in Figure 9.4. The problem is solved twice. In the first solution B_1 is taken as the redundant reaction. In the second solution A_2 is taken as the redundant reaction.

First Solution The beam is statically indeterminate to the first degree. The cut-back or primary structure corresponding to B_1 as the redundant reaction is the cantilever beam as shown in Figure 9.4b, in which u is the displacement at B, measured positive in the direction of X, due to the combined action of the applied load and the force X. The function of the redundant reaction is to enforce the boundary condition that the vertical displacement at support B of the given beam be zero. The mechanical states of the cut-back structure (Figure 9.4b) and the given structure (Figure 9.4a) will coincide if X is chosen so that the net displacement u at B of the cut-back structure is zero. By the principle of superposition, the displacement can be expressed in the form $u = u_0 + \mu X$, where u_0 is the displacement due to the applied load acting alone, and μ is the displacement when $X = 1$ acts alone. These displacements are easily evaluated using either the moment-area method or the principle of virtual work. The results are as indicated in Figure 9.4c and 9.4d, respectively. Therefore,

$$u = -\frac{wL^4}{8EI} + \frac{L^3}{3EI} X$$

Setting $u = 0$ gives $X = 3wL/8$. The remaining reactions are found from equilibrium. The results are displayed in Figure 9.4e.

Second Solution The couple A_2 in Figure 9.4a is taken as the redundant reaction. The corresponding cut-back structure is the simply supported beam shown in Figure 9.5a in which u is

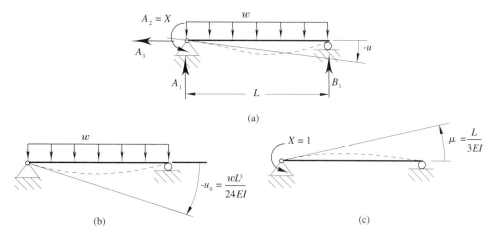

FIGURE 9.5 Second solution.

the rotation of the tangent at A, measured positive in the direction of the couple X, due to the combined action of the couple and the applied load. In this case the function of the redundant reaction is to enforce the boundary condition that the rotation of the tangent at A be zero. The mechanical states of the cut-back structure (Figure 9.5a) and the given structure (Figure 9.4a) will be the same if the couple X is chosen so that the rotation u of the tangent at A in Figure 9.5a is zero. The rotation u_0 due to the applied load acting alone is shown in Figure 9.5b, and the corresponding rotation u_0 due to a the unit couple $X = 1$ acting alone is shown in Figure 9.5c.

By the principle of superposition,

$$u = u_0 + \mu X = -\frac{wL^3}{24EI} + \frac{L}{3EI} X$$

Setting $u = 0$ gives $X = wL^2/8$, which is exactly as found in the first solution.

Example 9.2

Determine the reactions and the deflection at point C of the beam shown in Figure 9.6a. The beam is statically indeterminate to the first degree. With couple D_2 chosen as the redundant reaction, the corresponding cut-back structure is as shown in Figure 9.6b. The function of the redundant reactive couple (Figure 9.6a) is to enforce the condition that the rotation of the tangent at D be zero. Clearly, the mechanical states of the given beam and the cut-back structure will coincide if the couple of moment X is chosen so that the rotation u of the tangent at D of the cut-back structure is zero. Displacements of the cut-back structure when $X = 0$ and when $X = 1$ is the only load on the structure are shown in Figure 9.6c and Figure 9.6d, respectively. By the principle of superposition, for the cut-back structure,

$$u = -\frac{7wL^3}{24EI} + \frac{L}{EI} X, \quad v = \frac{7wL^4}{24EI} - \frac{2L^2}{3EI} X$$

Setting $u = 0$ yields $X = 7wL^2/24$. Substitution of this result into the expression for the deflection at C gives $v = 7wL^4/(72EI)$. The remaining reactions are found from statics. The results are summarized in Figure 9.6e.

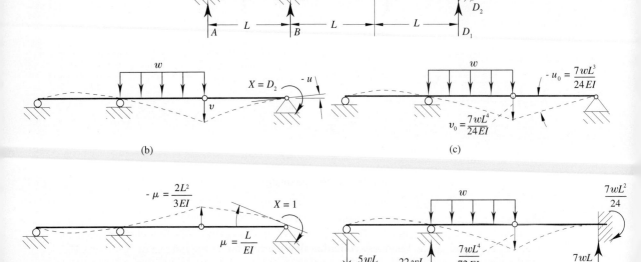

FIGURE 9.6 Compound beam.

Example 9.3

Determine the reactions for the continuous beam shown in Figure 9.7. Evaluate the reactions and construct the shear and bending moment diagrams for $w = 2$ k/ft and $L = 14$ ft. The beam is statically indeterminate to the second degree. The problem is solved twice. In the first solution, the reactions at A and B are taken as the redundant forces. In the second solution, the bending moment in the beam over the support at B and the bending moment at C are taken as the redundant forces.

First Solution The cut-back structure corresponding to $X_1 = A$ and $X_2 = B$ as the redundant reactions is the cantilever beam shown in Figure 9.7b in which u_1 and u_2 are displacements measured positive in the directions of X_1 and X_2, respectively. Clearly, the mechanical states of the cut-back structure and the given structure will coincide if X_1 and X_2 are chosen so that $u_1 = 0$ and $u_2 = 0$. The displacements at A and B due to the applied load with $X_1 = X_2 = 0$ are u_{01} and u_{02}, respectively, as shown in Figure 9.7c. With $X_1 = 1$ as the only load on the cut-back structure, the displacement at A is μ_{11} and the displacement at B is μ_{21} as shown in Figure 9.7d. As shown in Figure 9.7e, when $X_2 = 1$ is the only load on the cut-back structure, the corresponding displacements are μ_{12} and μ_{22}. μ_{11}, μ_{21}, μ_{12}, and μ_{22} are the influence (or flexibility) coefficients for displacements as discussed in Chapter 8. Observe that $\mu_{12} = \mu_{21}$, as required by Maxwell's reciprocal theorem. By superposition the displacements are

$$u_1 = u_{01} + \mu_{11}X_1 + \mu_{12}X_2 = -\frac{2wL^4}{EI} + \frac{8L^3}{3EI}X_1 + \frac{5L^3}{6EI}X_2$$

$$u_2 = u_{02} + \mu_{21}X_1 + \mu_{22}X_2 = -\frac{17wL^4}{24EI} + \frac{5L^3}{6EI}X_1 + \frac{L^3}{3EI}X_2$$

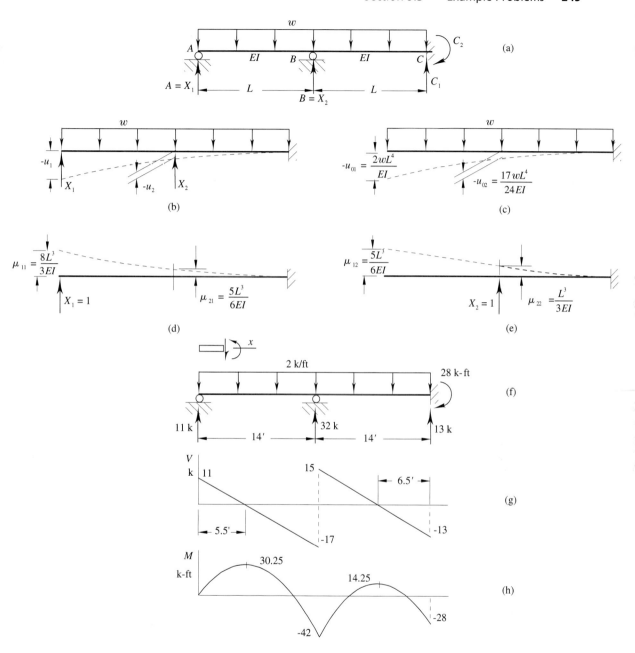

FIGURE 9.7 Continuous beam.

The solution of the equations for $u_1 = u_2 = 0$ is $X_1 = 11wL/28$, $X_2 = 32wL/28$ and from statics $C_2 = wL^2/14$. The reactions and the shear and bending moment diagrams for $w = 2$ k/ft and $L = 14$ ft are as shown in Figure 9.7f, g, and h, respectively.

Second Solution With the bending moments at B and C taken to be the redundant forces, the cut-back structure consists of a pair of simply supported beams with a common support at B as shown in Figure 9.8b. In Figure 9.8b u_1 is the rotation of the tangent $t_{B[AB]}$ at the B-end of beam AB with respect to the rotation of the tangent $t_{B[BC]}$ at the B-end of BC. For the given beam the tangent at B is continuous and the tangent at C is horizontal. The discontinuity u_1 in the tangent at B and the rotation u_2 of the tangent at C in the cut-back structure are controlled by the couples X_1 and X_2. The mechanical states of the given beam and the cut-back structure can be made to coincide by choosing X_1 and X_2 so that $u_1 = 0$ and $u_2 = 0$ simultaneously. Expressions for u_1 and u_2 are formulated with the aid of the principle of superposition and the results displayed in Figure 9.8c, d, and e. Thus,

$$u_1 = \frac{wL^3}{12EI} + \frac{2L}{3EI}X_1 + \frac{L}{6EI}X_2$$

$$u_2 = \frac{wL^3}{24EI} + \frac{L}{6EI}X_1 + \frac{L}{3EI}X_2$$

The solution of the equations for $u_1 = u_2 = 0$ is $X_1 = -3wL^2/28$, $X_2 = -wL^2/14$. For $w = 2$ k/ft and $L = 14$ ft, $X_1 = -42$ k-ft and $X_2 = -28$ k-ft. These results coincide with those obtained for the bending moments at B and C in the first solution. It should be observed that the flexibilities μ_{12} and μ_{21} shown in Figure 9.8d and e are equal as required by Maxwell's reciprocal theorem.

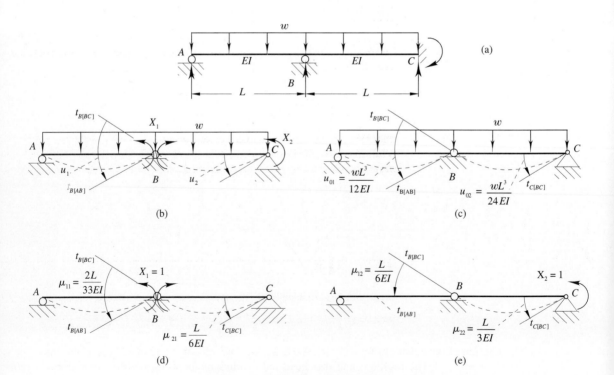

FIGURE 9.8 Internal forces as redundant forces.

Example 9.4

Determine the reactions, the horizontal component of displacement of joint C, and the rotation of joint C of the frame shown in Figure 9.9a. Assume the members are axially rigid. The frame is statically indeterminate to the first degree. Reaction components A_1 and D_1 can be determined from statics alone. Either A_2 or D_2 is a logical choice for the redundant reactive force; $D_2 = X$ is selected in this example. The cut-back structure corresponding to D_2 as the redundant reaction is shown in Figure 9.9b. The mechanical states of the given structure and the cut-back structure will coincide if the redundant reaction is chosen so that for the cut-back structure the displacement u at D is zero. Superposition is used to obtain the desired quantities.

The geometry of the structure, the condition of inextensional deformation, and the displacements to be calculated suggest that displacements be evaluated by use of (small-angle) geometry and the moment-area theorems. Displacements corresponding to $X = 0$ with only the applied load acting are shown in Figure 9.9c. Displacements due to the loading $X = 1$ and no other loads are shown in Figure 9.9d. The results shown in Figure 9.9c and d were obtained using

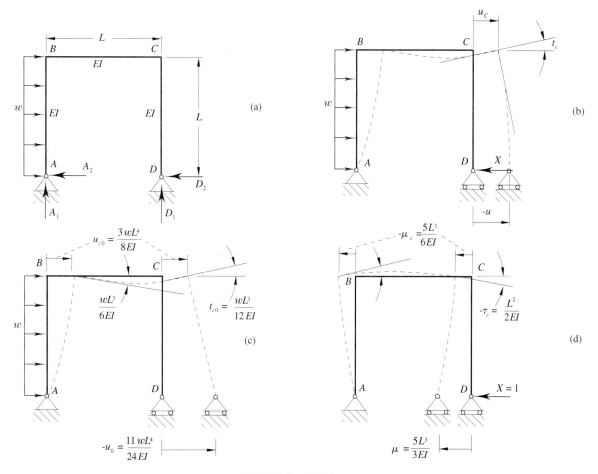

FIGURE 9.9 Rigid frame.

the moment-area method. In each case, the condition of inextensional deformation together with the boundary conditions at A and D, the compatibility conditions at B and C, and the moment-area theorems were used to establish the tangents at the ends of member BC to within a horizontal rigid translation. The translation of beam BC was determined from the conditions of compatibility at joint B, geometry, the moment-area theorems, and the boundary conditions at A. The displacement of D was determined with the aid of geometry, the moment-area theorems, and the conditions of compatibility at joint C. By superposition,

$$u = -\frac{11wL^4}{24EI} + \frac{5L^3}{3EI}X, \quad t_c = \frac{wL^3}{12EI} - \frac{L^2}{2EI}X, \quad u_c = \frac{3wL^4}{8EI} - \frac{5L^3}{6EI}X$$

The boundary condition $u = 0$ yields $X = 11wL/40$. Back substitution into the expressions for t_c and u_c gives

$$t_c = -\frac{13wL^3}{240EI}, \quad u_c = \frac{7wL^4}{48EI}$$

Example 9.5

The frame in Example 9.4 is modified by adding a tie bar AD connecting joints A and D and replacing the pin connection at D by a roller support as shown in Figure 9.10a. The frame is externally statically determinate and internally statically indeterminate one time. The tie bar has cross sectional area A and modulus of elasticity E. Calculate the tension in the tie bar.

Although not essential, the tension in the tie bar is taken as the redundant internal force. The cut-back mechanical system corresponding to the tension in the tie bar as the redundant force consists of the frame and the tie bar as shown in Figure 9.10b. The mechanical states of the cut-back system and the given frame will coincide if the tension X is chosen so that the increase in the distance between joints A and D of the frame and the increase ΔL in the length of the tie bar (shown in Figure 9.10b) are equal. Because joint A cannot displace the increase in the distance between joints A and D is just the displacement u of joint D as shown in Figure 9.10b. The elongation ΔL is given by $\Delta L = XL/(EA)$. With the aid of the principle of superposition and the results shown in Figure 9.10c and d, the compatibility condition $\Delta L = u$ becomes

$$\Delta L = \frac{L}{EA}X = \frac{11wL^4}{24EI} - \frac{5L^3}{3EI}X$$

from which

$$X = \frac{11wL^4}{24EI}\left(\frac{L}{EA} + \frac{5L^3}{3EI}\right)^{-1} = \frac{11wL}{40}\left(1 + \frac{3I}{5AL^2}\right)^{-1}$$

It is clear from the expression for X that the tension approaches $11wL/40$ as AL^2 becomes large relative to I. The implication of this result is that when $AL^2 \gg I$ the mechanical states of members AB, BC, and BD of the frame with the pinned support at D and the mechanical states of these members in the case of the tied frame are essentially the same. The inequality can be put into a more useful form by expressing the common area moment of inertia, I, of the beam and the columns in terms of the cross sectional area A_f and radius of gyration r. With the aid of the relation $I = A_f r^2$ the inequality reduces to

$$A \gg (r/L)^2 A_f$$

Ordinarily, $r/L \ll 1$ and, therefore, the tied frame can be made to perform as a frame with pinned supports by employing a tie of relatively small cross-sectional area.

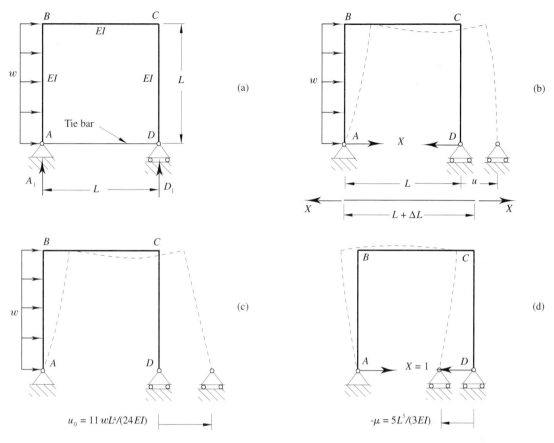

FIGURE 9.10 Tied frame.

Example 9.6

Calculate the reactions for the frame shown in Figure 9.11a. Disregard axial deformation in the analysis.

The horizontal and vertical components of the reaction at D are taken as the redundant forces and the corresponding cut-back structure is shown in Figure 9.11b. In Figure 9.11b u_1 is the component of displacement at D in the direction of X_1 and u_2 is the component of displacement at D in the direction of X_2. Clearly, to have the mechanical state of the cut-back structure match that of the given structure, X_1 and X_2 must be such that the boundary conditions $u_1 = u_2 = 0$ are satisfied. As in the previous examples, the required displacements are evaluated with the aid of the principle of superposition. Because the cut-back structure is built-in at point A, it is a relatively easy task to calculate the displacements by the moment-area method. Displacements corresponding to the basic load conditions employed in the superposition process are displayed in Figure 9.c, d, and e. It is observed that in agreement with Maxwell's reciprocal theorem, $\mu_{12} = \mu_{21}$. By the principle of superposition,

$$EIu_1 = -5238 + 1584X_1 + 1026X_2$$
$$EIu_2 = -1822.5 + 1026X_1 + 2952X_2$$

250 Chapter 9 Force Methods

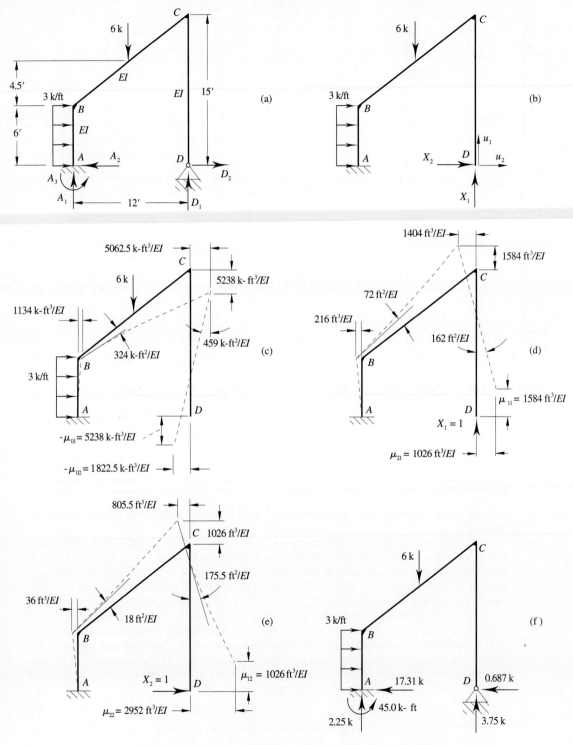

FIGURE 9.11 Frame is statically indeterminate to the second degree.

Example Problems 251

For $u_1 = u_2 = 0$, the solution of the equations is $X_1 = 3.7515$ k, $X_2 = -0.6865$ k. The remaining reactions are determined from equations of equilibrium. The results are displayed in Figure 9.11f.

Example 9.7

Calculate the reactions, the bar tensions for all members, and the vertical component of displacement of the joint at b of the truss shown in Figure 9.12a. The modulus of elasticity is $E = 29 \times 10^3$ ksi.

The force at the roller support at c is taken as the redundant reaction. The cut-back structure corresponding to this choice for the redundant reaction is shown in Figure 9.12b. For the mechanical state of the cut-back structure to be the same as that of the given truss, X must be chosen to satisfy the displacement condition $u = 0$. The displacement u is evaluated with the aid of the principle of superposition. The individual loading cases to be considered are shown in Figure 9.12c and d. Direct geometric methods (analytical or graphical) can be used to evaluate displacements; however, here the method of virtual work is most convenient. Denote the bar forces due to the applied loads acting alone (Figure 9.12c) by N_0 and denote the bar forces when

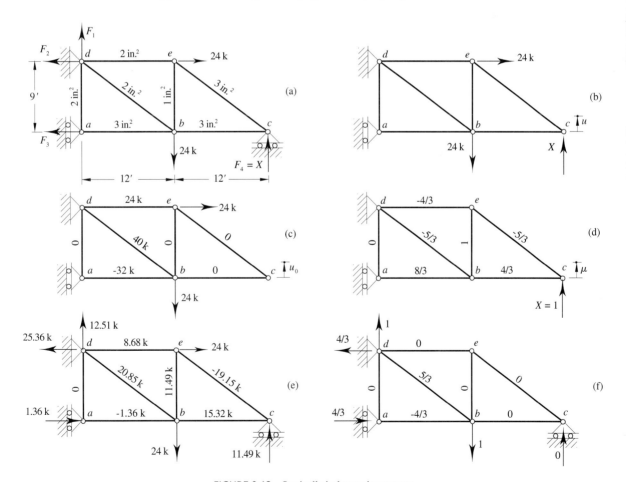

FIGURE 9.12 Statically indeterminate truss.

$X = 1$ acts alone (Figure 9.12d) by n. The bar forces are displayed on the line diagrams of the truss in Figure 9.12c and d, respectively.

By the principle of superposition, the tension N in any bar of the truss and the displacement u are

$$N = N_0 + nX, \quad u = u_0 + \mu X$$

where u_0 and μ are the displacements at c when the applied loads act alone (Figure 9.12c) and when $X = 1$ acts alone (Figure 9.12d), respectively. By the principle of virtual work,

$$u_0 = \sum_m \frac{N_0 nL}{EA}, \quad \mu = \sum_m \frac{n^2 L}{EA}$$

The calculations are arranged in tabular form as shown in columns 1 through 7 of Table 9.1. From the table,

$$u_0 = \sum_m \frac{N_0 nL}{EA} = \frac{-12400 \text{ k/in.}}{E}, \quad \mu = \sum_m \frac{n^2 L}{EA} = \frac{1079.33 \text{ in.}^{-1}}{E}$$

Therefore,

$$Eu = -12400 + 1079.33X = 0 \Rightarrow X = 11.49 \text{ k}$$

Superposition is used to calculate the bar tensions, which are displayed in column 8 of Table 9.1 and in Figure 9.12e. The vertical component of displacement of joint b is calculated with the aid of the principle of virtual work. A set of bar forces n^* that satisfy the equilibrium requirements for application of the principle of virtual work is displayed in Figure 9.12f (note: the forces shown constitute one set of an infinite number of sets of forces that may be employed). By the principle of virtual work,

$$u_b = \sum_m \frac{n^* NL}{EA}$$

The necessary data for the evaluation are displayed in columns 9 and 10 of Table 9.1. From the tabulated data,

$$u_b = 3214.5/(29 \times 10^3) = 0.1108 \text{ in.}$$

TABLE 9.1 Reaction and displacement calculations

Bar	L ft	A in²	N_0 kip	n —	$nN_0 L/A$ k/in	$n^2 L/A$ in⁻¹	$N_0 + nX$ kip	n^* —	$n^* NL/A$ k/in
(1)	(2)	(3)	(4)	(5)	(6)	(7)	(8)	(9)	(10)
ab	12	3	−32	8/3	−4096	341.33	−1.36	−4/3	87.0
ad	9	2	0	0	0	0	0	0	0
bc	12	3	0	4/3	0	85.33	15.32	0	0
bd	15	2	40	−5/3	−6000	250.00	20.85	5/3	3127.5
be	9	1	0	1	0	108.00	11.49	0	0
ce	15	3	0	−5/3	0	166.67	−19.15	0	0
de	12	2	24	−4/3	−2304	128.00	8.68	0	0
					−12400	1079.33			3214.5

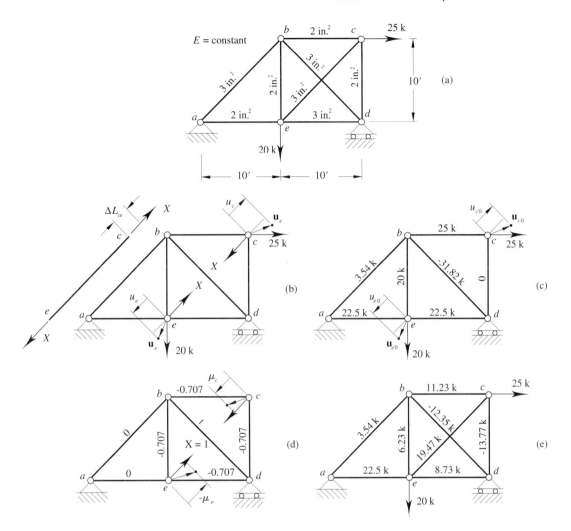

FIGURE 9.13 Internally statically indeterminate truss.

Example 9.8

Calculate the bar forces for the truss shown in Figure 9.13a. Bars ec and bd are not connected at their crossing point. The truss is internally statically indeterminate.

The force in any bar except for bars ab and ae may be taken as the redundant force. In this example, the force in bar ce is taken as the redundant force. The corresponding cut-back mechanical system consists of the statically determinate truss and bar ce as shown in Figure 9.13b. In Figure 9.13b, $\mathbf{u}_c$ and $\mathbf{u}_e$ are the displacements of joints c and e, respectively; u_c is the projection of the displacement at c on the direction from e toward c; and u_e is the projection of the displacement at e on the direction from c toward e. The increase in the distance between joints e and c, denoted by $u = \Delta d_{ec}$, is the sum of u_c and u_e, i.e., $u = u_c + u_e$. The increase in the length of bar ec is denoted by ΔL_{ce}. To have the mechanical state of the cut-back system coincide with that of the given truss, the redundant force must be such that the compatibility condition $\Delta d_{ec} = u = \Delta L_{ce}$ is satisfied. Let u_0 be the increase in the distance between joints c and e of the cut-back truss when the applied loads act alone, and let μ be the increase in the distance when $X = 1$ is the only load acting on the cut-back truss. Then, the compatibility condition is

$$\Delta d_{ce} = u_0 + \mu X = \Delta L_{ce}$$

u_0 and μ are readily evaluated with the aid of the principle of virtual work. Let N_0 denote a bar force for the loading shown in Figure 9.13c and let n denote a bar force for the loading shown in Figure 9.13d. Then, because the necessary requirements on forces and deformations are satisfied, the virtual work δW_e of the external forces shown in Figure 9.13d on the displacements shown in Figure 9.13c must be equal to the virtual work δW_σ of the tensions n (due to the loading shown in Figure 9.13d) on the deformations induced by the tensions N_0 (due to the loading shown in Figure 9.13c). Therefore, by the principle of virtual work,

$$\delta W_e = -1 u_{c0} - 1 u_{e0} = -u_0 = \sum_{m'} \frac{n N_0 L}{AE} = \delta W_\sigma$$

where the notation m' is employed to indicate that the summation extends only over all members of the cut-back truss (which does not include bar ce). The internal and external forces and the deformations and displacements for the cut-back truss under the loading shown in Figure 9.13d satisfy the necessary conditions of the principle of virtual work, and by the principle of virtual work, it follows that

$$-\mu = \sum_{m'} \frac{n^2 L}{AE}$$

The elongation of bar ce is given by

$$\Delta L_{ce} = \left(\frac{L}{AE}\right)_{ce} X$$

Substituting the foregoing relations into the compatibility equation and rearranging the terms gives

$$-\sum_{m'} \frac{n N_0 L}{AE} = \left[\left(\frac{L}{AE}\right)_{ce} + \sum_{m'} \frac{n^2 L}{AE}\right] X$$

Because $N_0 = 0$ and $n = 1$ for bar ce, the foregoing relation can also be written as

$$-\sum_{m} \frac{n N_0 L}{AE} = \left(\sum_{m} \frac{n^2 L}{AE}\right) X$$

in which the notation m is used to indicate that the summations extend over all members of the cut-back system (i.e., over all members of the original truss). The calculations are arranged in tabular form as displayed in Table 9.2. From the data in columns 6 and 7 in Table 9.2,

$$\sum_{m} \frac{n N_0 L}{A} = -4344.9 \text{ k/in.}, \quad \sum_{m} \frac{n^2 L}{A} = 223.12 \text{ in.}^{-1}$$

Therefore, the compatibility equation becomes $4344.9 = 223.12X$ from which $X = 19.47$ k. Superposition is used to obtain the bar forces. The results are given in column 8 of Table 9.2 and are displayed in Figure 9.13e.

TABLE 9.2 Internally statically indeterminate truss calculations

Bar	L ft	A in^2	N_0 kip	n	nN_0L/A k/in	n^2L/A in^{-1}	$N_0 + nX$ kip
(1)	(2)	(3)	(4)	(5)	(6)	(7)	(8)
ae	10	2	22.5	0	0	0	22.5
bc	10	2	25	−0.707	−1060.5	30	11.23
bd	14.14	3	−31.82	1	−1799.7	56.56	−12.35
be	10	2	20	−0.707	−848.4	30	6.23
cd	10	2	0	−0.707	0	30	−13.77
ce	14.14	3	0	1	0	56.56	19.47
de	10	3	22.5	−0.707	−636.3	20	8.73
					−4344.9	223.12	

Example 9.9

Calculate the bar tensions for the two times internally statically indeterminate truss shown in Figure 9.14a.

The cut-back system corresponding to the selection of the tensions in bars bf and df as the redundant forces is shown in Figure 9.14b. In Figure 9.14b u_{1b} is the projection of the displacement vector at joint b onto the direction from f toward b as indicated by the arrow labeled u_{1b}; u_{1f}, u_{2f}, and u_{2d} are projections of the same type. As indicated in the sketch, u_1 is the increase in the distance between joints b and f of the cut-back truss and u_2 is the increase in the distance between joints d and f. To have the mechanical state of the cut-back system coincide with that of the given truss, the redundant forces must be chosen to satisfy the compatibility conditions that the increase in the distance between joints b and f be equal to the elongation of bar bf, and the increase in the distance between joints d and f be equal to the elongation of bar df. These compatibility conditions are formulated as

$$u_1 = \Delta d_{bf} = \Delta L_{bf}, \quad u_2 = \Delta d_{df} = \Delta L_{df}$$

u_1 and u_2 are conveniently evaluated with the aid of the principle of superposition and the principle of virtual work. By the principle of superposition, the tension N in any bar of the cut-back truss is given by

$$N = N_0 + n_1 X_1 + n_2 X_2$$

where N_0, n_1, and n_2 are the tensions in the bar due to the loadings shown in Figures. 9.14c, d, and e, respectively. The virtual work of the external forces shown in Figure 9.14d on the displacements of the cut-back truss under the loading shown in Figure 9.14b must be equal to the virtual work of the tensions n_1 on the deformations due to the loading shown in Figure 9.14b. Because $u_1 = u_{1b} + u_{1f}$, the work equation yields

$$-u_1 = \sum_{m'} n_1 \left[(N_0 + n_1 X_1 + n_2 X_2) \frac{L}{AE} \right]$$

where the summations are over all members of the cut-back truss (which does not include bars bf and df). In the same way, u_2 is obtained by applying the principle of virtual work, using forces from Figure 9.14e and deformations and displacements from Figure 9.14b. Thus,

$$-u_2 = \sum_{m'} n_2 \left[(N_0 + n_1 X_1 + n_2 X_2) \frac{L}{AE} \right]$$

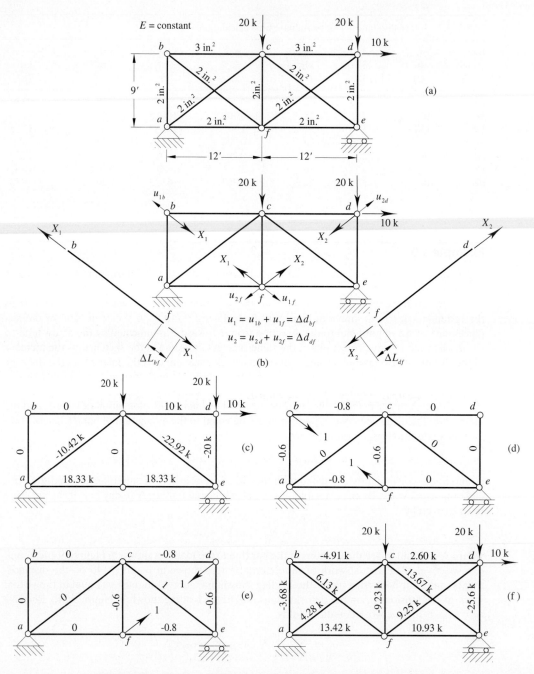

FIGURE 9.14 Truss, internally statically indeterminate to the second degree.

The elongations of bars bf and df are

$$\Delta L_{bf} = \left(\frac{L}{AE}\right)_{bf} X_1, \quad \Delta L_{df} = \left(\frac{L}{AE}\right)_{df} X_2$$

After substituting the foregoing expressions for $u_1, u_2, \Delta L_{bf}$, and ΔL_{df} in terms of X_1 and X_2 into the compatibility equations and rearranging terms, one obtains

$$\left[\left(\frac{L}{AE}\right)_{bf} + \sum_{m'}\frac{n_1^2 L}{AE}\right]X_1 + \left[\sum_{m'}\frac{n_1 n_2 L}{AE}\right]X_2 = -\sum_{m'}\frac{N_0 n_1 L}{AE}$$

$$\left[\sum_{m'}\frac{n_1 n_2 L}{AE}\right]X_1 + \left[\left(\frac{L}{AE}\right)_{df} + \sum_{m'}\frac{n_2^2 L}{AE}\right]X_2 = -\sum_{m'}\frac{N_0 n_2 L}{AE}$$

Because $N_0 = 0$ for bf and df, $n_1 = 1$ for bf and $n_1 = 0$ for df, and $n_2 = 0$ for bf and $n_2 = 1$ for df, the foregoing equations can be written in the more compact form

$$\left(\sum_m \frac{n_1^2 L}{AE}\right)X_1 + \left(\sum_m \frac{n_1 n_2 L}{AE}\right)X_2 = -\sum_m \frac{N_0 n_1 L}{AE}$$

$$\left(\sum_m \frac{n_1 n_2 L}{AE}\right)X_1 + \left(\sum_m \frac{n_2^2 L}{AE}\right)X_2 = -\sum_m \frac{N_0 n_2 L}{AE}$$

in which the notation m is employed to indicate that the summations now extend over all members of the original truss. Observe that the coefficient of X_2 in the first equation and the coefficient of X_1 in the second equation are equal. Maxwell's reciprocal theorem guarantees the equality. Calculation of the coefficients is arranged in tabular form as displayed in Table 9.3. From the data in Table 9.3,

$$\sum_m \frac{n_1^2 L}{A} = 295.68 \text{ in.}^{-1}, \quad \sum_m \frac{n_1 n_2 L}{A} = 19.44 \text{ in.}^{-1}, \quad \sum_m \frac{n_2^2 L}{A} = 295.68 \text{ in.}^{-1}$$

$$\sum_m \frac{N_0 n_1 L}{A} = -1993.6 \text{ k/in.}, \quad \sum_m \frac{N_0 n_2 L}{A} = -2854.6 \text{ k/in.}$$

Because E is a constant, the compatibility equations become

$$295.68 X_1 + 19.44 X_2 = 1993.6$$
$$19.44 X_1 + 295.68 X_2 = 2854.6$$

the solution of which is $X_1 = 6.13$ k, $X_2 = 9.25$ k. The superposition relation $N = N_0 + n_1 X_1 + n_2 X_2$ is used to evaluate the bar tensions. The results are given in Table 9.3 and a summary is displayed in Figure 9.14f.

9.4 PRESCRIBED DISPLACEMENTS

At any point in a structure, either the force or the displacement may be prescribed. Where the force is prescribed, the displacement is considered to be an unknown quantity to be determined, and where the displacement is prescribed, the force is an unknown quantity to be determined. Supports for a structure are understood to be particular points at which certain components of displacement are prescribed. Ordinarily, the prescribed value for the displacement at all supports is zero. The analysis of a structure for which the prescribed value for a displacement at a support is non-zero is

TABLE 9.3 Calculations for a two times internally statically indeterminate truss

Bar	L	A	N_0	n_1	n_2	$\dfrac{N_0 n_1 L}{A}$	$\dfrac{N_0 n_2 L}{A}$	$\dfrac{n_1^2 L}{A}$	$\dfrac{n_2^2 L}{A}$	$\dfrac{n_1 n_2 L}{A}$	$N = N_0 +$ $n_1 X_1 + n_2 X_2$
	ft	in²	k	—	—	k/in	k/in	in⁻¹	in⁻¹	in⁻¹	k
(1)	(2)	(3)	(4)	(5)	(6)	(7)	(8)	(9)	(10)	(11)	(12)
ab	9	2	0	−0.6	0	0	0	19.44	0	0	−3.68
ac	15	2	−10.42	1	0	−937.8	0	90.00	0	0	−4.28
af	12	2	18.33	−0.8	0	−1055.8	0	46.08	0	0	13.42
bc	12	3	0	−0.8	0	0	0	30.72	0	0	−4.91
bf	15	2	0	1	0	0	0	90.00	0	0	6.13
cd	12	3	10	0	−0.8	0	−384.0	0	30.72	0	2.60
ce	15	2	−22.92	0	1	0	−2062.8	0	90.00	0	−13.67
cf	9	2	0	−0.6	−0.6	0	0	19.44	19.44	19.44	−9.23
de	9	2	−20	0	−0.6	0	648	0	19.44	0	−25.55
df	15	2	0	0	1	0	0	0	90.00	0	9.25
ef	12	2	18.33	0	−0.8	0	−1055.8	0	46.08	0	10.93
						−1993.6	−2854.6	295.68	295.68	19.44	

described as a problem dealing with settlement of supports. Because some settlement of the supports of a structure is inevitable, the effect of settlement on the internal forces in a structure is of considerable practical importance.

Clearly, non-zero values for displacements of the supports play no role in the determination of the reactions or the internal forces for a statically determinate structure. Moreover, for the statically determinate structure, all reactions and internal forces are necessarily zero for all prescribed displacements of the supports. The situation is significantly different in the case of a statically indeterminate structure. For the statically indeterminate structure subject to any applied load, the reactions and the internal forces always depend on the assigned values for the displacements of the supports. In addition, as will be shown by specific examples, the reactions and internal forces in a statically indeterminate structure under no load are independent of support displacements only when those displacements define a rigid-body displacement of the structure.

Because the principle of superposition can be used to determine the combined effects of applied loads and prescribed displacements, only the effects of prescribed displacements are treated in the examples. The basic ideas of analysis based on the method of consistent deformation are the same for all structures. Accordingly, the examples are limited to beams.

Example 9.10

Determine the reactions of the propped cantilever beam due to settlement Δ of the tip as shown in Figure 9.15a. The problem is solved twice. In the first solution, the reaction at B is taken as the redundant force. In the second solution, the reaction couple at A is taken as the redundant force.

First Solution The cut-back structure corresponding to the reaction at B as the redundant force is the statically determinate cantilever beam shown in Figure 9.15b. The mechanical states of the cut-back structure and the given beam will coincide if the force X is chosen so

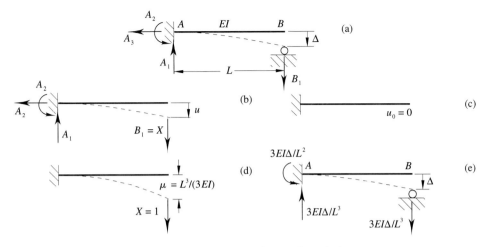

FIGURE 9.15 Support settlement; first solution.

that $u = \Delta$. By the principle of superposition, $u = u_0 + \mu X$, where u_0 is the displacement at B when $X = 0$ and μ is the displacement when $X = 1$, as shown in Figures 9.15c and d, respectively. Because $u_0 = 0$, $u = L^3 X/(3EI)$ and the compatibility condition yields $X = 3EI\Delta/L^3$. The remaining reactions are calculated from statics. The results are displayed in Figure 9.15e.

Second Solution With the reactive couple at A taken as the redundant force, the cut-back structure is a simply supported beam as shown in Figure 9.16b. To have the mechanical states of the cut-back structure and the given beam coincide, the moment X of the couple at A must be chosen so that the rotation of the tangent at A is zero. By superposition, $u = u_0 + \mu X$, where u_0 and μ are the rotations as identified in Figures 9.16c and d, respectively. Thus,

$$u = -\frac{\Delta}{L} + \frac{L}{3EI} X$$

Setting $u = 0$ yields $X = 3EI\Delta/L^2$ which coincides with the value of A_2 as obtained from the first solution.

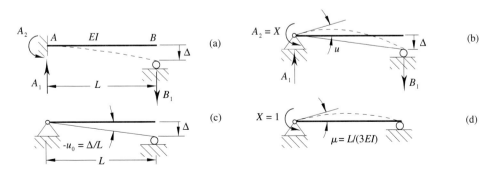

FIGURE 9.16 Support settlement; second solution.

Example 9.11

The beam shown in Figure 9.7 and reproduced in Figure 9.17a undergoes a support settlement in form of a rotation of the built-in support at C. Calculate the reactions. The solution is obtained using two different choices for the redundant reactions. In the first solution, the reaction forces at supports A and B are taken as the redundant quantities. In the second solution, the bending moment over support B and the bending moment at C are taken as the redundant forces.

First Solution The cut-back structure is the cantilever beam shown in Figure 9.17b. Because there is no settlement at support A or support B, the forces X_1 and X_2 must be chosen so that the displacements u_1 and u_2 are zero. With the aid of the information contained in Figures 9.17c, d, and e, and the principle of superposition,

$$u_1 = 2L\theta + \frac{8L^3}{3EI} X_1 + \frac{5L^3}{6EI} X_2$$

$$u_2 = L\theta + \frac{5L^3}{6EI} X_1 + \frac{L^3}{3EI} X_2$$

The solution of the equations for $u_1 = u_2 = 0$ is

$$X_1 = \frac{6EI}{7L^2}\theta, \quad X_2 = -\frac{36EI}{7L^2}\theta$$

The bending moments at B and C are

$$M_B = \frac{6EI}{7L}\theta, \quad M_C = -\frac{24EI}{7L}\theta$$

Second Solution The cut-back structure corresponding to the choices of the bending moments at B and at C as the redundant forces is a pair of simply supported beams as shown in Figure 9.18b.

To satisfy the compatibility conditions for the given beam, the couples X_1 and X_2 must be chosen to satisfy the condition that the tangent at B be continuous and that the rotation u_2 of the

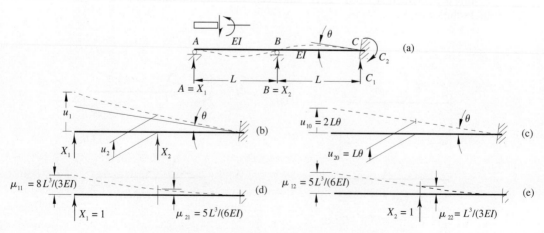

FIGURE 9.17 Continuous beam; first solution.

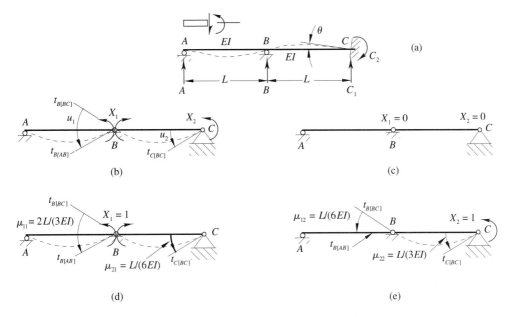

FIGURE 9.18 Continuous beam, second solution.

tangent at C be equal to $-\theta$. As indicated in Figure 9.18c, because there are no loads applied to the cut-back structure when $X_1 = X_2 = 0$, all rotations for this case of loading are zero. With the aid of the results shown in Figures 9.18c, d, and e and the principle of superposition,

$$u_1 = \frac{2L}{3EI} X_1 + \frac{L}{6EI} X_2$$

$$u_2 = \frac{L}{6EI} X_1 + \frac{L}{3EI} X_2$$

The solution of the equation for $u_1 = 0$ and $u_2 = -\theta$ is

$$X_1 = \frac{6EI}{7L} \theta, \qquad X_2 = -\frac{24EI}{7L} \theta$$

which coincides with the values found for M_B and M_C found in the first solution.

9.5 THE THREE-MOMENT EQUATION

A beam is said to be continuous if the deflection $v(x)$ and the slope $\theta(x) = v'(x)$ are continuous over the full length of the beam. Examples of beams that are said to be continuous over interior supports are shown in Figure 9.19. In the analysis of such beams, it is commonly assumed that translational displacements of the supports are prescribed and that the tangents over the interior supports are free to rotate. At an exterior support, either the bending moment or the rotation of the tangent to the beam at the support is

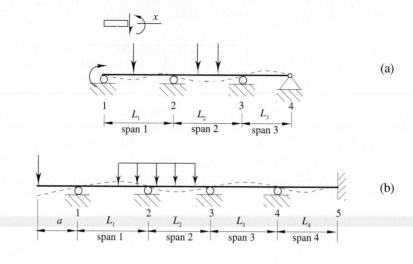

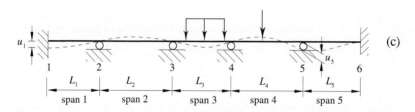

FIGURE 9.19 Continuous beams.

prescribed. The degree of indeterminacy of a beam that is continuous over several interior supports is determined by the number of interior supports and the prescribed conditions at the end or exterior supports. For a beam that is continuous over n interior supports the degree of indeterminacy is n if the bending moments at both exterior supports are prescribed; the degree of indeterminacy is $n + 1$ if the bending moment at one of the exterior supports is prescribed; and the degree of indeterminacy is $n + 2$ if the bending moment at neither of the exterior supports is prescribed. Because the bending moments at supports 1 and 4 can be calculated from statics, the beam in Figure 9.19a has a degree of indeterminacy $d = 2$. The degree of indeterminacy of the beam shown in Figure 9.19b is $d = 4$. This follows from the fact that the bending moment at support 1 can be determined from statics and the fact that the rotation of the tangent at support 5 is prescribed (the rotation is zero). Because the rotations at supports 1 and 6 are prescribed, the beam shown in Figure 9.19c has a degree of indeterminacy $d = 6$.

The bending moments over the supports are among the possible choices for the redundant forces in the analysis of continuous beams by the method of consistent deformation. One important advantage of choosing these moments as the redundant forces is that after they have been determined, construction of the bending moment diagram for the entire beam can be accomplished quickly without computing reactions.

Another very important advantage is that the associated compatibility conditions can be cast in a generic form that can be applied easily and systematically to derive a complete set of equations, the solution of which consists of the unknown bending moments. The generic compatibility equation is called the *three-moment equation*.

The cut-back structure corresponding to the selection of the bending moments over the supports as the redundant forces is a system of simply supported beams as shown in Figure 9.20.

For the mechanical states of the given beam and the cut-back structure to coincide, the bending moments acting on the ends of the beams in the cut-back structure must be chosen to satisfy the compatibility and boundary conditions for the given beam. Because the given beam is continuous, the compatibility condition to be satisfied at each interior support is that the rotations of the tangents at the ends of the members at the support be equal. The condition, expressed in terms of bending moments, loads, and displacements of the beams on each side of the support, is the three-moment equation.

Castigliano's theorem of least work could be used to derive the three-moment equation without explicit reference to compatibility conditions. (Recall that by its derivation the theorem takes into account compatibility conditions automatically.) The traditional approach is employed herein to emphasize that force methods of analysis are under consideration and that satisfaction of compatibility conditions are at the core of force methods. The derivation is carried out with the aid of relationships between loads, displacements, and the rotations of the tangents at the ends of a generic, simply supported beam. The notations and the sign conventions employed in the analysis are illustrated in Figure 9.21. In Figure 9.21a, φ_a is the rotation of the tangent drawn at a from the chord ab and θ_a is the rotation of the tangent drawn at a from the reference axis. The angles of rotation φ_a and θ_a are measured positive in the clockwise direction. Angle φ_b is the rotation of the tangent drawn at b from the chord ab and angle θ_b is the rotation of the tangent drawn at b from the reference axis. Rotations φ_b and θ_b are measured positive in the counterclockwise direction. ρ is the rotation of chord ab from the reference axis, measured positive in the clockwise direction.

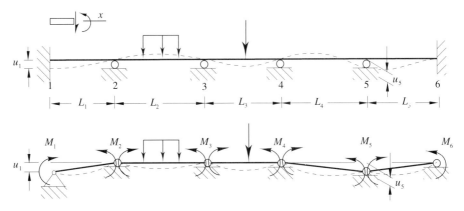

FIGURE 9.20   Cut-back structure.

264 Chapter 9 Force Methods

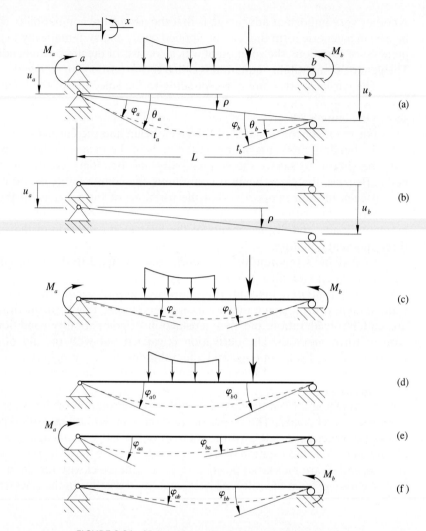

FIGURE 9.21 Nomenclature for the three-moment equation.

The rotations θ_a and θ_b are evaluated with the aid of the principle of superposition. From the geometry shown in Figures 9.21a, b, and c,

$$\theta_a = \varphi_a + \rho = \varphi_a + \frac{(u_b - u_a)}{L}$$

$$\theta_b = \varphi_b - \rho = \varphi_b - \frac{(u_b - u_a)}{L}$$

By the principle of superposition, the rotation $\varphi_a(\varphi_b)$ in Figure 9.21c is the sum of the rotations at the end $a(b)$ for the loadings shown in Figures 9.21d, e, and f. Thus,

$$\theta_a = \varphi_{aa} + \varphi_{ab} + \varphi_{a0} + \frac{(u_b - u_a)}{L}$$

$$\theta_b = \varphi_{ba} + \varphi_{bb} + \varphi_{b0} - \frac{(u_b - u_a)}{L}$$

For a uniform beam,

$$\varphi_{aa} = \frac{L}{3EI} M_a, \quad \varphi_{bb} = \frac{L}{3EI} M_b$$

$$\varphi_{ab} = \frac{L}{6EI} M_b, \quad \varphi_{ba} = \frac{L}{6EI} M_a$$

The equality of the coefficients of M_a and M_b in the expressions for φ_{aa} and φ_{bb} follows from the symmetry of the beam. The equality of the coefficients of M_a and M_b in the expressions for φ_{ab} and φ_{ba} follow from the reciprocal theorem. The rotations may now be expressed as

$$\theta_a = \frac{L}{3EI} M_a + \frac{L}{6EI} M_b + \varphi_{a0} + \frac{(u_b - u_a)}{L}$$

$$\theta_b = \frac{L}{6EI} M_a + \frac{L}{3EI} M_b + \varphi_{b0} - \frac{(u_b - u_a)}{L}$$

Before going on to the derivation of the three-moment equation, consider φ_{a0} and φ_{b0}. As shown in Figure 9.21d, these are the rotations of the tangents at the ends of the beam due to the loading between the ends when $u_a = u_b = 0$. With the aid of the method of virtual work or the moment-area method, it can be shown that these rotations are given by

$$\varphi_{a0} = \int_0^L \frac{M_0(x)}{EI}\left(1 - \frac{x}{L}\right) dx, \quad \varphi_{b0} = \int_0^L \frac{M_0(x)}{EI}\left(\frac{x}{L}\right) dx$$

where $M_0(x)$ is the bending moment in the simply supported beam due to the loads between the ends. The rotations are readily evaluated for elementary cases of loading. Complex load cases can be handled with the aid of the principle of superposition and results for elementary loadings. Rotations for three loading cases of practical importance are displayed in Figure 9.22.

To apply the foregoing results to the derivation of the three-moment equation, consider two adjacent beam segments lc and cr which span three consecutive supports l, c, and r as shown in Figure 9.23.

The rotations θ_{bl} and θ_{ar} are given by

$$\theta_{bl} = \frac{L_l}{6EI_l} M_l + \frac{L_l}{3EI_l} M_c + \varphi_{bo(l)} + \frac{u_l - u_c}{L_l}$$

$$\theta_{ar} = \frac{L_r}{3EI_r} M_c + \frac{L_r}{6EI_r} M_r + \varphi_{ao(r)} + \frac{u_r - u_c}{L_r}$$

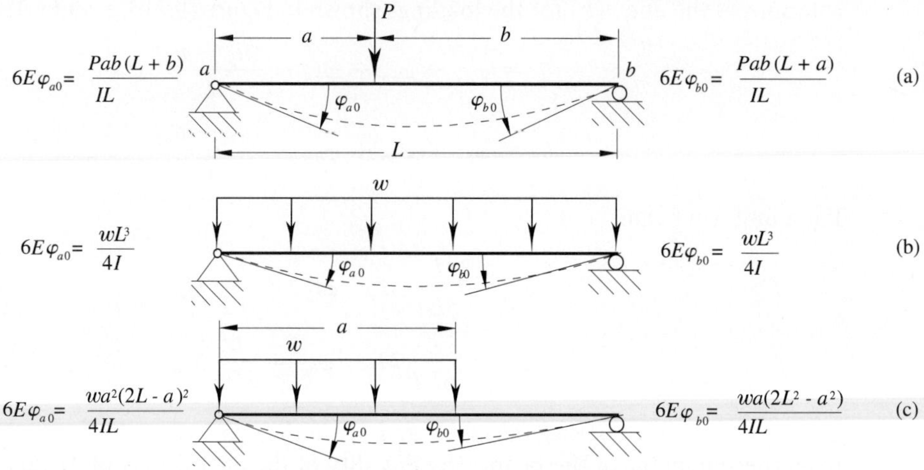

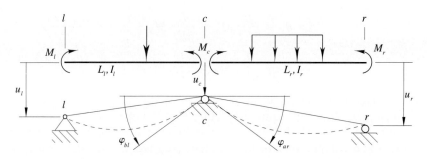

FIGURE 9.22 End rotations.

FIGURE 9.23 Discontinuity in rotation.

Now, $\theta_{bl} + \theta_{ar}$ is the discontinuity in the rotations of the tangents at the ends of the beams at support c. If c is an interior support of a continuous beam the tangent must be continuous over that support, and the compatibility condition to be satisfied is $\theta_{bl} + \theta_{ar} = 0$. Thus, after adding the expressions for the rotations, setting the result equal to zero and rearranging the terms, one obtains

$$\frac{L_l}{I_l} M_l + 2\left(\frac{L_l}{I_l} + \frac{L_r}{I_r}\right) M_c + \frac{L_r}{I_r} M_r$$
$$= -6E\varphi_{bo(l)} - 6E\varphi_{ao(r)} - 6E\left[\frac{1}{L_l} u_l - \left(\frac{1}{L_l} + \frac{1}{L_r}\right) u_c + \frac{1}{L_r} u_r\right] \quad (I)$$

which is the three-moment equation. The equation applies at all interior supports of a continuous beam. Evaluation of the right side of the equation offers no difficulty because the displacements at the ends and the loads between the ends of the members are assigned. To solve a problem, the three-moment equation must be supplemented with relations that reflect the boundary conditions. Let the supports be numbered from left to right starting with 1 and ending with n. Then at the exterior support 1, which is the a-end of the first span,

$$2\frac{L_1}{I_1}M_1 + \frac{L_1}{I_1}M_2 = -6E\varphi_{ao(1)} - 6E\left(\frac{u_2 - u_1}{L_1}\right) + 6E\theta_{a(1)} \quad \text{(E1)}$$

in which either M_1 or $\theta_{a(1)}$ is prescribed. If M_1 is prescribed the equation is used to calculate $\theta_{a(1)}$ *after* M_2 has been found. If $\theta_{a(1)}$ is prescribed, M_1 is unknown and Equation (E1) is used as a supplement to the three-moment equation. At exterior support n, which is the b-end of the last span (span $n - 1$),

$$\frac{L_{n-1}}{I_{n-1}}M_{n-1} + 2\frac{L_{n-1}}{I_{n-1}}M_n = -6E\varphi_{bo(n-1)} - 6E\left(\frac{u_{n-1} - u_n}{L_{n-1}}\right) + 6E\theta_{b(n-1)} \quad \text{(En)}$$

in which either M_n or $\theta_{b(n-1)}$ is prescribed. Equations (I), (E1), and (En) may be rewritten to meet specialized conditions. Such specializations are not of interest here.

It is to be observed that Equations (I), (E1), and (En) are linear in the prescribed displacements of the supports and the loads (as measured by the terms $6E\varphi$). Thus, the combined effects of applied loads and prescribed displacements can be determined with the aid of the principle of superposition. In other words, the effects of applied loads and of prescribed displacements can be evaluated independently and the results superimposed to obtain the combined effect. As noted earlier, the effects of any combination of applied loads can also be determined with the aid of the principle of superposition.

9.6 APPLICATION PROCEDURE

The analysis of a continuous beam with the aid of the three-moment equation can be reduced to an organized sequence of operations as outlined below.

1. Generate the load terms $6E\varphi_{a(0)}$ and $6E\varphi_{b(0)}$ for each span. For any combination of uniform and concentrated loads, use the results displayed in Figure 9.22 and the principle of superposition to determine the net values of $6E\varphi_{a(0)}$ and $6E\varphi_{b(0)}$.
2. If $\theta_{a(1)}$ is prescribed (M_1 is unknown), generate Equation (E1). Skip this step if M_1 is prescribed.
3. Generate Equation (I) at interior supports $2, 3, \ldots, (n - 1)$, using the known values of M_1 and M_n if these quantities are prescribed.
4. If $\theta_{b(n-1)}$ is prescribed (M_n is unknown), generate Equation (En). Skip this step if M_n is prescribed.
5. Solve the equations generated in Steps 2–4.

Example 9.12

Calculate the bending moments over the supports of the continuous beam shown in Figure 9.24. This is the same beam considered in Example 9.3 to illustrate the basic method of consistent deformation.

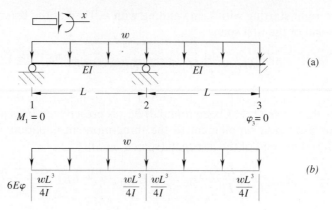

FIGURE 9.24 One interior support.

The prescribed conditions at the exterior supports are $M_1 = 0$ and $\theta_{b(2)} = 0$. Accordingly, only Equation (I) at the interior support (support 2) and Equation (En) at support 3 need be generated. The load terms are displayed in Figure 9.24b. The equations to be solved are

$$2\left(\frac{L}{I} + \frac{L}{I}\right)M_2 + \frac{L}{I}M_3 = -\frac{wL^3}{2I}$$

$$\frac{L}{I}M_2 + 2\frac{L}{I}M_3 = -\frac{wL^3}{4I}$$

The solution values are $M_2 = -3wL^2/28$, $M_3 = -wL^2/14$, which coincide with the results obtained previously.

Example 9.13

For the continuous beam with an overhang shown in Figure 9.25: (a) calculate the bending moments over the supports, (b) construct the shear and bending moment diagrams, (c) calculate the deflection at the point of application of the 20-k load.

The bending moment over support 3 is calculated from statics with the aid of a free-body diagram of span 3-4. Therefore, the boundary conditions at supports 1 and 3 are $\theta_{a(1)} = 0$ and $M_3 = -200$ k-ft, respectively. Only Equation (El) and Equation (I) at support 2 need be generated to solve the problem. The load terms, evaluated with the aid of the results given in Figure 9.22, are displayed in Figure 9.25b. The equations to be solved are

$$2\left(\frac{20}{I_0}\right)M_1 + \frac{20}{I_0}M_2 = -\frac{2400}{I_0}$$

$$\frac{20}{I_0}M_1 + 2\left(\frac{20}{I_0} + \frac{30}{2I_0}\right)M_2 + \frac{30}{2I_0}(-200) = -\frac{7400}{I_0}$$

The solution values are $M_1 = -33.33$ k-ft, $M_2 = -53.33$ k-ft.

With the values of the bending moments over the supports known, the shear and bending moment functions and the corresponding diagrams for each span can be constructed by starting with equilibrium of the free-body diagrams shown in Figure 9.25c. The shear and bending moment diagrams are as shown in Figure 9.25d and Figure 9.25e, respectively.

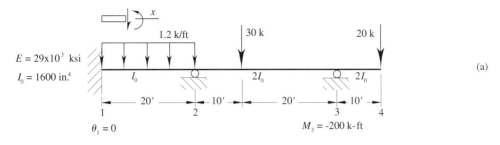
(a)

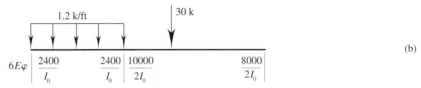
(b)

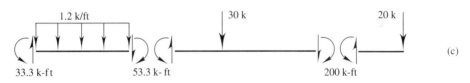
(c)

(d)

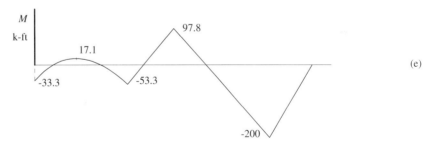
(e)

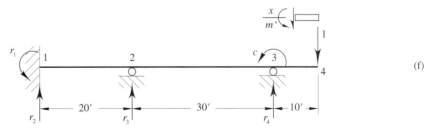
(f)

FIGURE 9.25 Continuous beam with an overhang.

Calculation of the deflection at point 4 is facilitated by first observing that all of the information that is needed to calculate the rotation of the tangent at support 3 from Equation (En) is available. Thus, with $n = 3$, Equation (En) becomes

$$\left(\frac{30}{2I_0}\right)(-53.33) + 2\left(\frac{30}{2I_0}\right)(-200) = -\frac{8000}{2I_0} + 6E\theta_{b(2)}$$

from which $EI_0\theta_{b(2)} = -466.7$ k-ft^2. In the derivation of the three-moment equation, the rotation of the tangent at the b-end of a beam is measured positive in the counterclockwise direction. In view of this and the continuity of the tangent over a support, the rotation of the tangent at support 3 is obtained from $EI_0\theta_3 = EI_0\theta_{b(2)} = -466.7$ k-ft^2, where θ_3 is measured positive in the counterclockwise sense. With θ_3 known and knowing that $u_3 = 0$, either the moment-area method or the principle of virtual work can be applied to obtain the displacement at the tip of the beam. The method of virtual work is applied here. A set of forces that will do the job is displayed in Figure 9.25f. Any choice of values for the couple c and the reactions r_1, r_2, r_3, r_4 such that equilibrium of the entire beam is satisfied will result in a set of internal bending moments m^* that satisfy the requirements for the application of the principle of virtual work. The choice $r_1 = r_2 = r_3 = 0$ requires $c = 10', r_4 = 1$. The advantage of this choice is that it yields $m^* = 0$ for spans 12 and 23. With the sign convention indicated in Figure 9.25f, the principle of virtual work yields

$$10\theta_3 + u_4 = \int_0^{10} x\left(\frac{20x}{2EI_0}\right) dx$$

from which $EI_0 u_4 = 8000$ k-ft^3. For the given values of E and I_0, $u_4 = 0.298$ inches.

Example 9.14

The beam shown in Figure 9.26 undergoes support settlements in the form of displacement $u_1 = \Delta$, and rotation θ of support at 3 as indicated in the figure. Calculate the reactions due to these settlements.

The reactions are easy to determine once the bending moments over the supports are known. Thus, attention is focussed on the bending moments over the supports which are conveniently determined by the three-moment equation method. An obvious advantage in this approach is that it sidesteps the awkward elements of formulating compatibility equations in a direct attack using the basic method of consistent deformation. Because $M_1 = 0$, and $\theta_{b(2)} = -\theta$ are prescribed, all one need do is generate Equation (I) at support 2 and generate Equation (En) for span 2. By direct substitution into the equations

$$2\left(\frac{L}{I} + \frac{L}{I}\right)M_2 + \frac{L}{I}M_3 = -6E\frac{\Delta}{L}$$

$$\frac{L}{I}M_2 + 2\frac{L}{I}M_3 = -6E\theta$$

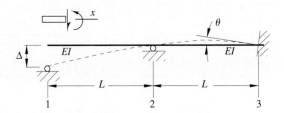

FIGURE 9.26 Support settlements.

the solution values of which are

$$M_2 = -\frac{12EI\Delta}{7L^2} + \frac{6EI\theta}{7L}, \quad M_3 = \frac{6EI\Delta}{7L^2} - \frac{24EI\theta}{7L}$$

For $\Delta = 0$ the expressions for M_2 and M_3 reduce to the values obtained for the corresponding quantities in Example 9.11 in the section on prescribed displacements.

9.7 COMBINED METHODS

Not infrequently one comes across a situation in which the three-moment equation cannot be applied and the traditional approach used in analysis by the method of consistent deformation is not very appealing or efficient. The following example demonstrates how the methods can be combined to achieve an efficient analysis. The compound beam shown in Figure 9.27a is statically indeterminate to the third degree. It is required to calculate the bending moments over the supports and the deflection at D. The three-moment equation cannot be used in the analysis of the beam as it stands because the necessary continuity conditions are not satisfied at the pin. However, a solution can be constructed with the aid of the three-moment equation method, the principle of superposition, and the basic concepts of the method of consistent deformation.

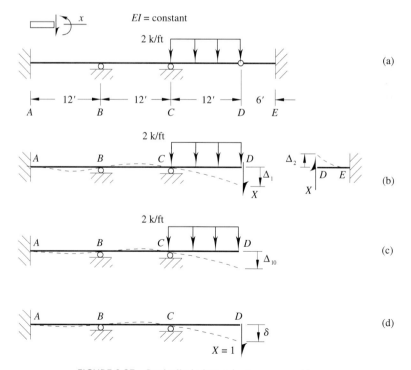

FIGURE 9.27 Statically-indeterminate compound beam.

Consider the cut-back system consisting of the continuous beam $ABCD$ and the cantilever beam DE as shown in Figure 9.27b. The mechanical states of the given beam and the cut-back system will coincide if the force pair denoted by X is chosen so that the discontinuity in the displacements at the point D is zero, i.e., the mechanical states will coincide if X is chosen so that $\Delta_1 + \Delta_2 = 0$. When this is done, the force pair satisfies exactly the same conditions as do the forces of interaction at the pin in the given structure.

Let Δ_{10} and δ be the deflection of beam $ABCD$ for the load conditions shown in Figure 9.27c and Figure 9.27d, respectively. These deflections can be determined with relative ease by an analysis of beam $ABCD$ with the aid of the three-moment equation to find the bending moments over the supports, and either virtual work or the moment-area method to calculate the deflections. Let the bending moment at any location for the conditions shown in Figure 9.27c be denoted by M_0 and let the corresponding bending moments for the conditions shown in Figure 9.27d be denoted by m. Then, by the principle of superposition the deflection Δ_1 and the bending moment M at any location in beam $ABCD$ is given by

$$\Delta_1 = \Delta_{10} + \delta X, \quad M = M_0 + mX$$

Because $3EI\Delta_2 = 6^3 X$ the compatibility condition $\Delta_1 + \Delta_2 = 0$ becomes

$$0 = \Delta_{10} + \left(\delta + \frac{72}{EI}\right)X$$

For the loading shown in Figure 9.27c, $M_{C0} = -144$ k-ft. The equations for the determination of M_{A0} and M_{B0} are

$$2\left(\frac{12}{I}\right)M_{A0} + \frac{12}{I}M_{B0} = 0$$

$$\frac{12}{I}M_{A0} + 2\left(\frac{12}{I} + \frac{12}{I}\right)M_{B0} + \frac{12}{I}(-144) = 0$$

The solution values are $M_{A0} = -20.57$ k-ft, $M_{B0} = 41.14$ k-ft. The deflection at D is calculated to be $EI\Delta_{10} = 11109$ k-ft^3. For the loading shown in Figure 9.27d, $m_{C0} = -12$ ft and the equations to be solved for m_{A0} and m_{B0} are

$$2\left(\frac{12}{I}\right)m_{A0} + \frac{12}{I}m_{B0} = 0$$

$$\frac{12}{I}m_{A0} + 2\left(\frac{12}{I} + \frac{12}{I}\right)m_{B0} + \frac{12}{I}(-12) = 0$$

from which $m_{A0} = -1.714$ ft, $m_{B0} = 3.429$ ft. For the deflection at D, $EI\delta = 1069.9$ ft^3. The compatibility equation now becomes

$$0 = \frac{11109}{EI} + \left(\frac{1069.9}{EI} + \frac{72}{EI}\right)X$$

from which $X = -9.73$ k. By superposition, $M_A = -3.89$ k-ft, $M_B = 7.78$ k-ft, and

$M_C = -27.24$ k-ft. From equilibrium of beam DE, $M_E = -58.4$ k-ft. For the deflection at the pin, $EI\Delta_1 = 700.5$ k-ft^3.

9.8 METHOD OF LEAST WORK

Discussion is initiated with a review of the key elements in the theorem. For a plane-frame structure made of a material that obeys Hooke's law, the energy function Φ in Castigliano's theorem of least work has the particular formulation

$$\Phi = \sum_m \int_0^L \left(\frac{M^2}{2EI} + \frac{N^2}{2EA} \right) dx - \sum_r R u_R$$

in which u_R is a prescribed component of displacement and R is the force required at the point where u_R is prescribed, measured in the direction of u_R. If u_R is a prescribed rotation, R is understood to be the moment of the required couple. M and N in the formulation are understood to define any distribution of internal forces such that the equations of equilibrium are satisfied throughout the structure. The *theorem of least work* (developed in Chapter 7) asserts that of all the internal force distributions that satisfy the equilibrium conditions the true distribution has the property that it makes Φ a minimum. The theorem is trivially true and has no practical application in the case of a statically determinate structure. The theorem is particularly useful in the force method of analysis of a statically indeterminate structure when formulation of the compatibility conditions is a problem. Its practical utility follows from the fact that among all of the internal force distributions that satisfy the equations of equilibrium is the force distribution that satisfies the force-deformation relations and the conditions of compatibility. In other words, among these force distributions is that distribution which is the object of the force analysis of a statically indeterminate structure. The key to successful application of the theorem is to formulate expressions for M and N that include *all* possible distributions of these quantities that satisfy the equations of equilibrium. As will be seen by reference to specific examples, construction of the required functions is easy and the overall process contains considerable flexibility. Because specific examples serve to illustrate the ideas involved, no attempt is made to go into generalized formulations.

Example 9.15

Consider the force analysis of the propped cantilever beam of Figure 9.4, reproduced here in Figure 9.28. The problem is to calculate the reactions. The beam is statically indeterminate to the first degree. Any one of A_1, A_2, B_1 can be chosen as the redundant reaction.

With $X = B_1$ as the redundant reaction, all bending moment distributions satisfying the equilibrium conditions are given by

$$M = X(L - x) - \frac{w(L - x)^2}{2}$$

Because $N = 0$ and all prescribed displacements have value zero, the energy function Φ reduces to

$$\Phi = \Phi(X) = \int_0^L \frac{M^2}{2EI} dx$$

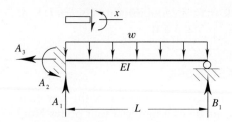

FIGURE 9.28 Propped cantilever beam.

which, in view of the expression for M, is a function of X. According to the theorem of least work, the true value of X makes $\Phi(X)$ a minimum. The necessary condition for $\Phi(X)$ to be a minimum is that its first derivative with respect to X be zero. Whether the differentiation is carried out before or after the integration is inconsequential. However, because less work is involved, good practice is to differentiate with respect to X first. Thus,

$$\frac{d\Phi}{dX} = \int_0^L \frac{M}{EI}\frac{\partial M}{\partial X}dx = \int_0^L \frac{1}{EI}\left[X(L-x) - \frac{w(L-x)^2}{2}\right](L-x)\,dx$$

$$= \frac{1}{EI}\left(\frac{L^3}{3}X - \frac{wL^4}{8}\right)$$

The necessary condition for Φ to be a minimum yields $X = 3wL/8$, which coincides with the result obtained by direct application of the basic method of consistent deformation. The obvious advantage here is that no role is played by the sketches employed in the derivation of the compatibility condition in the basic method of consistent deformation.

Example 9.16

Use the theorem of least work to calculate the reactions for the compound beam shown in Figure 9.29. The beam is that of Example 9.2 used to illustrate the basic method of consistent deformation. The beam is statically indeterminate to the first degree. There are no displacements of the supports and the axial force is zero throughout the beam. Therefore, the function Φ will contain only the bending moment terms. Let X identify the redundant reactive or interactive force of choice. Then $\Phi = \Phi(X)$ and the necessary condition for Φ to be a minimum takes the form

$$\Phi'(X) = \frac{d\Phi}{dX} = \sum_m \int_0^L \frac{M}{EI}\frac{\partial M}{\partial X}dx = 0$$

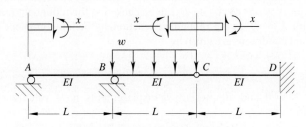

FIGURE 9.29 Compound beam.

The choice for the redundant force is at the option of the analyst. The results are independent of the sign convention adopted in formulating the expressions for the bending moments. Accordingly, the sign convention to be used is optional.

In this example, the bending moment over the support at B is taken to be the redundant force, and the sign conventions for each member are as shown in Figure 9.29. For member AB

$$M = \frac{x}{L}X, \quad \frac{\partial M}{\partial X} = \frac{x}{L}, \quad \int_0^L \frac{M}{EI}\frac{\partial M}{\partial X}\,dx = \frac{LX}{3EI}$$

For member BC

$$M = \frac{x}{L}X + \frac{w}{2}(Lx - x^2), \quad \frac{\partial M}{\partial X} = \frac{x}{L}, \quad \int_0^L \frac{M}{EI}\frac{\partial M}{\partial X}\,dx = \frac{LX}{3EI} + \frac{wL^3}{24EI}$$

For member CD

$$M = \frac{x}{L}X + \frac{wLx}{2}, \quad \frac{\partial M}{\partial X} = \frac{x}{L}, \quad \int_0^L \frac{M}{EI}\frac{\partial M}{\partial X}\,dx = \frac{LX}{3EI} + \frac{wL^3}{6EI}$$

Insertion of the foregoing results into the expression for $\Phi'(X)$ yields

$$\Phi'(X) = \frac{LX}{EI} + \frac{5wL^3}{24EI}$$

Therefore, the necessary condition for Φ to be a minimum is satisfied by $X = -5wL^2/24$, which is consistent with the results obtained in Example 9.2 used to illustrate the basic method of consistent deformation. It is to be observed that the problem has been solved without use of the sketches involved in the solution by the method of consistent deformation.

Example 9.17

The frame shown in Figure 9.11 is reproduced below as Figure 9.30a. Axial deformation is negligibly small. Use the theorem of least work to calculate the reactions.

The frame is statically indeterminate to the second degree. To illustrate the flexibility that is available in analysis by the method of least work, the bending moments at the midpoint and at the C end of member BC are taken as the redundant reaction forces. The bending moments in the given frame are exactly the same as those in the frame shown in Figure 9.30b. By the principle of superposition, the bending moment at any location in the frame can be expressed in the form

$$M = M_0 + m_1 X_1 + m_2 X_2$$

where M_0 is the bending moment when the applied loads act alone, m_1 is the bending moment due to $X_1 = 1$ acting alone, and m_2 is the bending moment due to $X_2 = 1$ acting alone. Because there are no non-zero prescribed displacements, the energy function Φ reduces to

$$\Phi = \Phi(X_1, X_2) = \sum_m \int_0^L \frac{1}{2EI}(M_0 + m_1 X_1 + m_2 X_2)^2 \, dx$$

The necessary conditions for the function Φ to be a minimum are

$$0 = \frac{\partial \Phi}{\partial X_1} = \sum_m \int_0^L \frac{1}{EI}(M_0 + m_1 X_1 + m_2 X_2) m_1 \, dx$$

$$0 = \frac{\partial \Phi}{\partial X_2} = \sum_m \int_0^L \frac{1}{EI}(M_0 + m_1 X_1 + m_2 X_2) m_2 \, dx$$

276　Chapter 9　Force Methods

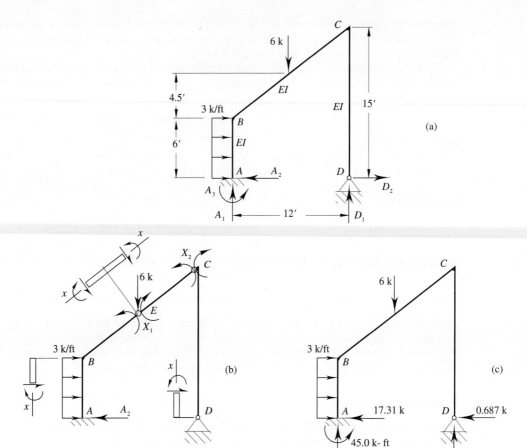

FIGURE 9.30 Flexibility in choosing redundant reaction forces.

which after expanding and rearranging terms become

$$0 = \sum_m \int_0^L \frac{1}{EI} M_0 m_1 \, dx + \left(\sum_m \int_0^L \frac{1}{EI} m_1^2 \, dx \right) X_1 + \left(\sum_m \int_0^L \frac{1}{EI} m_1 m_2 \, dx \right) X_2$$

$$0 = \sum_m \int_0^L \frac{1}{EI} M_0 m_2 \, dx + \left(\sum_m \int_0^L \frac{1}{EI} m_1 m_2 \, dx \right) X_1 + \left(\sum_m \int_0^L \frac{1}{EI} m_2^2 \, dx \right) X_2$$

a pair of simultaneous equations to be solved for X_1 and X_2 With coordinates and sign conventions as shown in Figure 9.30b, the bending moments are

AB:　$M_0 = -36 - 1.5x^2$,　　$m_1 = 2$,　　　　　$m_2 = -1 - x/15$　　$0 \leq x \leq 6$
BE:　$M_0 = -4.8x$,　　　　　$m_1 = 1 + x/7.5$,　$m_2 = -x/7.5$　　　$0 \leq x \leq 7.5$
EC:　$M_0 = 0$,　　　　　　　$m_1 = 1 - x/7.5$,　$m_2 = x/7.5$　　　　$0 \leq x \leq 7.5$
CD:　$M_0 = 0$,　　　　　　　$m_1 = 0$,　　　　　$m_2 = -x/15$　　　　$0 \leq x \leq 15$

and the sums of the integrals are

$$\sum_m \int_0^L \frac{1}{EI} M_0 m_1 \, dx = -\frac{873}{EI} \text{ k-ft}^2, \qquad \sum_m \int_0^L \frac{1}{EI} M_0 m_2 \, dx = \frac{489.6}{EI} \text{ k-ft}^2,$$

$$\sum_m \int_0^L \frac{1}{EI} m_1^2 \, dx = \frac{44}{EI} \text{ ft} \qquad \sum_m \int_0^L \frac{1}{EI} m_1 m_2 \, dx = -\frac{19.4}{EI} \text{ ft},$$

$$\sum_m \int_0^L \frac{1}{EI} m_2^2 \, dx = \frac{18.72}{EI} \text{ ft}$$

Insertion of the foregoing results into the necessary conditions for Φ to be a minimum gives

$$44 X_1 - 19.40 X_2 = 873$$
$$-19.40 X_1 + 18.72 X_2 = -489.6$$

the solution of which is $X_1 = 15.31$ k-ft, $X_2 = -10.30$ k-ft. The corresponding values for the reactions at the supports are found from statics and have values as displayed in Figure 9.30c. As it must be, the results are the same as those obtained in the analysis of the frame by the basic method of consistent deformation.

Example 9.18

Determine the bending moments over the supports due to the support settlements of the beam shown in Figure 9.31. This is the same beam considered in Example 9.14 (shown in Figure 9.26) to illustrate application of the three-moment equation.

Let R be the force required at support 1 in the direction of Δ and let C be the couple required at support 3 in the direction of θ. Then, because there are no axial forces, the function Φ reduces to

$$\Phi = \sum_m \int_0^L \frac{M^2}{2EI} \, dx - R\Delta - C\theta$$

Let M_2 and M_3 be the bending moments over the supports at 2 and 3, respectively. From equilibrium, $RL = -M_2$ and $C = -M_3$. By the principle of superposition the bending moment at any location in the beam, for all possible values of M_2 and M_3, can be expressed as $M = m_2 M_2 + m_3 M_3$, where m_2 is the bending moment when $M_2 = 1$ and $M_3 = 0$, and m_3 is the bending moment when $M_3 = 1$ and $M_2 = 0$. In terms of M_2 and M_3 the energy function Φ is

$$\Phi = \Phi(M_2, M_3) = \sum_m \int_0^L \frac{1}{EI}(m_2 M_2 + m_3 M_3)^2 \, dx + \frac{M_2}{L}\Delta + M_3\theta$$

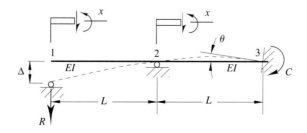

FIGURE 9.31 Support settlements.

By the theorem of least work, the true values of M_1 and M_2 make the energy function Φ a minimum. The necessary conditions for Φ to be a minimum are

$$0 = \frac{\partial \Phi}{\partial M_2} = \sum_m \int_0^L \frac{1}{EI}(m_2 M_2 + m_3 M_3) m_2 \, dx + \frac{\Delta}{L}$$

$$0 = \frac{\partial \Phi}{\partial M_3} = \sum_m \int_0^L \frac{1}{EI}(m_2 M_2 + m_3 M_3) m_3 \, dx + \theta$$

With the coordinates and sign conventions as shown in Figure 9.31 the bending moments are:

span 1–2: $\quad M = \dfrac{x}{L} M_2 \quad\quad\quad 0 \leq x \leq L$

span 2–3: $\quad M = \left(1 - \dfrac{x}{L}\right) M_2 + \dfrac{x}{L} M_3 \quad\quad 0 \leq x \leq L$

Substitution of the foregoing relations into the necessary conditions for Φ to be a minimum yields

$$0 = \frac{2L}{3EI} M_2 + \frac{L}{6EI} M_3 + \frac{\Delta}{L}$$

$$0 = \frac{L}{6EI} M_2 + \frac{L}{3EI} M_3 + \theta$$

These equations may be rearranged to read

$$\frac{4L}{I} M_2 + \frac{L}{I} M_3 = -6E \frac{\Delta}{L}$$

$$\frac{L}{I} M_2 + \frac{2L}{I} M_3 = -6E\theta$$

which are identical to the equations derived in solving the same problem by the three-moment equation method.

Example 9.19

The following example illustrates the ease with which the theorem of least work can be used to formulate problems for which the conventional treatment of the compatibility conditions is involved. Use the theorem of least work to derive the equations to be solved for the tensions X_1 and X_2 in any two appropriately selected redundant members of the internally statically indeterminate truss shown in Figure 9.32.

The truss was considered earlier as a specific example to illustrate application of the basic method of consistent deformation to the force analysis of internally statically indeterminate structures (see Example 9.9). To be suitable as choices for the redundant members, the members must have the property that the truss be capable of being in equilibrium under arbitrary loads when the tensions in the selected members are set to zero. In other words, the cut-back structure corresponding to the removal of the selected bars must be statically determinate. Because the loads are applied only to the joints, the bending moments are zero everywhere. The members are uniform and all prescribed displacement are zero so the function Φ is simply

$$\Phi = \sum_m \frac{N^2 L}{2EA}$$

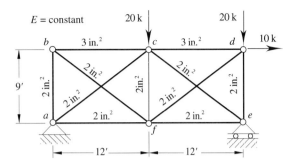

FIGURE 9.32 Truss, internally statically indeterminate to the second degree.

By the principle of superposition, the tension in any bar can be expressed in the form

$$N = N_0 + n_1 X_1 + n_2 X_2$$

where N_0 is the tension due to the applied loads with $X_1 = X_2 = 0$, n_1 is the tension when $X_1 = 1$, $X_2 = 0$ and there are no applied loads, and n_2 is the tension when $X_2 = 1$, $X_1 = 0$ and there are no applied loads. In terms of X_1 and X_2

$$\Phi = \Phi(X_1, X_2) = \sum_m \frac{L}{EA}(N_0 + n_1 X_1 + n_2 X_2)^2$$

Application of the necessary conditions for Φ to be a minimum yields

$$0 = \frac{\partial \Phi}{\partial X_1} = \sum_m \frac{N_0 n_1 L}{EA} + \left(\sum_m \frac{n_1^2 L}{EA}\right) X_1 + \left(\sum_m \frac{n_1 n_2 L}{EA}\right) X_2$$

$$0 = \frac{\partial \Phi}{\partial X_2} = \sum_m \frac{N_0 n_2 L}{EA} + \left(\sum_m \frac{n_1 n_2 L}{EA}\right) X_1 + \left(\sum_m \frac{n_2^2 L}{EA}\right) X_2$$

which are the desired equations. They are of exactly the same form as derived in Example 9.9; however, here, the derivation has been accomplished with much less effort.

KEY POINTS

1. Success in the force method of analysis of a statically indeterminate structure is tied to the ability to recognize and properly formulate the compatibility equations in terms of redundant reaction forces.

2. In the method of consistent deformation, attention is focussed directly on formulation of the compatibility equations in terms of forces.

3. The three-moment equation is a generic statement of the condition of continuity of slope over an interior support of a continuous beam in terms of bending moments over the supports.

4. In the method of least work, the compatibility conditions appropriate to the chosen redundant reaction forces are taken into account automatically.

PROBLEMS

9.1–9.12 General instructions: Each of the beams is statically indeterminate to the first degree. a) Construct the cut-back structure corresponding to the assigned redundant force. b) Use the basic method of consistent deformation to determine the redundant force. c) Construct the shear and bending moment diagrams. d) Calculate the assigned displacement.

9.1 Take the reaction at B as the redundant force. Calculate the deflection at C. Follow the general instructions for problems 9.1–9.12.

9.2 Take the reaction moment at D as the redundant force. Calculate the deflection at A. Follow the general instructions for problems 9.1–9.12.

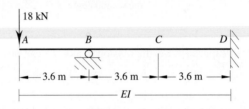

PROBLEMS 9.1 and 9.2

9.3 Take the reaction at A as the redundant force. Calculate the deflection at C. Follow the general instructions for problems 9.1–9.12.

9.4 Take the reaction at B as the redundant force. Calculate the rotation at C. Follow the general instructions for problems 9.1–9.12.

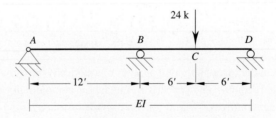

PROBLEMS 9.3 and 9.4

9.5 Take the reaction at A as the redundant force. Calculate the rotation at A. Follow the general instructions for problems 9.1–9.12.

9.6 Take the bending moment over the support at B as the redundant force. Calculate the rotation at C. Follow the general instructions for problems 9.1–9.12.

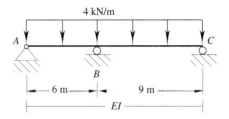

PROBLEMS 9.5 and 9.6

9.7 Take the reaction at A as the redundant force. Calculate the deflection at B. Follow the general instructions for problems 9.1–9.12.

9.8 Take the reaction at C as the redundant force. Calculate the deflection at D. Follow the general instructions for problems 9.1–9.12.

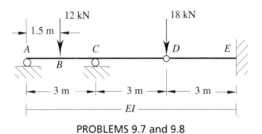

PROBLEMS 9.7 and 9.8

9.9 Take the reaction at C as the redundant force. Calculate the deflection at B. Follow the general instructions for problems 9.1–9.12.

9.10 Take the reaction moment at D as the redundant force. Calculate the deflection at D. Follow the general instructions for problems 9.1–9.12.

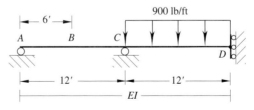

PROBLEMS 9.9 and 9.10

9.11 Take the reaction at B as the redundant force. Calculate the rotation at B. Follow the general instructions for problems 9.1–9.12.

9.12 Take the reaction at D as the redundant force. Calculate the deflection at C. Follow the general instructions for problems 9.1–9.12.

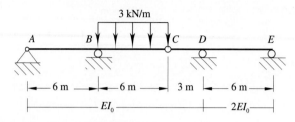

PROBLEMS 9.11 and 9.12

9.13–9.24 General instructions: Each of the beams is statically indeterminate to the second degree.
a) Construct the cut-back structure corresponding to the assigned redundant forces.
b) Use the basic method of consistent deformation to determine the redundant forces.
c) Construct the shear and bending moment diagrams.

9.13 Take the reaction at B and at D as the redundant forces. Follow the general instructions for problems 9.13–9.24.

9.14 Take the reaction moment at A and the reaction at D as the redundant forces. Follow the general instructions for problems 9.13–9.24.

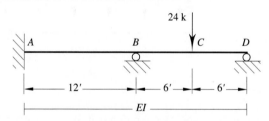

PROBLEMS 9.13 and 9.14

9.15 Take the reaction at C and the reactive moment at E the redundant forces. Follow the general instructions for problems 9.13–9.24.

9.16 Take the reaction at A and the reaction force at E as the redundant forces. Follow the general instructions for problems 9.13–9.24.

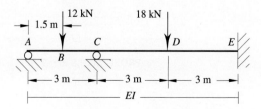

PROBLEMS 9.15 and 9.16

9.17 Take the reaction at A and at D as the redundant forces. Follow the general instructions for problems 9.13–9.24.

9.18 Take the reactions at B and at D as the redundant forces. Follow the general instructions for problems 9.13–9.24.

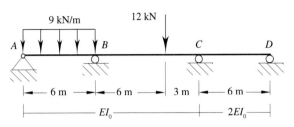

PROBLEMS 9.17 and 9.18

9.19 Take the reaction moment at A and the reaction force at C as the redundant forces. Follow the general instructions for problems 9.13–9.24.

9.20 Take the reaction force at A and at C as the redundant forces. Follow the general instructions for problems 9.13–9.24.

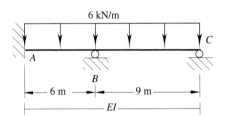

PROBLEMS 9.19 and 9.20

9.21 Take the reaction moment at A and the reaction moment at C as the redundant forces. Follow the general instructions for problems 9.13–9.24.

9.22 Take the reaction moment at A and the reaction force at C as the redundant forces. Follow the general instructions for problems 9.13–9.24.

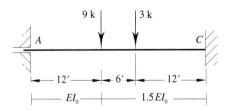

PROBLEMS 9.21 and 9.22

9.23 Take the reaction moment at A and the reaction force at C as the redundant forces. Follow the general instructions for problems 9.13–9.24.

9.24 Take the reaction moment at D and the reaction force at C as the redundant forces. Follow the general instructions for problems 9.13–9.24.

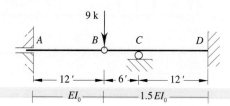

PROBLEMS 9.23 and 9.24

9.25–9.36 General instructions: Each of the frames is statically indeterminate to the first degree. The deformation is inextensional. a) Construct the cut-back structure corresponding to the assigned redundant force. b) Use the basic method of consistent deformation to determine the redundant force. c) Calculate the assigned displacement.

9.25 Take the reaction at D as the redundant force. Calculate the vertical component of displacement at C. Follow the general instructions for problems 9.25–9.36.

9.26 Take the reaction moment at A as the redundant force. Calculate the horizontal component of displacement at B. Follow the general instructions for problems 9.25–9.36.

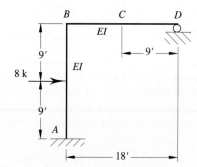

PROBLEMS 9.25 and 9.26

9.27 Take the horizontal reaction at A as the redundant force. Calculate the vertical component of displacement at B. Follow the general instructions for problems 9.25–9.36.

9.28 Take the vertical reaction at E as the redundant force. Calculate the horizontal component of displacement at D. Follow the general instructions for problems 9.25–9.36.

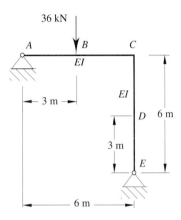

PROBLEMS 9.27 and 9.28

9.29 Take the vertical reaction at C as the redundant force. Calculate the horizontal component of displacement at A. Follow the general instructions for problems 9.25–9.36.

9.30 Take the reaction moment at A as the redundant force. Calculate the rotation at C. Follow the general instructions for problems 9.25–9.36.

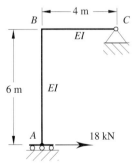

PROBLEMS 9.29 and 9.30

9.31 Take the horizontal reaction at *A* as the redundant force. Calculate the horizontal component of displacement at *D*. Follow the general instructions for problems 9.25–9.36.

9.32 Take the vertical reaction *E* as the redundant force. Calculate the horizontal component of displacement at *B*. Follow the general instructions for problems 9.25–9.36.

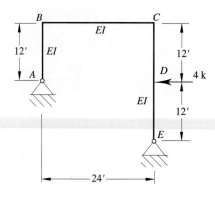

PROBLEMS 9.31 and 9.32

9.33 Take the horizontal reaction at *A* as the redundant force. Calculate the horizontal component of displacement at *C*. Follow the general instructions for problems 9.25–9.36.

9.34 Take the vertical reaction at *D* as the redundant force. Calculate the vertical component of displacement at *C*. Follow the general instructions for problems 9.25–9.36.

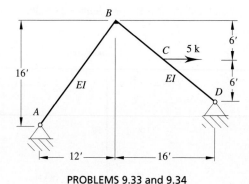

PROBLEMS 9.33 and 9.34

9.35 Take the reactive moment at A as the redundant force. Calculate the vertical component of displacement at C. Follow the general instructions for problems 9.25–9.36.

9.36 Take the vertical reaction at D as the redundant force. Calculate the horizontal component of displacement at B. Follow the general instructions for problems 9.25–9.36.

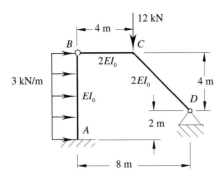

PROBLEMS 9.35 and 9.36

9.37–9.46 General instructions: Each of the frames is statically indeterminate to the second degree. Construct the cut-back structure corresponding to the assigned redundant forces. Use the basic method of consistent deformation to determine the redundant forces and use statics to determine the remaining reactive forces. The members are axially rigid.

9.37 Take the vertical reaction at A and the horizontal reaction at D as the redundant forces. Follow the general instructions for problems 9.37–9.46.

9.38 Take the horizontal reaction at A and the horizontal reaction at E as the redundant forces. Follow the general instructions for problems 9.37–9.46.

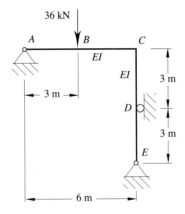

PROBLEMS 9.37 and 9.38

9.39 Take the reactive moment at A and the vertical reaction at E as the redundant forces. Follow the general instructions for problems 9.37–9.46.

9.40 Take the reaction at D and the reaction at E as the redundant forces. Follow the general instructions for problems 9.37–9.46.

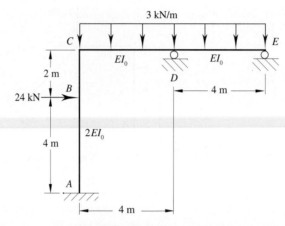

PROBLEMS 9.39 and 9.40

9.41 Take the reactive moment at E and the horizontal reaction at A as the redundant forces. Follow the general instructions for problems 9.37–9.46.

9.42 Take the reactive moment at A and the horizontal reaction at E as the redundant forces. Follow the general instructions for problems 9.37–9.46.

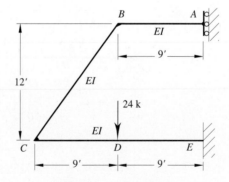

PROBLEMS 9.41 and 9.42

9.43 Take the horizontal reaction at D and the vertical reaction at D as the redundant forces. Follow the general instructions for problems 9.37–9.46.

9.44 Take the reactive moment at A and the vertical reaction at D as the redundant forces. Follow the general instructions for problems 9.37–9.46.

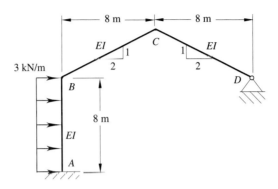

PROBLEMS 9.43 and 9.44

9.45 Take the horizontal reaction at D and the vertical reaction at E as the redundant forces. Follow the general instructions for problems 9.37–9.46.

9.46 Take the horizontal reaction at E and the vertical reaction at E as the redundant forces. Follow the general instructions for problems 9.37–9.46.

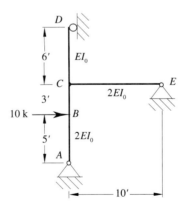

PROBLEMS 9.45 and 9.46

9.47 Determine the reactions by the method of consistent deformation.

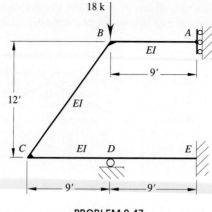

PROBLEM 9.47

9.48 Determine the reactions by the method of consistent deformation.

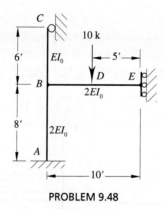

PROBLEM 9.48

9.49 Determine the reactions by the method of consistent deformation.

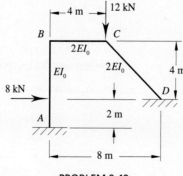

PROBLEM 9.49

9.50–9.59 General instructions: Each truss is statically indeterminate to the first degree. a) Construct the cut-back structure corresponding to the assigned redundant force. b) Use the basic method of consistent deformation to determine the redundant force. c) Determine the bar forces. d) calculate the assigned displacement. Arrange the calculations in tabular form. Display the bar forces for each case of loading on a line diagram of the truss.

9.50 Take the reaction at d as the redundant force. Calculate the vertical component of displacement of joint e. $E = 30{,}000$ ksi.

9.51 Take the horizontal reaction at a as the redundant force. Calculate the horizontal component of displacement of joint c. $E = 30{,}000$ ksi.

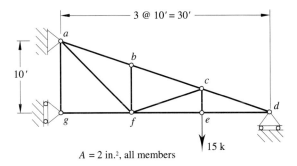

PROBLEMS 9.50 and 9.51

9.52 Take the vertical reaction at c as the redundant force. Calculate the vertical component of displacement of joint d. $E = 200$ GPa.

9.53 Take the horizontal reaction at a as the redundant force. Calculate the horizontal component of displacement of joint b. $E = 200$ GPa.

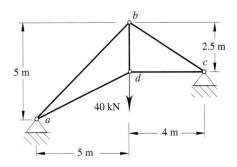

PROBLEMS 9.52 and 9.53

9.54 Take the reaction at h as the redundant force. Calculate the vertical component of displacement of joint g. $E = 30,000$ ksi.

9.55 Take the reaction at f as the redundant force. Calculate the vertical component of displacement of joint i. $E = 30,000$ ksi.

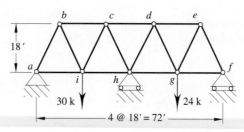

$A_{bc} = A_{cd} = A_{de} = 3$ in.^{2}
$A_{ai} = A_{ih} = A_{hg} = A_{gf} = 2.5$ in.^{2}
$A_{ba} = A_{bi} = A_{ci} = A_{ch} = A_{dh} = A_{dg} = A_{eg} = A_{ef} = 2$ in.^{2}

PROBLEMS 9.54 and 9.55

9.56 Take the reaction at a as the redundant force. Calculate the vertical component of displacement of joint e. $E = 30,000$ ksi.

9.57 Take the reaction at d as the redundant force. Calculate the vertical component of displacement of joint b. $E = 30,000$ ksi.

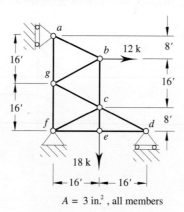

$A = 3$ in.^{2}, all members

PROBLEMS 9.56 and 9.57

9.58 Take the vertical reaction at h as the redundant force. Calculate the vertical component of displacement of joint f. $E = 200$ GPa.

9.59 Take the horizontal reaction at h as the redundant force. Calculate the vertical component of displacement of joint b. $E = 200$ GPa.

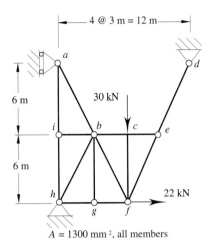

PROBLEMS 9.58 and 9.59

9.60–9.65 General instructions: Each truss is internally statically indeterminate to the first degree.
a) Construct the cut-back structure corresponding to the assigned redundant bar force.
b) Use the basic method of consistent deformation to determine the redundant force.
c) Determine the bar forces. Arrange the calculations in tabular form. Display the bar forces for each case of loading on a line diagram of the truss.

9.60 Take the tension force in bc as the redundant quantity. Follow the general instructions for problems 9.60–9.65.

9.61 Take the tension force in ed as the redundant quantity. Follow the general instructions for problems 9.60–9.65.

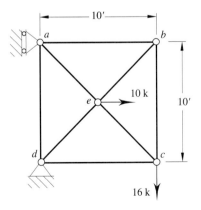

PROBLEMS 9.60 and 9.61

9.62 Take the tension force in *ad* as the redundant quantity. Follow the general instructions for problems 9.60–9.65.

9.63 Take the tension force in *ec* as the redundant quantity. Follow the general instructions for problems 9.60–9.65.

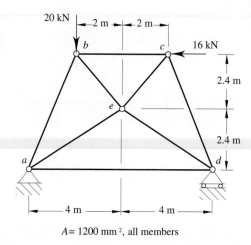

$A = 1200$ mm², all members

PROBLEMS 9.62 and 9.63

9.64 Take the tension force in *bc* as the redundant quantity. Follow the general instructions for problems 9.60–9.65.

9.65 Take the tension force in *ed* as the redundant quantity. Follow the general instructions for problems 9.60–9.65.

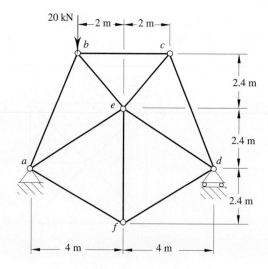

$A = 1000$ mm², all members

PROBLEMS 9.64 and 9.65

9.66 Use the method of consistent deformation to determine the reactions and all bar tensions. Display all results on a line diagram of the truss.

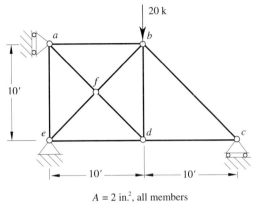

$A = 2$ in.^{2}, all members

PROBLEM 9.66

9.67 Use the method of consistent deformation to determine the reactions. $I = 1728$ in.^{4}, $A = 2$ in.^{2}

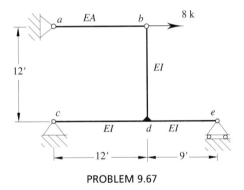

PROBLEM 9.67

9.68 Use the method of consistent deformation to determine the tension in the tie bar. The joints at d and c are rigid. $I = 675 \times 10^6$ mm.^{4}, $A = 1300$ mm^2.

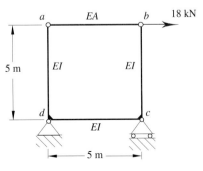

PROBLEM 9.68

9.69 Use the method of consistent deformation to determine the reactions due to a 0.36 inches downward settlement of support C. $EI = 360{,}000$ k-ft².

9.70 Use the method of consistent deformation to determine the reactions due to a downward settlement of 0.24 inches of support A and a downward settlement of 0.36 inches of support B. $EI = 360{,}000$ k-ft².

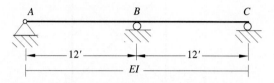

PROBLEMS 9.69 and 9.70

9.71 Use the method of consistent deformation to determine the reactions due to a 12 mm heaving (upward displacement) of support B. $EI = 166$ MN·m².

9.72 Use the method of consistent deformation to determine the reactions due to a downward settlement of 16 mm of support A and a heaving (upward displacement) of 10 mm of support B. $EI = 166$ MN·m².

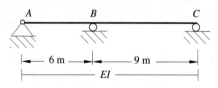

PROBLEMS 9.71 and 9.72

9.73 Use the method of consistent deformation to determine the reactions due to a 12 mm downward settlement of support D. $EI = 180$ MN·m².

9.74 Use the method of consistent deformation to determine the reactions due to a downward settlement of 8 mm of support A and a downward settlement of 16 mm. of support D. $EI = 180$ MN·m².

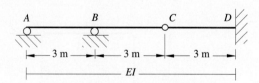

PROBLEMS 9.73 and 9.74

9.75 Use the method of consistent deformation to determine the reactions due to a 0.36 inches downward settlement of support C. $EI = 300{,}000$ k-ft².

9.76 Use the method of consistent deformation to determine the reactions due to a downward settlement of 0.24 inches of support A and a downward settlement of 0.36 inches of support B. $EI = 300{,}000$ k-ft^2.

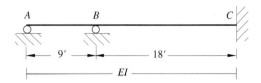

PROBLEMS 9.75 and 9.76

9.77 Use the method of consistent deformation to determine the reactions due to a 10 mm downward settlement and a displacement to the left of 8 mm of support C. $EI = 180$ MN·m^2.

9.78 Use the method of consistent deformation to determine the reactions due to a 15 mm downward settlement of support A. $EI = 180$ MN·m^2.

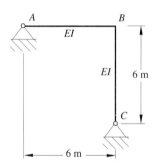

PROBLEMS 9.77 and 9.78

9.79 Use the method of consistent deformation to determine the reactions due to a downward settlement of 0.30 inches of support A. $EI = 320{,}000$ k-ft^2.

9.80 Use the method of consistent deformation to determine the reactions due to a downward settlement of 0.36 inches of support D. $EI = 320{,}000$ k-ft^2.

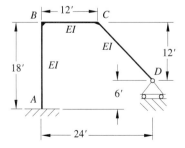

PROBLEMS 9.79 and 9.80

9.81 a) Use the three-moment equation to determine the bending moments over the supports, b) construct the shear and bending moment diagrams, and c) determine the deflection at C.

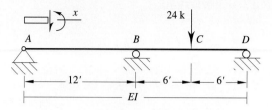

PROBLEM 9.81

9.82 a) Use the three-moment equation to determine the bending moments over the supports, b) construct the shear and bending moment diagrams, and c) determine the rotation at C.

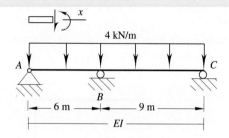

PROBLEM 9.82

9.83 a) Use the three-moment equation to determine the bending moments over the supports, b) construct the shear and bending moment diagrams, and c) determine the deflection at E.

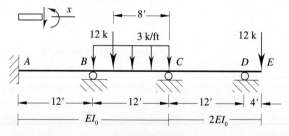

PROBLEM 9.83

9.84 a) Use the three-moment equation to determine the bending moments over the supports, b) construct the shear and bending moment diagrams, and c) determine the deflection at C.

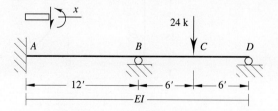

PROBLEM 9.84

9.85 a) Use the three-moment equation to determine the bending moments over the supports, b) construct the shear and bending moment diagrams, and c) determine the deflection at B.

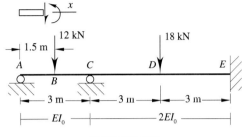

PROBLEM 9.85

9.86 a) Use the three-moment equation to determine the bending moments over the supports, b) construct the shear and bending moment diagrams, and c) determine the deflection at C.

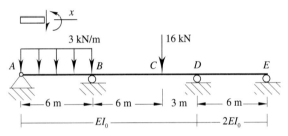

PROBLEM 9.86

9.87 a) Use the three-moment equation to determine the bending moments over the supports, b) construct the shear and bending moment diagrams, and c) determine the deflection at C.

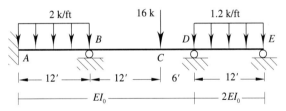

PROBLEM 9.87

9.88 a) Use the three-moment equation to determine the bending moment over the supports, b) construct the shear and bending moment diagrams, and c) determine the rotation of the tangent at B.

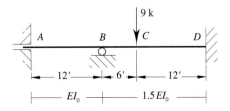

PROBLEM 9.88

9.89 Support B heaves upward 12 mm. Use the three-moment equation to determine the bending moment over support B and then calculate the reactions. $EI = 166$ MN · m².

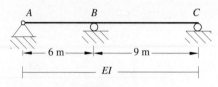

PROBLEM 9.89

9.90 Support A settles 0.36 inches. Use the three-moment equation to determine the bending moment over supports B and C and then calculate the reactions. $EI = 180{,}000$ k-ft².

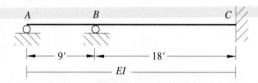

PROBLEM 9.90

9.91 Support A settles 0.24 inches and support D heaves upward 0.12 inches. Use the three-moment equation to determine the bending moments over the supports and then calculate the reactions. $EI_0 = 240{,}000$ k-ft².

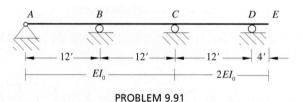

PROBLEM 9.91

9.92 Use Castigliano's theorem of least work and statics to determine the reactions. Take the reaction at B as the redundant force.

9.93 Use Castigliano's theorem of least work and statics to determine the reactions. Take the reaction moment at C as the redundant force.

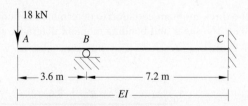

PROBLEMS 9.92 and 9.93

9.94 Use Castigliano's theorem of least work and statics to determine the reactions. Take the reaction at B as the redundant force.

9.95 Use Castigliano's theorem of least work and statics to determine the reactions. Take the reaction at A as the redundant force.

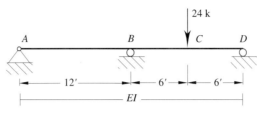

PROBLEMS 9.94 and 9.95

9.96 Use Castigliano's theorem of least work and statics to determine the reactions. Take the reaction at A as the redundant force.

9.97 Use Castigliano's theorem of least work and statics to determine the reactions. Take the reaction at B as the redundant force.

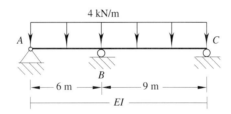

PROBLEMS 9.96 and 9.97

9.98 Use Castigliano's theorem of least work and statics to determine the reactions.

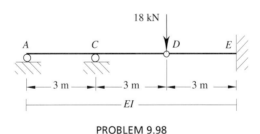

PROBLEM 9.98

9.99 Use Castigliano's theorem of least work and statics to determine the reactions.

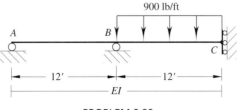

PROBLEM 9.99

9.100 Use Castigliano's theorem of least work and statics to determine the reactions.

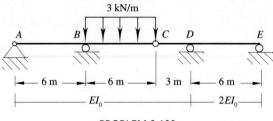

PROBLEM 9.100

9.101 Use Castigliano's theorem of least work and statics to determine the reactions.

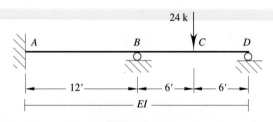

PROBLEM 9.101

9.102 Use Castigliano's theorem of least work and statics to determine the reactions.

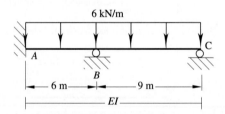

PROBLEM 9.102

9.103 Use Castigliano's theorem of least work and statics to determine the reactions.

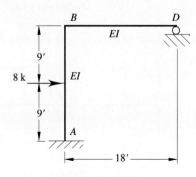

PROBLEM 9.103

9.104 Use Castigliano's theorem of least work and statics to determine the reactions.

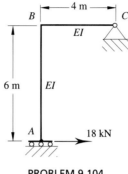

PROBLEM 9.104

9.105 Use Castigliano's theorem of least work and statics to determine the reactions.

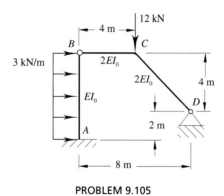

PROBLEM 9.105

9.106 Use Castigliano's theorem of least work and statics to determine the reactions.

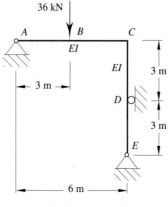

PROBLEM 9.106

9.107 Use Castigliano's theorem of least work and statics to determine the reactions.

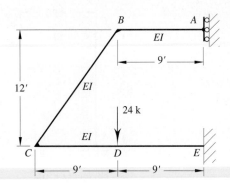

PROBLEM 9.107

9.108 Use Castigliano's theorem of least work and statics to determine the reactions due to a 12 mm heaving (upward) displacement of support B. Take the reaction at B as the redundant force. $EI = 166$ MN·m².

9.109 Use Castigliano's theorem of least work and statics to determine the reactions due to a 12 mm heaving (upward) displacement of support B. Take the reaction at A as the redundant force. $EI = 166$ MN·m².

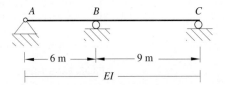

PROBLEMS 9.108 and 9.109

9.110 Use Castigliano's theorem of least work and statics to determine the reactions due to a downward settlement of 8 mm of support A and a downward settlement of 16 mm of support D. Take the reaction at A as the redundant force. $EI = 180$ MN·m².

9.111 Use Castigliano's theorem of least work and statics to determine the reactions due to a downward settlement of 8 mm of support A and a downward settlement of 16 mm of support D. Take the reaction at B as the redundant force. $EI = 180$ MN·m².

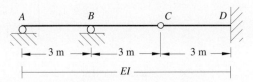

PROBLEMS 9.110 and 9.111

9.112 Use Castigliano's theorem of least work and statics to determine the reactions due to a 0.24 inch downward settlement of support C. Take the vertical reaction at C as the redundant force. $EI = 240{,}000$ k-ft^2.

9.113 Use Castigliano's theorem of least work and statics to determine the reactions due to a 0.24 inch downward settlement of support C. Take the vertical reaction at A as the redundant force. $EI = 240{,}000$ k-ft^2.

9.114 Use Castigliano's theorem of least work and statics to determine the reactions due to a 0.24 inch downward settlement of support C. Take the horizontal reaction at C as the redundant force. $EI = 240{,}000$ k-ft^2.

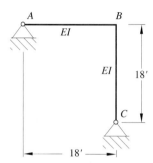

PROBLEMS 9.112, 9.113 and 9.114

9.115 Use Castigliano's theorem of least work, statics, and the principle of superposition to determine the reactions and all bar tensions. $E = 30{,}000$ ksi.

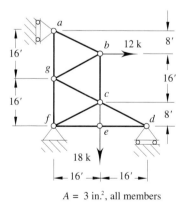

$A = 3$ in.^{2}, all members

PROBLEM 9.115

9.116 Use Castigliano's theorem of least work, statics, and the principle of superposition to determine the reactions and all bar tensions. $E = 200$ GPa.

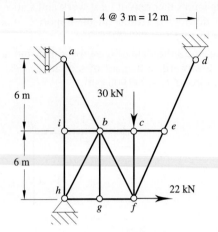

$A = 1300$ mm², all members

PROBLEM 9.116

9.117 Use Castigliano's theorem of least work, statics, and the principle of superposition to determine the reactions and all bar tensions. $E = 30{,}000$ ksi.

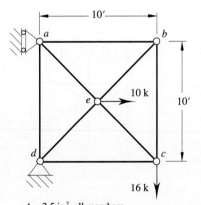

$A = 2.5$ in.², all members

PROBLEM 9.117

9.118 Use Castigliano's theorem of least work, statics, and the principle of superposition to determine the reactions and all bar tensions. $E = 200$ GPa.

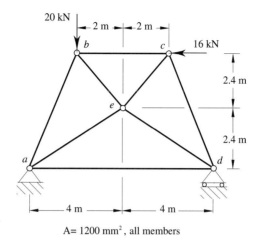

A= 1200 mm², all members

PROBLEM 9.118

9.119 Use Castigliano's theorem of least work, statics, and the principle of superposition to determine the reactions and all bar tensions. $E = 30,000$ ksi.

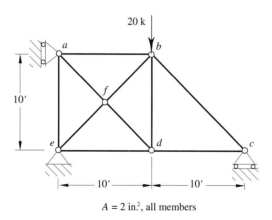

$A = 2$ in.², all members

PROBLEM 9.119

CHAPTER 10

Displacement Methods of Analysis

In the displacement method of analysis, displacements are taken as the fundamental mechanical variables. The compatibility equations, the displacement-deformation equations, and the force-deformation equations are employed to express equations of equilibrium in terms of displacements. Corresponding to each independent position or displacement variable (i.e., for each kinematic degree of freedom) there is one independent equation of equilibrium. Solution of the equations of equilibrium for the displacements followed by back substitution into the deformation-displacement relations, force-deformation equations, and the remaining equations of equilibrium yields the forces of reaction and interaction.

How one goes about deriving the independent equations of equilibrium corresponding to the independent position variables is of no consequence. Usually, the appropriate equations, or at least some of the equations, can be written down by inspection. However, the equations can always be derived with the aid of the principle of virtual work or the principle of stationary potential energy.

A fact of central importance in the displacement method of analysis of plane frames is that each member in such a frame can be treated as a six degree-of freedom mechanical system (five degree-of-freedom system in the case of inextensional deformation) for which the mechanical state is defined in terms of the loads between the ends of the member, the displacements of the ends of the member and the rotations of the tangents at the ends of the member. The purpose of the following abbreviated analysis is to establish this fact when the material obeys Hooke's law and the deformations and displacements are small. Assume that the beam shown in Figure 10.1a is in equilibrium under prescribed loads $w(x)$ and $p(x)$ and that the end-displacements (i.e., $v_A, \theta_A, u_A, v_B, \theta_B$, and u_B) are prescribed. Then, the end-forces M_A, V_A, N_A, M_B, V_B, and N_B and the displacement functions $v(x)$ and $u(x)$ comprise the list of unknown quantities to be determined.

For simplicity, assume the beam is uniform. The differential equations of equilibrium are

$$EIv'''' = -w(x), \quad EAu'' = -p(x)$$

Let $v_0(x)$ and $u_0(x)$ be displacement functions that satisfy the following conditions:

$$EIv_0'''' = -w(x), \quad v_0(0) = v_0(L) = v_0'(0) = v_0'(L) = 0$$
$$EAu_0'' = -p(x), \quad u_0(0) = u_0(L) = 0$$

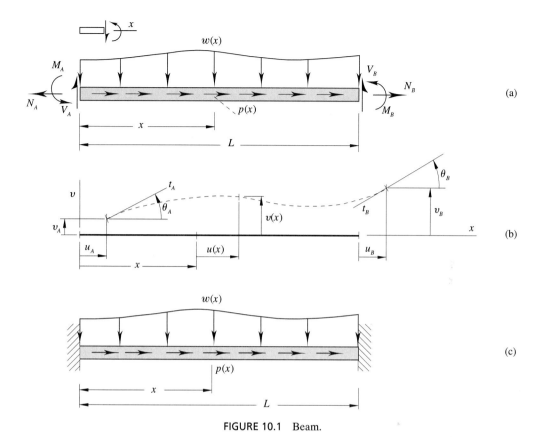

FIGURE 10.1 Beam.

The displacement functions $v_0(x)$ and $u_0(x)$ are seen to be nothing more than the displacement functions due to the loads $w(x)$ and $p(x)$ when the ends of the beam are built in to rigid supports as indicated in Figure 10.1c. With the aid of the functions $v_0(x)$ and $u_0(x)$, the original differential equations can now be written in the forms

$$v'''' = v_0'''', \quad u'' = u_0''$$

Successive integrations of the equations yields

$$v''' = A_1 + v_0''',$$
$$v'' = xA_1 + A_2 + v_0'',$$
$$v' = \frac{x^2}{2}A_1 + xA_2 + A_3 + v_0',$$
$$v = \frac{x^3}{6}A_1 + \frac{x^2}{2}A_2 + xA_3 + A_4 + v_0$$

$$u' = B_1 + u_0'$$
$$u = xB_1 + B_2 + u_0$$

Equations to be solved for the constants of integration are derived with the aid of the boundary conditions $v(0) = v_A$, $v(L) = v_B$, $v'(0) = \theta_A$, $v'(L) = \theta_B$, $u(0) = u_A$,

$u(L) = u_B$, and the properties of the functions $v_0(x)$ and $u_0(x)$. The equations are

$$\theta_A = A_3, \qquad u_A = B_2$$

$$\theta_B = \frac{L^2}{2} A_1 + L A_2 + A_3, \qquad u_B = L B_1 + B_2$$

$$v_A = A_4$$

$$v_B = \frac{L^3}{6} A_1 + \frac{L^2}{2} A_2 + L A_3 + A_4,$$

Therefore, the constants of integration are linear functions of the displacements of the ends of the member. From the expressions for v and u, it follows that for a given loading defined by $w(x)$ and $p(x)$ the displacements at the ends of a member define all equilibrium configurations of the member to within a constant function of x.

10.1 SLOPE-DEFLECTION EQUATIONS

The slope-deflection equations form the basis of a primary displacement method of analysis of plane frames. The equations can be derived by completing the analysis outlined in the previous section. However, the traditional method of derivation is presented herein. Within the framework of the small displacement theory of structures, axial loads do not affect the bending deformations of a beam. Accordingly, axial loads are omitted in the derivation.

Sign conventions are implicit in any formulation of the slope-deflection equations. Here, the sign conventions are set out with reference to local member-coordinate axes. As shown in Figure 10.2, the local x-axis has its origin at the A-end of the member and is directed from the A-end toward the B-end. The local v-axis is chosen so that the x-v coordinate system is right-handed (i.e., $x \otimes y \Rightarrow a$ z-axis upward out of the plane of

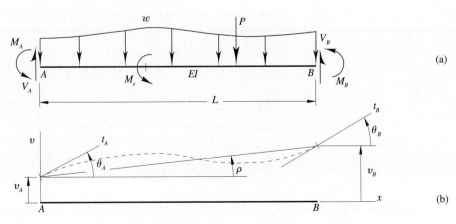

FIGURE 10.2 Sign conventions.

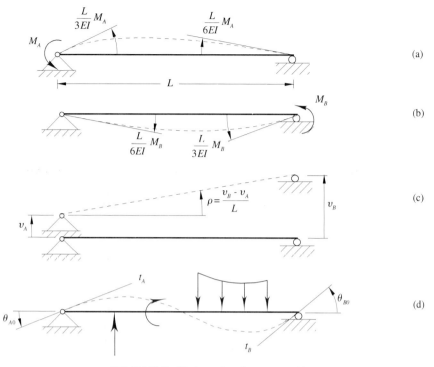

FIGURE 10.3 End rotations by superposition.

the paper). $v'(x) > 0$ is understood to define the positive sense for rotation. Clearly, with the selected coordinates, rotations are measured positive in the counterclockwise sense.

End-moments M_A and M_B are measured positive in the positive sense of the end-rotations θ_A and θ_B, respectively. End-shears V_A and V_B measured positive in the direction of the end-displacements v_A and v_B, respectively. Transverse loads between the ends of the member are taken positive in the sense opposite to the v-axis. Applied couples such as M_e are taken positive in the counterclockwise sense. In Figure 10.2b, ρ is the rotation of the chord measured positive in the counterclockwise sense.

For a uniform beam, the end-rotation at A and at B can be obtained with the aid of the principle of superposition, small angle geometry, and the results shown in Figure 10.3. Thus,

$$\theta_A = \frac{L}{3EI} M_A - \frac{L}{6EI} M_B + \frac{(v_B - v_A)}{L} + \theta_{A0}$$

$$\theta_B = -\frac{L}{6EI} M_A + \frac{L}{3EI} M_B + \frac{(v_B - v_A)}{L} + \theta_{B0}$$

The values for the rotations θ_{A0} and θ_{B0} due to any combination of loads between the ends of the beam can also be obtained with the aid of the principle of superposition.

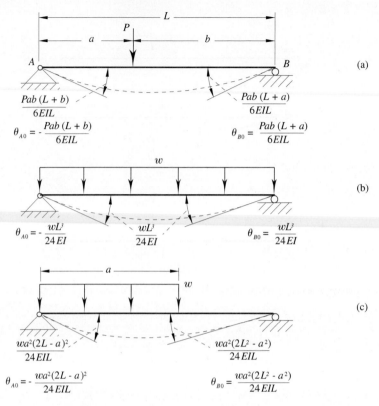

FIGURE 10.4 End rotations due to transverse loads.

The rotations for some elementary cases of loading are displayed in Figure 10.4. The equations for the rotations may be solved to obtain M_A and M_B in terms of the remaining variables. The equations of equilibrium may then be used to obtain expressions for the end-shears V_A and V_B. The solution of the equations and the expressions for the end-shears are given by

$$M_A = \frac{EI}{L}\left(4\theta_A + 2\theta_B + \frac{6}{L}v_A - \frac{6}{L}v_B\right) + M_{FA}$$

$$M_B = \frac{EI}{L}\left(2\theta_A + 4\theta_B + \frac{6}{L}v_A - \frac{6}{L}v_B\right) + M_{FB}$$

$$V_A = \frac{EI}{L}\left(\frac{6}{L}\theta_A + \frac{6}{L}\theta_B + \frac{12}{L^2}v_A - \frac{12}{L^2}v_B\right) + V_{FA}$$

$$V_B = \frac{EI}{L}\left(-\frac{6}{L}\theta_A - \frac{6}{L}\theta_B - \frac{12}{L^2}v_A + \frac{12}{L^2}v_B\right) + V_{FB}$$

(a)

in which

$$M_{FA} = -\frac{EI}{L}(4\theta_{A0} + 2\theta_{B0}), \quad M_{FB} = -\frac{EI}{L}(2\theta_{A0} + 4\theta_{B0})$$

Equations (a) are the slope-deflection equations.

Section 10.1   Slope-Deflection Equations

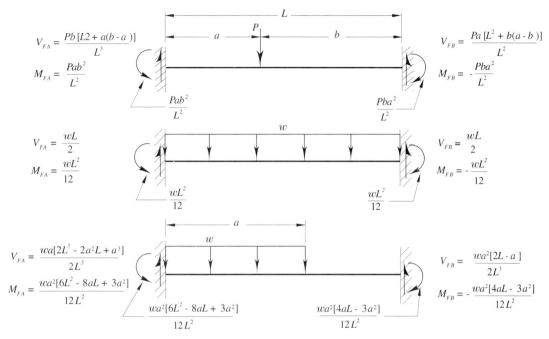

FIGURE 10.5   Fixed-end forces.

The terms $M_{FA}, M_{FB}, V_{FA}, V_{FB}$ can be given interpretations. The interpretations follow directly from Equations (a) which define $M_A, M_B, V_A,$ and V_B in terms of the loads between the ends of the member and all assignments of the end-displacements and end-rotations. With $\theta_A = \theta_B = v_A = v_B = 0$, Equations (a) yield $M_A = M_{FA}$, $M_B = M_{FB}, V_A = V_{FA}, V_B = V_{FB}$. However, the conditions $\theta_A = \theta_B = v_A = v_B = 0$ are the boundary conditions for a beam with built in ends. Accordingly, $M_{FA}, M_{FB}, V_{FA}, V_{FB}$ are the end-moments and end-shears due to the load applied between the ends of the beam when the ends of the beam are built in or *fixed*. This interpretation is the basis for referring to M_{FA} and M_{FB} as *fixed-end moments* and for referring to V_{FA} and V_{FB} as *fixed-end shears*. The symbolism of the interpretation and values of the fixed-end moments and fixed end-shears for some elementary loads of practical importance are illustrated in Figure 10.5.

The form of the slope-deflection equations shows that the total end-force is the superposition of the fixed end-forces and the end forces due to prescribed end-displacements. With $\theta_A \neq 0$ and $\theta_B = v_B = v_A = 0$ and no loads between the ends of the beam, the end-forces are as shown in Figure 10.6a; with $\theta_B \neq 0$ and $\theta_A = v_B = v_A = 0$ and no loads between the ends of the beam, the end-forces are as shown in Figure 10.6b; and, with $\theta_A = \theta_B = 0$ and $(v_B - v_A) \neq 0$ and no loads between the ends of the beam, the end forces are as shown in Figure 10.6c.

The slope-deflection equations, which are member force-deformation equations for bending, together with the compatibility conditions, bending displacement-deformation relations, and equilibrium equations are sufficient for the analysis framed structures in which the effects of axial deformation are negligibly small. The great

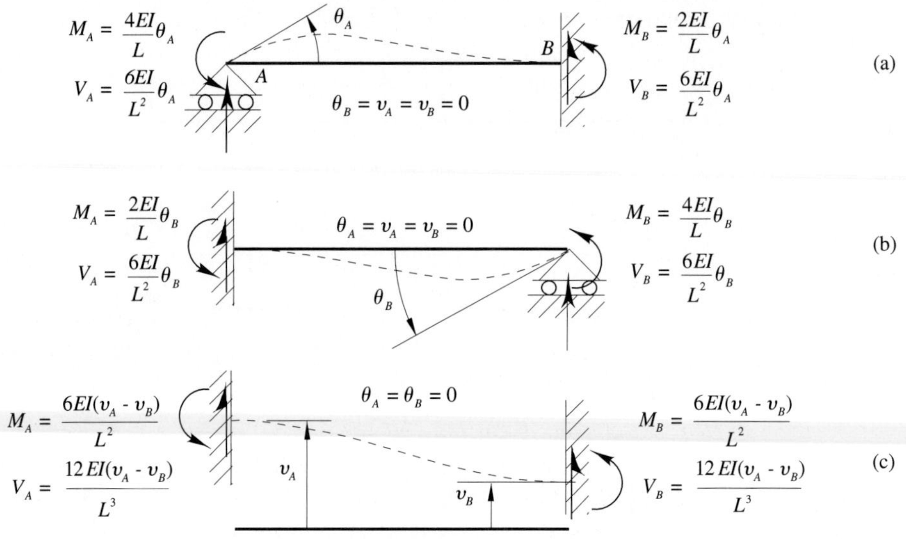

FIGURE 10.6 End-displacements and end-forces.

majority of practical beam and rigid frame structures are such that the effects of axial deformation are negligible. In the analysis of such structures, the members are assumed to be axially rigid. The assumption is employed because it is reasonable and it reduces the total effort required to complete the analysis. The assumption is not reasonable for truss type structures in which axial deformation is important and the effects of bending deformation are small. To carry out an accurate analysis of truss type structures which contain rigid joints, the slope-deflection equations must be augmented with axial force-deformation relations.

10.2 APPLICATIONS

The independent position or displacement variables that define the configuration of a framed structure may be placed into two groups. The first group contains the variables that define the rotations of the ends of the members. The second group consists of the variables that define the translational displacements of the ends of the members. Ordinarily the group that defines the rotations will consist of the independent rotations of the ends of the members that form the joints. Considerable flexibility is available in setting out the position variables that define the translations of the ends of the members. This point will be discussed in somewhat greater detail after application of the slope-deflection equations to the analysis of beams and frames for which joint displacements are prescribed. A simple procedure for analysis is set out below.

10.2.1 Procedure

1. Identify the independent displacement variables and expresses all member end-rotations and end-displacements in terms of the independent variables.

2. Use the relationships developed in Step 1 and the slope-deflection equations to express all end-forces in terms of fixed-end forces and displacements.
3. Formulate the independent equations of equilibrium corresponding to the independent displacement variables.
4. Substitute the expressions developed in Step 2 into the equations of equilibrium formulated in Step 3.
5. Solve the equations obtained in Step 4 for the unknown displacement variables and calculate the end-forces by back substitution into the slope-deflection equations.

10.3 NOTE ON THE EQUATIONS OF EQUILIBRIUM

By the mode of derivation, the slope-deflections equations are such that transverse force and moment equilibrium of each member in a frame are satisfied. However, for a frame to be in equilibrium moment equilibrium must be satisfied for each joint. In addition, the axial forces at the ends of each member must be such that axial force equilibrium for each member is satisfied and such that force equilibrium of each joint is satisfied. Moment equations of equilibrium for joints are easy to formulate; in fact, the moment equation for a joint can be expressed in a standard form. If the members are axially rigid and there are no joint displacements, the moment equations of equilibrium for the joints are sufficient to determine the rotations of the joints. If the members are treated as being axially rigid and joint translation is not prevented, derivation of the equations of equilibrium corresponding to the displacement variables that define the translational joint displacements can be a challenge. The challenge is met easily with the aid of the principle of virtual work and elementary operations from matrix algebra. (This point will be expanded upon later in this chapter.)

10.4 RESTATEMENT OF THE SLOPE-DEFLECTION EQUATIONS

It is advantageous to rewrite the generic slope-deflections equations, Equations (a), in a way that is more suitable for practical applications. The desired form is obtained by the simple operation of replacing the end identifier $A(B)$ by the new identifier $ab\,(ba)$. When this is done, the equations read

$$M_{ab} = \frac{EI}{L}\left(4\theta_{ab} + 2\theta_{ba} + \frac{6}{L}v_{ab} - \frac{6}{L}v_{ba}\right) + M_{Fab}$$

$$M_{ba} = \frac{EI}{L}\left(2\theta_{ab} + 4\theta_{ba} + \frac{6}{L}v_{ab} - \frac{6}{L}v_{ba}\right) + M_{Fba}$$

$$V_{ab} = \frac{EI}{L}\left(\frac{6}{L}\theta_{ab} + \frac{6}{L}\theta_{ba} + \frac{12}{L^2}v_{ab} - \frac{12}{L^2}v_{ba}\right) + V_{Fab}$$

$$V_{ba} = \frac{EI}{L}\left(-\frac{6}{L}\theta_{ab} - \frac{6}{L}\theta_{ba} - \frac{12}{L^2}v_{ab} + \frac{12}{L^2}v_{ba}\right) + V_{Fba}$$

(A)

With the foregoing notation, the letter combination ab is understood to identify the

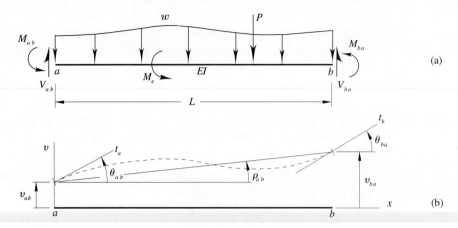

FIGURE 10.7 Notation.

member and the ordering of the letters identify the end of the member. Thus, *ab* identifies the *a*-end of member *ab* and *ba* identifies the *b*-end of the member. The physical entities corresponding to the symbols appearing in Equations (A) are illustrated in Figure 10.7.

10.5 END-ROTATIONS

For certain applications, it is advantageous to have available expressions for the end-rotations in terms of the end-displacements, end-moments, and fixed end-moments. The desired relations are obtained by treating the first two of Equations (A) as equations to be solved for the joint rotations as unknown quantities. The solution of the equations is

$$\theta_{ab} = \frac{L}{6EI}[(2M_{ab} - M_{ba}) - (2M_{Fab} - M_{Fba})] + \frac{v_{ba} - v_{ab}}{L}$$

$$\theta_{ba} = \frac{L}{6EI}[(2M_{ba} - M_{ab}) - (2M_{Fba} - M_{Fab})] + \frac{v_{ba} - v_{ab}}{L}$$

(B)

10.6 EXAMPLES OF BEAMS AND FRAMES WITH PRESCRIBED JOINT DISPLACEMENTS

When the joints of a frame undergo non-zero displacements the frame is said to have experienced sidesway. In the examples that follow the sidesway is prescribed.

Example 10.1

Determine the end-moments for the propped cantilever beam shown in Figure 10.8. The beam is statically indeterminate to the first degree and kinematically indeterminate to the first degree. Accordingly, the effort required to complete the analysis in a displacement method of analysis is essentially the same as in a force method of analysis.

Section 10.6 Examples of Beams and Frames with Prescribed Joint Displacements

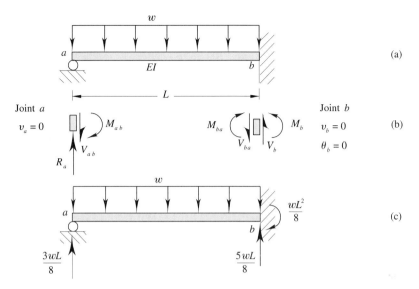

FIGURE 10.8 Propped cantilever beam.

As indicated by the free-body diagrams shown in Figure 10.8b, infinitely thin slices of the beam at its ends are understood to define the joints at the ends of the beam. The forces acting on a slice consist of the reaction forces exerted by the supports and the reversed end-forces exerted by the beam. Let v_a, v_b, θ_a, and θ_b be the displacements of the joints. The rotation of joint b is zero and the displacements of the joints are zero. Thus, from the compatibility equations, $v_{ab} = v_a = 0$, $v_{ba} = v_b = 0$, $\theta_{ba} = \theta_b = 0$, and $\theta_{ab} = \theta_a$. The fixed end-moments for the uniformly distributed load are obtained from Figure 10.5. Thus, the slope-deflection equations for the problem reduce to

$$M_{ab} = \frac{EI}{L}(4\theta_a) + \frac{wL^2}{12}, \quad M_{ba} = \frac{EI}{L}(2\theta_a) - \frac{wL^2}{12}$$

The equations for the end-shears are not essential to the solution the problem and, therefore, the equations are not written. In any case, in this problem, if the shears are required it is easiest to simply calculate them using equations of equilibrium after the end-moments have been found. The equation of equilibrium for joint a is simply $M_{ab} = 0$. Substituting the slope-deflection equations into the moment equilibrium equation yields

$$0 = \frac{EI}{L}(4\theta_a) + \frac{wL^2}{12} \Rightarrow \frac{EI\theta_a}{L} = -\frac{wL^2}{48}$$

Back substitution into the slope-deflection equations gives $M_{ab} = 0$, $M_{ba} = -wL^2/8$. The end-shears and corresponding reactions are obtained from statics. The results (which coincide with those obtained by force methods of analysis) are summarized in Figure 10.8c.

Example 10.2

Determine the end-moments for the beam shown in Figure 10.9. Construct the shear and bending moment diagrams for the case $L = 14'$ and $w = 2$ k/ft. This is same beam problem taken up in Example 9.3 which was used to illustrate the elements of the method of consistent deformation. The joints are located at/over the supports. Denote the displacements and rotations of the joints by v_1, v_2, v_3, θ_1, θ_2, θ_3. Then from the compatibility conditions and the boundary conditions, all

318 Chapter 10 Displacement Methods of Analysis

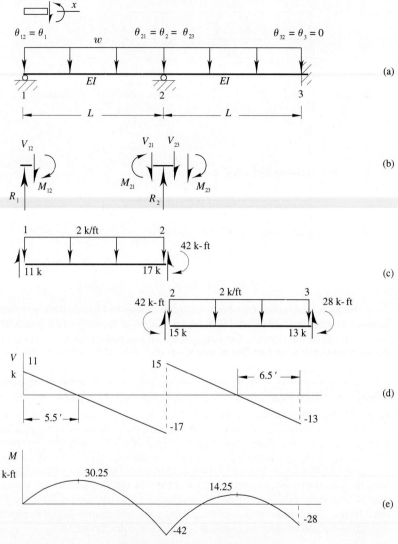

FIGURE 10.9 Continuous beam.

member end-displacements are zero, and the member end-rotations are $\theta_{12} = \theta_1$, $\theta_{21} = \theta_{23} = \theta_2$, $\theta_{32} = \theta_3 = 0$.

With the fixed end-moments as given in Figure 10.5, the slope-deflection equations reduce to

$$M_{12} = \frac{EI}{L}(4\theta_1 + 2\theta_2) + \frac{wL^2}{12}$$

$$M_{21} = \frac{EI}{L}(2\theta_1 + 4\theta_2) - \frac{wL^2}{12}$$

$$M_{23} = \frac{EI}{L}(4\theta_2) + \frac{wL^2}{12}$$

$$M_{32} = \frac{EI}{L}(2\theta_2) - \frac{wL^2}{12}$$

Section 10.6   Examples of Beams and Frames with Prescribed Joint Displacements   319

From the free-body diagrams of joints 1 and 2 as shown in Figure 10.9b, the moment equations of equilibrium are

$$0 = M_{12}$$
$$0 = M_{21} + M_{23}$$

which, after substituting from the slope-deflection equations, become

$$0 = 4\left(\frac{EI}{L}\right)\theta_1 + 2\left(\frac{EI}{L}\right)\theta_2 + \frac{wL^2}{12}$$

$$0 = 2\left(\frac{EI}{L}\right)\theta_1 + 8\left(\frac{EI}{L}\right)\theta_2$$

the solution values of which are

$$\frac{EI\theta_1}{L} = -\frac{wL^2}{42}, \quad \frac{EI\theta_2}{L} = \frac{wL^2}{168}$$

Back substitution into the slope-deflection equations yields

$$M_{12} = 0, \quad M_{21} = -\frac{3wL^2}{28}, \quad M_{23} = \frac{3wL^2}{28}, \quad M_{32} = -\frac{wL^2}{14}$$

For $w = 2$ k/ft and $L = 14'$, $M_{21} = -42$ k-ft, $M_{23} = 42$ k-ft, and $M_{32} = -28$ k-ft. Free-body diagrams for each of the members are shown in Figure 10.8c. With the sign conventions for shear and bending moment as shown in Figure 10.9a, the shear and bending moment diagrams are as shown in Figure 10.9d and e. It is to be noted that the sign conventions for the end-moments and the sign convention for bending moment are such that $M(0^+) = -M_{ab}$ and $M(L^-) = M_{ba}$.

Example 10.3

Calculate the end-moments for the beam shown in Figure 10.10. The beam is statically indeterminate to the fourth degree but kinematically indeterminate only to the second degree. Accordingly, a displacement method of analysis is indicated. The independent position variables are the joint rotations θ_2 and θ_3, both taken positive in the counterclockwise sense. All other joint rotations and displacements are zero. The fixed-end moments for each load and the total fixed

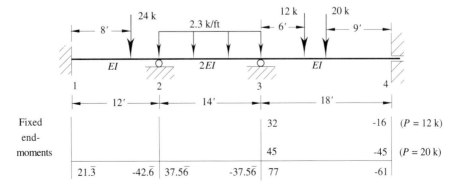

FIGURE 10.10 Beam, statically indeterminate to the fourth degree, kinematically indeterminate to the second degree.

end-moment at each end of each member are as shown in Figure 10.10. The formulas shown in Figure 10.5 were used to evaluate the fixed end-moments.
The slope-deflection equations are

$$M_{12} = \frac{EI}{12}(2\theta_2) + 21.\overline{3}, \qquad M_{21} = \frac{EI}{12}(4\theta_2) - 42.\overline{6}$$

$$M_{23} = \frac{2EI}{14}(4\theta_2 + 2\theta_3) + 37.5\overline{6}, \qquad M_{32} = \frac{2EI}{14}(2\theta_2 + 4\theta_3) - 37.5\overline{6}$$

$$M_{34} = \frac{EI}{18}(4\theta_3) + 77, \qquad M_{43} = \frac{EI}{18}(2\theta_3) - 61$$

Moment equilibrium for joints 2 and 3 require that

$$0 = M_{21} + M_{23}$$
$$0 = M_{32} + M_{34}$$

After substituting for the end-moments from the slope-deflection equations, the equations of equilibrium become

$$0 = \frac{19}{21}(EI\theta_2) + \frac{2}{7}(EI\theta_3) - 5.1$$

$$0 = \frac{2}{7}(EI\theta_2) + \frac{50}{63}(EI\theta_3) + 39.43\overline{3}$$

The solution values are $EI\theta_2 = 24.06$ k-ft, $EI\theta_3 = -58.35$ k-ft. Back substitution into the slope-deflection equations yields

$$M_{12} = 25.34 \text{ k-ft}, \quad M_{21} = -34.65 \text{ k-ft}, \quad M_{23} = 34.65 \text{ k-ft}, \quad M_{32} = -64.03 \text{ k-ft}$$
$$M_{34} = 64.03 \text{ k-ft}, \quad M_{43} = -67.48 \text{ k-ft}$$

Example 10.4

Determine the end-moments due to support settlements of the beam as shown in Figure 10.11.

With the local coordinates as shown, the prescribed joint displacements are $v_1 = -\Delta$, $v_2 = v_3 = 0$ and the prescribed rotation of the joint at 3 is $\theta_3 = -\theta$. In terms of Δ, θ, and the unknown joint rotations θ_1 and θ_2, the slope-deflection equations become

$$M_{12} = \frac{EI}{L}\left(4\theta_1 + 2\theta_2 - 6\frac{\Delta}{L}\right), \qquad M_{21} = \frac{EI}{L}\left(2\theta_1 + 4\theta_2 - 6\frac{\Delta}{L}\right)$$

$$M_{23} = \frac{EI}{L}(4\theta_2 - 2\theta), \qquad M_{32} = \frac{EI}{L}(2\theta_2 - 4\theta)$$

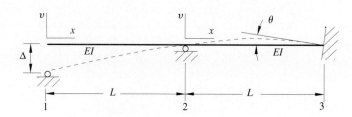

FIGURE 10.11 Support settlements.

Section 10.6 Examples of Beams and Frames with Prescribed Joint Displacements

The moment equations of equilibrium are

$$0 = M_{12}, \quad 0 = M_{21} + M_{23}$$

Substituting from the slope-deflection equations into the equations of equilibrium yields

$$0 = \frac{EI}{L}\left(4\theta_1 + 2\theta_2 - 6\frac{\Delta}{L}\right)$$

$$0 = \frac{EI}{L}\left(2\theta_1 + 8\theta_2 - 6\frac{\Delta}{L} - 2\theta\right)$$

the solution values of which are

$$\theta_1 = \frac{9\Delta}{7L} - \frac{\theta}{7}, \quad \theta_2 = \frac{3\Delta}{7L} + \frac{2\theta}{7}$$

Back substitution into the slope-deflection equations gives

$$M_{12} = 0, \qquad M_{21} = -\frac{12EI\Delta}{7L^2} + \frac{6EI\theta}{7L}$$

$$M_{23} = \frac{12EI\Delta}{7L^2} - \frac{6EI\theta}{7L}, \qquad M_{32} = \frac{6EI\Delta}{7L^2} - \frac{24EI\theta}{7L}$$

The results agree with those obtained previously by the three-moment equation.

Example 10.5

Determine the end-moments for all members and construct the shear and bending moment diagrams for member 1-2 of the rigid frame shown in Figure 10.12. The members are axially rigid.

Because the supports are built in, there is no end rotation or end or displacement at points 1, 4, or 5. From the conditions at the supports and the condition of axial rigidity of the members,

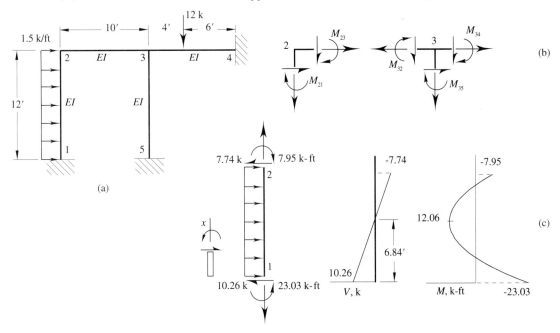

FIGURE 10.12 Two bay frame.

it follows that there is no displacement of joint 2 or joint 3. Accordingly, the deformed state of the frame is defined completely by the rotations of joints 2 and 3. It follows that the frame is kinematically indeterminate to the second degree whereas it is statically indeterminate to the sixth degree. Let θ_2 and θ_3 be the rotations of joints 2 and 3, respectively, measured positive in the counterclockwise sense. From the boundary conditions and conditions of compatibility at joints 2 and 3

$$\theta_{12} = \theta_{43} = \theta_{53} = 0, \quad \theta_{21} = \theta_{23} = \theta_2, \quad \theta_{32} = \theta_{34} = \theta_{35} = \theta_3$$

In terms of the joint rotations, the slope-deflection equations are

$$M_{12} = \frac{EI}{12}(2\theta_2) + 18, \qquad M_{21} = \frac{EI}{12}(4\theta_2) - 18$$

$$M_{23} = \frac{EI}{10}(4\theta_2 + 2\theta_3), \qquad M_{32} = \frac{EI}{10}(2\theta_2 + 4\theta_3)$$

$$M_{34} = \frac{EI}{10}(4\theta_3) + 17.28, \qquad M_{43} = \frac{EI}{10}(2\theta_3) - 11.52$$

$$M_{35} = \frac{EI}{12}(4\theta_3), \qquad M_{53} = \frac{EI}{12}(2\theta_3)$$

From the free-body diagrams for joints 2 and 3 shown in Figure 10.12 the joint moment equations of equilibrium are

$$0 = M_{21} + M_{23}$$
$$0 = M_{32} + M_{34} + M_{35}$$

After substituting for the end-moments from the slope-deflection equations, the equations of equilibrium become

$$0 = \frac{11}{15}(EI\theta_2) + \frac{2}{10}(EI\theta_3) - 18$$

$$0 = \frac{2}{10}(EI\theta_2) + \frac{17}{15}(EI\theta_3) + 17.28$$

which has the solution values $EI\theta_2 = 30.16$ k-ft, $EI\theta_3 = -20.57$ k-ft. The end-moments are obtained by back substitution into the slope-deflection equations. Thus

$M_{12} = 23.03$ k-ft, $M_{21} = -7.95$ k-ft, $M_{23} = 7.95$ k-ft, $M_{32} = -2.20$ k-ft
$M_{34} = 9.05$ k-ft, $M_{43} = -15.63$ k-ft, $M_{35} = -6.85$ k-ft, $M_{53} = -3.43$ k-ft

A free-body diagram of member 1-2 and the shear and bending moment diagrams for the member are shown in Figure 10.12c.

Example 10.6

Translation or sidesway displacement of the frame shown in Figure 10.13 is prevented by the horizontal link at joint 3. Calculate the force developed in the link. The members of the frame are axially rigid. With the link in place, the frame is statically indeterminate to the fourth degree. The frame is kinematically indeterminate to the second degree, the rotations of joints 2 and 3 being the only unknown displacement variables. A displacement analysis can be used to obtain the end-moments and end-shears from the (complete set of) slope-deflection equations. Because the members are axially rigid, their respective axial forces and the force in the link must be obtained from equations of equilibrium. The independent equations of equilibrium involving the axial forces are sufficient to complete the solution of the problem. In the example, the equa-

Section 10.6   Examples of Beams and Frames with Prescribed Joint Displacements   323

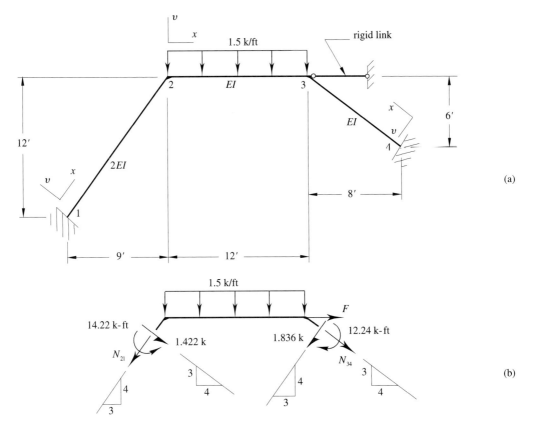

FIGURE 10.13   Calculation of restraining force.

tions of equilibrium are derived from a free body diagram. Subsequently, it will be shown how the principle of virtual work can be used to advantage to calculate the tie force independently of the axial forces in the members of the frame.

The slope-deflection equations in terms of the independent joint rotations θ_2 and θ_3 are

$$M_{12} = \frac{2EI}{15}(2\theta_2), \qquad M_{21} = \frac{2EI}{15}(4\theta_2)$$

$$M_{23} = \frac{EI}{12}(4\theta_2 + 2\theta_3) + 18, \qquad M_{32} = \frac{EI}{12}(2\theta_2 + 4\theta_3) - 18$$

$$M_{34} = \frac{EI}{10}(4\theta_3), \qquad M_{43} = \frac{EI}{10}(2\theta_3)$$

The moment equations of equilibrium are

$$0 = M_{21} + M_{23} \Rightarrow 0 = \frac{13}{15}(EI\theta_2) + \frac{1}{6}(EI\theta_3) + 18$$

$$0 = M_{32} + M_{34} \Rightarrow 0 = \frac{1}{6}(EI\theta_2) + \frac{11}{15}(EI\theta_3) - 18$$

from which $EI\theta_2 = -26.65$ k-ft, $EI\theta_3 = 30.60$ k-ft. Back substitution into the slope-deflection equations gives

$M_{12} = -7.11$ k-ft, $M_{21} = -14.22$ k-ft, $M_{23} = 14.22$ k-ft, $M_{32} = -12.24$ k-ft
$M_{34} = 12.24$ k-ft, $M_{43} = 6.12$ k-ft

The end-shears may be calculated from statics or from the slope-deflection equations. If the slope-deflection equations are used, it is important to keep in mind that in the local coordinates employed in the derivation of the slope-deflection equations, the end-shears are positive in the sense of the local v-coordinate axis. With local coordinates as shown in Figure 10.13a, the slope-deflection equations for the end-shears are

$$V_{12} = \frac{2EI}{15}\left(\frac{6}{15}\theta_2\right), \qquad V_{21} = \frac{2EI}{15}\left(-\frac{6}{15}\theta_2\right)$$

$$V_{23} = \frac{EI}{12}\left(\frac{6}{12}\theta_2 + \frac{6}{12}\theta_3\right) + 9, \qquad V_{32} = \frac{EI}{12}\left(-\frac{6}{12}\theta_2 - \frac{6}{12}\theta_3\right) + 9$$

$$V_{34} = \frac{EI}{10}\left(-\frac{6}{10}\theta_3\right), \qquad V_{43} = \frac{EI}{10}\left(\frac{6}{10}\theta_3\right)$$

Replacing the rotations by their known values gives

$V_{12} = -1.422$ k, $V_{21} = 1.422$ k, $V_{23} = 9.165$ k, $V_{32} = 8.835$ k
$V_{34} = -1.836$ k, $V_{43} = 1.836$ k

It is easy to verify that the end-shears are the same as would have been obtained from equilibrium of the free-body diagrams of each of the members. With the shears V_{21} and V_{34} known, the free-body diagram shown in Figure 10.13b can be used to calculate the axial forces N_{21} and N_{34} and the force in the link by statics. Summing moments about joint 2 gives $N_{34} = -17.17$ k, and summing moments about joint 3 yields $N_{21} = -12.52$ k. Finally, force equilibrium in the horizontal direction gives $F = 6.188$ k.

Example 10.7

The members of the frame shown in Figure 10.14 are axially rigid. Determine the force F required to maintain the frame in the position defined by the prescribed displacement u at joint 2.

Because the supports at 1 and 4 are built in, and because of the axial rigidity of the members, there is no vertical displacement at joint 2 or joint 3, and the horizontal component of displacement at joint 3 is u. The frame is kinematically indeterminate to the second degree. The rotations of the joints at 2 and 3 are taken as the independent position variables. The problem is solved in much the same way as in Example 10.6. The slope-deflection equations and moment equations of equilibrium for the joints are used to determine the rotations of the joints in terms of u. The end-moments are found by back substitution into the slope deflection equations, and force equations of equilibrium are then used to find F.

The slope-deflection equations for members 1-2 and 3-4 will include non-zero transverse end-displacement terms corresponding to v_{ab} and v_{ba} in the generic slope-deflection equations, Equations (A). It is essential that these terms be expressed correctly in terms of the displacement u. The key to a correct formulation is the assignment of local coordinates to the member under consideration that correspond to the local coordinates used in the derivation of the slope-deflection equations. The results are independent of which end of a member is selected for the origin of the local member x-axis. Once the coordinate assignment has been made determination of the end-displacements in terms of joint displacements is an uncomplicated matter of geometry.

From the force system shown in Figure 10.14a, for equilibrium in the horizontal direction

$$0 = F - V_{12} + V_{43}$$

Section 10.6 Examples of Beams and Frames with Prescribed Joint Displacements

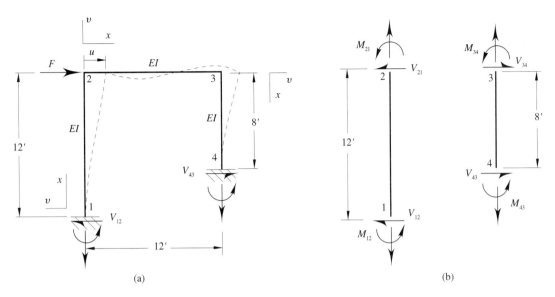

FIGURE 10.14 Prescribed displacement.

Two (superficially) different lines of attack can be employed to determine the shears V_{12} and V_{43}. In the first approach, the slope-deflection equations that define these shears are included in the analysis from the outset, and the shears are obtained by back substitution of known values of the displacement variables into the equations. In the second approach, the slope-deflection equations are used to calculate the end-moments for members 1-2 and 3-4. Then the equilibrium conditions for the members are used to calculate the shears. Both lines of approach are illustrated. With local coordinates selected as shown in Figure 10.14,

$$v_{12} = 0, \quad v_{21} = -u, \quad v_{23} = 0, \quad v_{32} = 0, \quad v_{34} = u, \quad v_{43} = 0$$

With the joint rotations θ_2 and θ_3 positive in the counterclockwise sense the slope-deflection equations are

$$M_{12} = \frac{EI}{12}\left(2\theta_2 + \frac{1}{2}u\right), \qquad M_{21} = \frac{EI}{12}\left(4\theta_2 + \frac{1}{2}u\right)$$

$$M_{23} = \frac{EI}{12}(4\theta_2 + 2\theta_3), \qquad M_{32} = \frac{EI}{12}(2\theta_2 + 4\theta_3)$$

$$M_{34} = \frac{EI}{8}\left(4\theta_3 + \frac{3}{4}u\right), \qquad M_{43} = \frac{EI}{8}\left(2\theta_3 + \frac{3}{4}u\right)$$

$$V_{12} = \frac{EI}{12}\left(\frac{1}{2}\theta_2 + \frac{1}{12}u\right), \qquad V_{21} = \frac{EI}{12}\left(-\frac{1}{2}\theta_2 - \frac{1}{12}u\right)$$

$$V_{34} = \frac{EI}{8}\left(\frac{3}{4}\theta_3 + \frac{3}{16}u\right), \qquad V_{43} = \frac{EI}{8}\left(-\frac{3}{4}\theta_3 - \frac{3}{16}u\right)$$

The moment equations of equilibrium at joints 2 and 3 are

$$0 = M_{21} + M_{23} \Rightarrow 0 = 8\theta_2 + 2\theta_3 + 0.5u$$

$$0 = M_{32} + M_{34} \Rightarrow 0 = 2\theta_2 + 10\theta_3 + 1.125u$$

from which $19\theta_2 = -0.6875u/\text{ft}$ and $19\theta_3 = -2u/\text{ft}$. Back substitution into the slope-deflection equations for end-moments and end-shears gives

$$456M_{12} = 16.25\,EIu/\text{ft}^2, \qquad 456M_{21} = 13.50\,EIu/\text{ft}^2$$
$$456M_{23} = -13.50\,EIu/\text{ft}^2, \qquad 456M_{32} = -18.75\,EIu/\text{ft}^2$$
$$456M_{34} = 18.75\,EIu/\text{ft}^2, \qquad 456M_{43} = 30.75\,EIu/\text{ft}^2$$
$$5472V_{12} = 29.75\,EIu/\text{ft}^3, \qquad 5472V_{21} = -29.75\,EIu/\text{ft}^3$$
$$5472V_{34} = 74.25\,EIu/\text{ft}^3, \qquad 5472V_{43} = -74.25\,EIu/\text{ft}^3$$

As noted above, an alternative to using the slope-deflection equations is to obtain the end-shears from equilibrium conditions of the individual members after the end-moment have been determined. From the free-body diagrams of members 1-2 and 3-4 as shown in Figure 10.14b,

$$V_{12} = \frac{M_{12} + M_{21}}{12}, \qquad V_{43} = -\frac{M_{34} + M_{43}}{8}$$

Substitution of the known values for the end-moments into these relations gives, as it must, the same results as obtained above. The required force is

$$F = \frac{13}{684}\,EIu/\text{ft}^3$$

10.7 FRAMES WITH SIDESWAY

Frames that undergo sidesway as a consequence of loading are inherently more difficult to analyze than frames for which joint displacements are prescribed. The moment equations of equilibrium for the joints must be supplemented with one independent force equation of equilibrium for each degree of freedom of joint translation. Derivation of the essential force equations of equilibrium for rectangular frames is a relatively easy task; however, the derivation can be demanding in the case of frames with sloping members. Because the points to be brought out cannot be treated in the abstract, two examples are used to bring into focus some of the characteristic difficulties of sidesway analysis and to provide motivation for methods that have been developed to reduce the burden of analysis.

Example 10.8

Determine the end-moments for the frame shown in Figure 10.15. Treat the members as being axially rigid. The frame is statically indeterminate to the first degree and kinematically indeterminate to the fifth degree. The displacement variables are the joint rotations $\theta_a, \theta_b, \theta_c, \theta_d$, and the common horizontal translation u of joints b and c. Clearly, a force analysis is indicated for greatest efficiency. However, a displacement analysis is carried out to illustrate ideas.

The slope-deflection equations for end-shears V_{ab} and V_{ba} as well as for the end-moments are required. With local coordinates as shown in Figure 10.15 the slope-deflection equations are

$$M_{ab} = \frac{EI}{10}(4\theta_a + 2\theta_b + 0.6u) + \frac{50}{3}, \qquad M_{ba} = \frac{EI}{10}(2\theta_a + 4\theta_b + 0.6u) - \frac{50}{3}$$

$$M_{bc} = \frac{EI}{10}(4\theta_b + 2\theta_c), \qquad M_{cb} = \frac{EI}{10}(2\theta_b + 4\theta_c)$$

$$M_{cd} = \frac{EI}{12}(4\theta_c + 2\theta_d + u), \qquad M_{dc} = \frac{EI}{12}(2\theta_c + 4\theta_d + u)$$

$$V_{ab} = \frac{EI}{10}(0.6\theta_a + 0.6\theta_b + 0.12u) + 10, \qquad V_{dc} = \frac{EI}{12}(\theta_c + \theta_d + 0.33\overline{3}u)$$

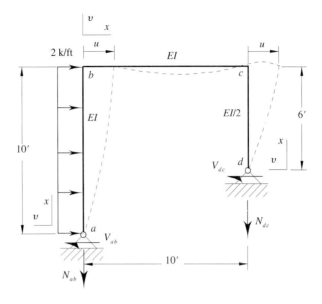

FIGURE 10.15 Sidesway induced by loads.

and the equations of equilibrium are

$$0 = M_{ab}$$
$$0 = M_{ba} + M_{bc}$$
$$0 = M_{cb} + M_{cd}$$
$$0 = M_{dc}$$
$$0 = V_{ab} + V_{dc} - 20$$

After substitution from the slope-deflection equations, the equations of equilibrium reduce to

$$-16.\overline{6} = 0.400(EI\theta_a) + 0.200(EI\theta_b) \hspace{3cm} + 0.0600(EIu)$$
$$16.\overline{6} = 0.200(EI\theta_a) + 0.800(EI\theta_b) + 0.2000(EI\theta_c) \hspace{1cm} + 0.0600(EIu)$$
$$0 = \hspace{2cm} 0.200(EI\theta_b) + 0.733\overline{3}(EI\theta_c) + 0.166\overline{6}(EI\theta_d) + 0.083\overline{3}(EIu)$$
$$0 = \hspace{4.5cm} 0.166\overline{6}(EI\theta_c) + 0.333\overline{3}(EI\theta_d) + 0.083\overline{3}(EIu)$$
$$10 = 0.060(EI\theta_a) + 0.060(EI\theta_b) + 0.083\overline{3}(EI\theta_c) + 0.083\overline{3}(EI\theta_d) + 0.039\overline{7}(EIu)$$

It is observed that the coefficients of the unknowns form a symmetric matrix. As will be shown later, the symmetry is not fortuitous. The solution values are

$$EI\theta_a = -298.3 \text{ k-ft}^2, \quad EI\theta_b = -7.55 \text{ k-ft}^2, \quad EI\theta_c = -109.0 \text{ k-ft}^2$$
$$EI\theta_d = -379.7 \text{ k-ft}^2, \quad EIu = 1736.4 \text{ k-ft}^3$$

The end-moments and end-shears are

$$M_{ab} = 0, \hspace{1cm} M_{ba} = 24.8 \text{ k-ft}, \hspace{1cm} V_{ab} = 12.5 \text{ k}$$
$$M_{bc} = -24.8 \text{ k-ft}, \hspace{1cm} M_{cb} = -45.1 \text{ k-ft}$$
$$M_{cd} = 45.1 \text{ k-ft}, \hspace{1cm} M_{dc} = 0 \hspace{1cm} V_{dc} = 7.5 \text{ k}$$

Example 10.9

Calculate the end-moments in the frame shown in Figure 10.16. This is the frame treated in Example 10.6, but with the rigid link at joint 3 removed. The members of the frame are axially rigid. As a consequence of the loading joints 2 and 3 will displace and the frame will deform somewhat as shown in Figure 10.16a to a greatly exaggerated scale. It is not known in which direction the joints will move. The amount and direction of displacement is determined in the course of solving the problem. The members undergo inextensional deformation and since there can be no displacement of joints 1 and 4, the displacement u_2 must be perpendicular to member 1-2 and the displacement u_3 must be perpendicular to member 3-4. Due to the axial rigidity of member 2-3 the rectangular component of u_2 in the direction from 2 to 3 must be equal to the rectangular component of u_3 in the same direction. The common value of these components is identified as u in Figure 10.16a. The frame has one degree of freedom with respect to joint displacement. Accordingly, any one parameter that completely defines the displacements of the joints may be chosen as the position variable. u, u_2, u_3, and the chord rotation of any one of the members are in the group of parameters from which the position variable may be chosen. u is selected for this example.

From the geometry shown in Figure 10.16a and for the local member coordinates shown,

$$u_2 = \frac{5}{4}u, \quad u_3 = \frac{5}{3}u, \quad v_{12} = 0, \quad v_{21} = -\frac{5}{4}u$$

$$v_{23} = -\frac{3}{4}u, \quad v_{32} = \frac{4}{3}u, \quad v_{34} = -\frac{5}{4}u, \quad v_{43} = 0$$

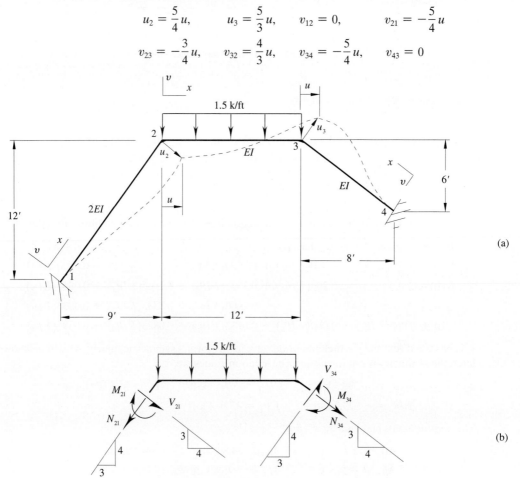

FIGURE 10.16 Frame with sloping members.

The essential slope-deflection equations are

$$M_{12} = \frac{2EI}{15}\left(2\theta_2 - \frac{6}{15}v_{21}\right), \qquad M_{21} = \frac{2EI}{15}\left(4\theta_2 - \frac{6}{15}v_{21}\right)$$

$$M_{23} = \frac{EI}{12}\left(4\theta_2 + 2\theta_3 + \frac{6}{12}v_{23} - \frac{6}{12}v_{32}\right) + 18,$$

$$M_{32} = \frac{EI}{12}\left(2\theta_2 + 4\theta_3 + \frac{6}{12}v_{23} - \frac{6}{12}v_{32}\right) - 18$$

$$M_{43} = \frac{EI}{10}\left(2\theta_3 - \frac{6}{10}v_{34}\right), \qquad M_{34} = \frac{EI}{10}\left(4\theta_3 - \frac{6}{10}v_{34}\right)$$

$$V_{21} = \frac{2EI}{15}\left(-\frac{6}{15}\theta_2 + \frac{12}{225}v_{21}\right), \qquad V_{34} = \frac{EI}{10}\left(-\frac{6}{10}\theta_3 + \frac{12}{100}v_{34}\right)$$

which, after replacing the member end-displacements by their equivalents in terms of u, become

$$M_{12} = \frac{2EI}{15}\left(2\theta_2 + \frac{1}{2}u\right), \qquad M_{21} = \frac{2EI}{15}\left(4\theta_2 + \frac{1}{2}u\right)$$

$$M_{23} = \frac{EI}{12}\left(4\theta_2 + 2\theta_3 - \frac{25}{24}u\right) + 18, \qquad M_{32} = \frac{EI}{12}\left(2\theta_2 + 4\theta_3 - \frac{25}{24}u\right) - 18$$

$$M_{43} = \frac{EI}{10}(2\theta_3 + u), \qquad M_{34} = \frac{EI}{10}(4\theta_3 + u)$$

$$V_{21} = \frac{2EI}{15}\left(-\frac{6}{15}\theta_2 - \frac{1}{15}u\right), \qquad V_{34} = \frac{EI}{10}\left(-\frac{6}{10}\theta_3 - \frac{1}{5}u\right)$$

Moment equilibrium for joints 2 and 3 provide two equations that relate θ_2, θ_3 and u. An additional independent equation of equilibrium that involves only θ_2, θ_3 and u as unknowns is required. The additional equation is derived with the aid of the free-body diagram shown in Figure 10.16b for which three independent equations of equilibrium are

$$0 = \Sigma M_2 \Rightarrow 0 = 108 - 9.6V_{34} + 7.2N_{34} + M_{21} + M_{34}$$

$$0 = \Sigma M_3 \Rightarrow 0 = -108 - 7.2V_{21} - 9.6N_{21} + M_{21} + M_{34}$$

$$0 = \Sigma F_h \Rightarrow 0 = 0.8V_{21} + 0.6V_{34} - 0.6N_{21} + 0.8N_{34}$$

The moment equations of equilibrium can be used to eliminate N_{21} and N_{34} from the force equation to obtain the force equation in the form

$$0 = -5.25 + \frac{5}{4}V_{21} + \frac{5}{3}V_{34} - \frac{25}{144}(M_{21} + M_{34})$$

The moment equations of equilibrium for joints 2 and 3 are

$$0 = M_{21} + M_{23}, \quad 0 = M_{32} + M_{34}$$

After substitution from the slope-deflection equations, the equations of equilibrium reduce to

$$0 = M_{21} + M_{23} = \frac{13}{15}(EI\theta_2) + \frac{1}{6}(EI\theta_3) - \frac{2.9}{144}(EIu) + 18$$

$$0 = M_{32} + M_{34} = \frac{1}{6}(EI\theta_2) + \frac{11}{15}(EI\theta_3) + \frac{1.9}{144}(EIu) - 18$$

$$0 = -\Sigma F_h = \frac{43}{270}(EI\theta_2) + \frac{12.2}{72}(EI\theta_3) + \frac{63.4}{864}(EIu) + 5.25$$

the solution values of which are

$$EI\theta_2 = -29.0 \text{ k-ft}^2, \quad EI\theta_3 = 32.65 \text{ k-ft}^2, \quad EIu = -84.0 \text{ k-ft}^3$$

Back substitution into the slope-deflection equations gives

$$M_{12} = -13.33 \text{ k-ft}, \quad M_{21} = -21.1 \text{ k-ft}, \quad M_{23} = 21.1 \text{ k-ft}, \quad M_{32} = -4.66 \text{ k-ft}$$
$$M_{43} = -1.88 \text{ k-ft}, \quad M_{34} = 4.66 \text{ k-ft}$$

10.8 MODIFIED SLOPE-DEFLECTION EQUATIONS

If an end-moment is known, as in the case of a beam with one end pinned, the associated slope-deflection equation can be used to eliminate the corresponding end-rotation as an independent position variable. The modified slope-deflection equations are the result of the elimination process. A reduction in effort required to complete an analysis is always obtained by employing the modified slope-deflection equations where appropriate; moreover, often the reduction is significant. In deriving the modified slope-deflection equations, the end-moment at the end of the member associated with the origin of the local coordinates is assumed to be known (i.e., M_{ab} is assumed to be known). The assumption does not limit the utility of the result since the origin of the local coordinates can be taken at either end of the member. The first of the slope-deflection equations is

$$M_{ab} = \frac{EI}{L}\left(4\theta_{ab} + 2\theta_{ba} + \frac{6}{L}v_{ab} - \frac{6}{L}v_{ba}\right) + M_{Fab}$$

from which

$$\theta_{ab} = -\frac{1}{2}\theta_{ba} - \frac{3}{2L}v_{ab} + \frac{3}{2L}v_{ba} + \frac{L}{4EI}(M_{ab} - M_{Fab})$$

Substitution of this expression for θ_{ab} into the slope-deflection equations for M_{ba}, V_{ab}, and V_{ba} yields the modified slope-deflection equations:

$$M_{ba} = \frac{EI}{L}\left(3\theta_{ba} + \frac{3}{L}v_{ab} - \frac{3}{L}v_{ba}\right) + M_{Fba} + \frac{1}{2}(M_{ab} - M_{Fab})$$

$$V_{ab} = \frac{EI}{L}\left(\frac{3}{L}\theta_{ba} + \frac{3}{L^2}v_{ab} - \frac{3}{L^2}v_{ba}\right) + V_{Fab} + \frac{3}{2L}(M_{ab} - M_{Fab})$$

$$V_{ba} = \frac{EI}{L}\left(-\frac{3}{L}\theta_{ba} - \frac{3}{L^2}v_{ab} + \frac{3}{L^2}v_{ba}\right) + V_{Fba} - \frac{3}{2L}(M_{ab} - M_{Fab})$$

10.9 APPLICATIONS

To provide comparative information on the benefits of using the modified slope-deflection equations in situations where they are applicable, three example problems that had been considered previously to illustrate basic ideas are solved again using the modified slope-deflection equations.

Example 10.10

Determine the end-moments for the propped cantilever beam shown in Figure 10.8, here reproduced as Figure 10.17. By inspection, $M_{ab} = 0$. Because $v_{ab} = v_{ba} = \theta_{ba} = 0$, and $M_{Fba} = -M_{Fab} = -wL^2/12$, the modified slope-deflection equation for M_{ba} reduces to

$$M_{ba} = M_{Fba} - \frac{1}{2}M_{Fab} = -\frac{wL^2}{8}$$

The result agrees, as it must, with that obtained previously.

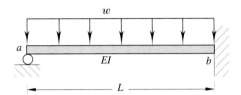

FIGURE 10.17 Propped cantilever beam.

Example 10.11

Determine the end-moments and the rotation $\theta_{12} = \theta_1$ of the tangent at the left end of the beam shown in Figure 10.18. The beam is the same as that of Example 10.2 used to illustrate basic applications of the slope-deflection equations. $M_{12} = 0$ so the modified slope-deflection equations are applicable to member 1-2. All of the end displacements are zero, so the slope-deflection equations are

$$M_{21} = \frac{EI}{L}(3\theta_2) - \frac{wL^2}{8}, \quad M_{23} = \frac{EI}{L}(4\theta_2) + \frac{wL^2}{12}, \quad M_{32} = \frac{EI}{L}(2\theta_2) - \frac{wL^2}{12}$$

The moment equation of equilibrium for joint 2 is

$$0 = M_{21} + M_{23} \Rightarrow 0 = 7\left(\frac{EI\theta_2}{L}\right) - \frac{wL^2}{24} \Rightarrow \frac{EI\theta_2}{L} = \frac{wL^2}{168}$$

Back substitution into the slope-deflection equations yields

$$M_{21} = -\frac{3wL^2}{28}, \quad M_{23} = \frac{3wL^2}{28}, \quad M_{32} = -\frac{wL^2}{14}$$

With $v_{12} = v_{21} = M_{12} = 0$ and $\theta_{21} = \theta_2$ the expression for the end-rotation $\theta_{12} = \theta_1$ yields

$$\theta_{12} = -\frac{1}{2}\theta_{21} - \frac{L}{4EI}M_{F21} = -\frac{wL^3}{336EI} - \frac{L}{4EI}\left(\frac{wL^2}{12}\right) = -\frac{wL^3}{42EI}$$

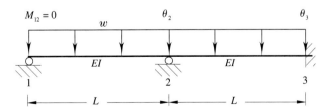

FIGURE 10.18 Continuous beam.

Example 10.12

Solve Example 10.8 with the aid of the modified slope-deflection equations. The frame is shown in Figure 10.19. Without the use of the modified slope-deflection equations the analysis involves the solution of a set of five simultaneous equations. With the aid of the modified equations the number of equations is reduced to three. $M_{ab} = 0$ and $M_{dc} = 0$, so the modified equations are used for members ab and dc. For direct application of the modified slope-deflection equations, the origin of the local coordinates for member ab is taken at a, and for member dc the origin is taken at d. The independent displacement variables are the joint rotations θ_b, and θ_c, and the joint translation u.

The slope-deflection equations are

$$M_{ba} = \frac{EI}{10}(3\theta_b + 0.3u) - 25, \qquad V_{ab} = \frac{EI}{10}(0.3\theta_b + 0.03u) + 7.5$$

$$M_{bc} = \frac{EI}{10}(4\theta_b + 2\theta_c), \qquad M_{cb} = \frac{EI}{10}(2\theta_b + 4\theta_c)$$

$$M_{cd} = \frac{EI}{12}(3\theta_c + 0.5u) \qquad V_{dc} = \frac{EI}{12}(0.5\theta_c + 0.083\overline{3}u)$$

and the equations of equilibrium are

$$0 = M_{ba} + M_{bc} \Rightarrow 0 = 0.70(EI\theta_b) + 0.20(EI\theta_c) + 0.03(EIu) - 25$$
$$0 = M_{cb} + M_{cd} \Rightarrow 0 = 0.20(EI\theta_b) + 0.65(EI\theta_c) + 0.041\overline{6}6(EIu)$$
$$0 = V_{ab} + V_{dc} - 20 \Rightarrow 0 = 0.03(EI\theta_b) + 0.041\overline{6}(EI\theta_c) + 0.0099\overline{4}(EIu) - 12.5$$

The solution values of the equations are

$$EI\theta_b = -7.55 \text{ k-ft}^2, \quad EI\theta_c = -109.0 \text{ k-ft}^2, \quad EIu = 1736.4 \text{ k-ft}^3$$

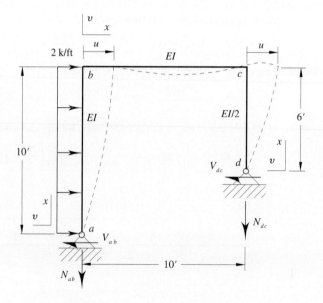

FIGURE 10.19 Sidesway induced by loads.

Substitution of these results back into the slope-deflection equations gives

$M_{ba} = 24.8$ k-ft, $V_{ab} = 12.5$ k, $M_{bc} = -24.8$ k-ft, $M_{cb} = -45.1$ k-ft

$M_{cd} = 45.1$ k-ft, $V_{dc} = 7.5$ k

10.10 SIDESWAY PROBLEMS AND SUPERPOSITION

The members of the frame shown in Figure 10.20 are axially rigid. If F is prescribed the frame has three kinematic degrees of freedom, two with respect to joint rotation and one with respect to joint translation. Let the joint rotation variables be θ_2 and θ_3 and let u be the joint translation variable. Then the basic displacement method of analysis requires the formulation and solution of one set of three simultaneous in the three unknown displacement variables one time. Clearly, perhaps the most critical and trying part of the analysis is formulation of the force equation of equilibrium corresponding to the sidesway displacement in terms of θ_2, θ_3, and u (see Example 10.9). If u is prescribed, the frame has two degrees of freedom with respect to joint rotation and no degrees of freedom with respect to joint translation. The joint rotations can be found by formulating and solving simultaneously the moment equations of equilibrium for joints 2 and 3 for θ_2 and θ_3. With the rotations known, the end-moments can be found by substitution into the slope-deflection equations. The required value for the force F can then obtained by direct application of statics using known, numerical values for the end-moments, and this can be done without the need for a general formulation in terms of θ_2 and θ_3 and u.

The foregoing observations together with the principle of superposition form the basis for a procedure which permits solution of the sidesway problem without the necessity of formulating force equations of equilibrium in terms of unknown joint rotations and sidesway displacement variables. An important secondary advantage of the procedure is that it reduces the order of the maximum number of equations to be solved simultaneously. For a frame with m degrees of freedom with respect to joint rotation and n degrees of freedom with respect to sidesway, the order of the maximum number of equations to be solved simultaneously without use of the procedure is $m + n$. With the procedure, the order is reduced to m. However, the m equations must

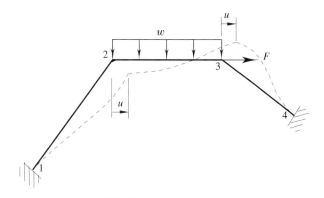

FIGURE 10.20   Sidesway.

be solved $n + 1$ times, and one additional solution of n simultaneous equations is required. Basic ideas are set out with the aid of specific examples. The generalization of the procedure can be related to a specific technique for solving simultaneous equations (see Appendix B).

For the frame shown in Figure 10.20, temporarily treat w as having any prescribed value and treat u as having any prescribed value. Under these conditions, the force F is an unknown to be determined. By the principle of superposition, the mechanical state of the frame is obtained as the superposition of the state that exists when $u = 0$ and the prescribed loads act and the state that exists under the prescribed value of u with the applied loads removed. The two states are illustrated in Figure 10.21. As indicated in the figure, subscript zero is used to identify the value of a mechanical variable when $u = 0$ and the prescribed loads acts, and subscript u is used to identify the corresponding value of the variable due to the assigned value of u with the prescribed loads removed. By the principle of superposition

$$F = F_0 + F_u, \quad M = M_0 + M_u, \quad V = V_0 + V_u, \quad N = N_0 + N_u$$

Again, by the principle of superposition, the value of any mechanical variable for any assigned value of u is u times the value of the variable when $u = 1$.

Let k, m, v, n, be the specific values of F_u, M_u, V_u, and N_u, respectively when $u = 1$. Then

$$F_u = ku, \quad M_u = mu, \quad V_u = vu, \quad N = nu$$

and for any u

$$F = F_0 + ku, \quad M = M_0 + mu, \quad V = V_0 + vu, \quad N = N_0 + nu$$

The relations are valid for any u. Accordingly, the first equation can be used to determine the value of u that is required when F rather than u is prescribed. The corresponding values of M, V, and N are calculated from the remaining equations after u has been determined. Clearly, for a frame with one degree of freedom with respect to sidesway, the procedure is to:

1. Calculate F_0, M_0, etc., for zero sidesway displacement.
2. Calculate k, m, etc., for a unit value of the sidesway displacement variable.
3. Calculate the sidesway displacement variable from the relation $F = F_0 + ku$, where F is prescribed.
4. Evaluate the remaining mechanical variables by superposition.

Example 10.13

Use superposition to calculate the end-moments for the frame shown in Figure 10.22 on page 336. The mechanical state corresponding to the prescribed displacement $u = 0$ with the applied loads acting on the frame is represented in Figure 10.22b. The end-moments, essential end-shears, and the force F_0 corresponding to the applied loads with the prescribed value of $u = 0$ were determined in the solution of Example 10.6 in the section on prescribed displacements. The values of

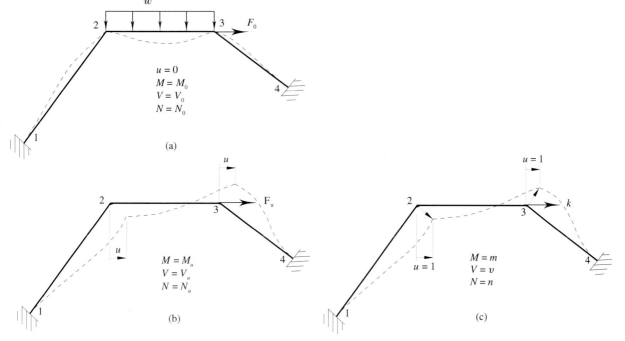

FIGURE 10.21 Notation for superposition.

these quantities are:

$M_{(0)12} = -7.11$ k-ft, $\quad M_{(0)21} = -14.22$ k-ft, $\quad V_{(0)12} = -1.422$ k, $\quad V_{(0)21} = 1.422$ k

$M_{(0)23} = 14.22$ k-ft, $\quad M_{(0)32} = -12.24$ k-ft,

$M_{(0)34} = 12.24$ k-ft, $\quad M_{(0)43} = 6.12$ k-ft, $\quad V_{(0)34} = -1.836$ k, $\quad V_{(0)43} = 1.836$ k

$F_0 = 6.188$ k

The free-body diagram used to calculate the value of F_0 is shown in Figure 10.22c.

Figure 10.22d represents the mechanical state of the frame when the assigned displacement is $u = 1$ and the applied loads are removed. With $u = 1$

$$u_2 = 5/4, \quad u_3 = 5/3, \quad v_{21} = -5/4, \quad v_{23} = -3/4, \quad v_{32} = 4/3, \quad v_{34} = -5/3$$

The slope-deflection equations are

$$M_{12} = \frac{2EI}{15}\left(2\theta_2 + \frac{1}{2}\right), \qquad M_{21} = \frac{2EI}{15}\left(4\theta_2 + \frac{1}{2}\right)$$

$$M_{23} = \frac{EI}{12}\left(4\theta_2 + 2\theta_3 - \frac{25}{24}\right), \qquad M_{32} = \frac{EI}{12}\left(2\theta_2 + 4\theta_3 - \frac{25}{24}\right)$$

$$M_{43} = \frac{EI}{10}(2\theta_3 + 1), \qquad M_{34} = \frac{EI}{10}(4\theta_3 + 1)$$

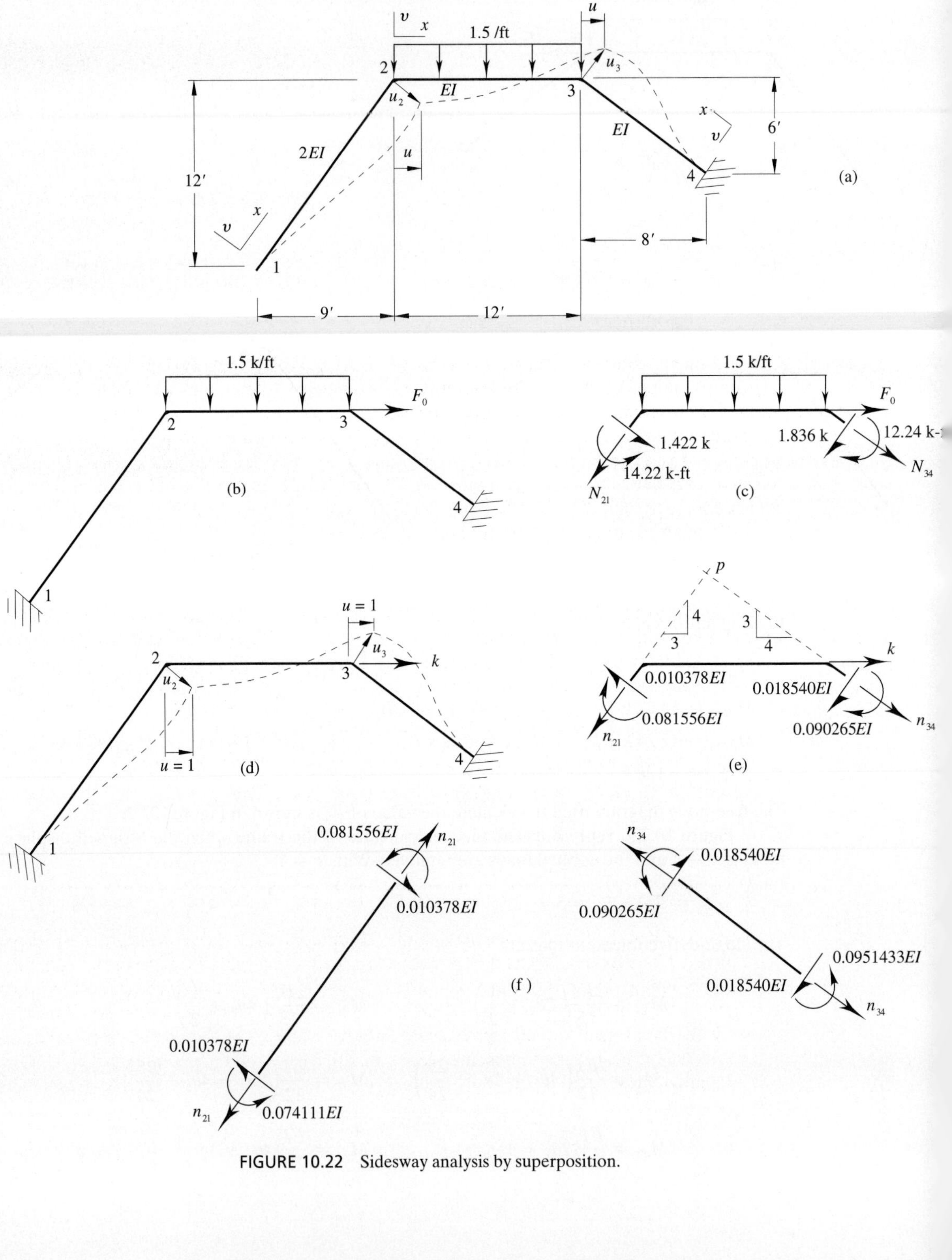

FIGURE 10.22 Sidesway analysis by superposition.

The moment equations of equilibrium for joints 2 and 3 are

$$0 = M_{21} + M_{23} \Rightarrow 0 = \frac{13}{15}\theta_2 + \frac{1}{6}\theta_3 - \frac{2.9}{144}$$

$$0 = M_{32} + M_{34} \Rightarrow 0 = \frac{1}{6}\theta_2 + \frac{11}{15}\theta_3 + \frac{1.9}{144}$$

with solution values

$$\theta_2 = 0.0279174, \quad \theta_3 = -0.0243373$$

Back substitution into the slope-deflection equations gives

$m_{12} = 0.074111EI, \quad m_{21} = 0.081556EI, \quad m_{23} = -0.081556EI, \quad m_{32} = -0.090265EI$

$m_{43} = 0.095133EI, \quad m_{34} = 0.090265EI$

The end-shears could be obtained by substitution of the values for θ_2, θ_3 and $u = 1$ into the appropriate slope-deflection equations. It is easier, however, to simply make free-body diagrams of the members and make direct use of the equations of equilibrium. The essential free-body diagrams and the calculated values for the end-shears are displayed in Figure 10.22f. The force k is obtained from equilibrium of the free-body diagram shown in Figure 10.22e. The most direct solution is obtained by summing moments about the point of intersection of the line of action of the axial force n_{21} and the line of action of n_{34}. The result is $k = 0.073703EI$.

By superposition, the horizontal force at joint 3 is

$$F = F_0 + ku = 6.188 + 0.073703EIu$$

which, for the given loading shown in Figure 10.22a, must be zero. Therefore, $EIu = -83.96$ k-ft^3, which agrees with the value obtained in the basic slope-deflection method of analysis of the problem (see Example 10.9 in the section Frames with Sidesway). By superposition, the final values for the end-moments are

$$M_{12} = -7.11 + 0.074111 \times (-83.96) = -13.33 \text{ k-ft}$$

$$M_{21} = -14.22 + 0.081556 \times (-83.96) = -21.10 \text{ k-ft} = -M_{23}$$

$$M_{32} = -12.24 - 0.090265 \times (-83.96) = -4.66 \text{ k-ft} = -M_{34}$$

$$M_{43} = 6.12 + 0.095133 \times (-83.96) = -1.87 \text{ k-ft}$$

Example 10.14

The support at d of the frame shown in Figure 10.23 prevents rotation of the tangent at the d end of member cd. The forces P, Q, and R, as well as the load w are prescribed. Determine the end-moments, assuming that the deformation is inextensional. The frame has two degrees of freedom with respect to joint rotation and two degrees of freedom with respect to joint translation or sidesway. By the principle of superposition, the mechanical state of the frame is the superposition of the mechanical states represented in Figures 10.23b, c, and d. Accordingly, the displacements u_1 and u_2 must be such that

$$P = F_{(0)1} + k_{11}u_1 + k_{12}u_2$$
$$R = F_{(0)2} + k_{21}u_1 + k_{22}u_2$$

Displacements u_1 and u_2 are prescribed for each of the loading cases represented by Figures 10.23b, c, and d. From earlier discussions concerning stiffness of linearly elastic bodies, it is clear that k_{11}, k_{12}, k_{21}, and k_{22} are stiffness coefficients with respect to displacements at points 1 and 2. By the reciprocal theorem, the cross-stiffnesses k_{12} and k_{21} must be equal. All of the variables that characterize the respective mechanical states can be determined by means of a slope-deflection equation analysis in which the only unknown position variables are the rotations of joints b and c. Because the procedures to be followed have been well documented by means of examples, only the results of the analyses are given.

Case 1 $u_1 = 0, u_2 = 0$, and prescribed loads act.

$$\theta_{(0)b} = \frac{wL^3}{90EI}, \qquad \theta_{(0)c} = -\frac{wL^3}{360EI}, \qquad M_{(0)ab} = \frac{19wL^2}{180}, \qquad M_{(0)ba} = -\frac{7wL^2}{180}$$

$$M_{(0)bc} = \frac{7wL^2}{180}, \qquad M_{(0)cb} = \frac{wL^2}{90}, \qquad M_{(0)cd} = -\frac{wL^2}{90}, \qquad M_{(0)dc} = -\frac{wL^2}{180}$$

$$F_{(0)1} = -Q - \frac{9wL}{20}, \qquad F_{(0)2} = \frac{wL}{60}$$

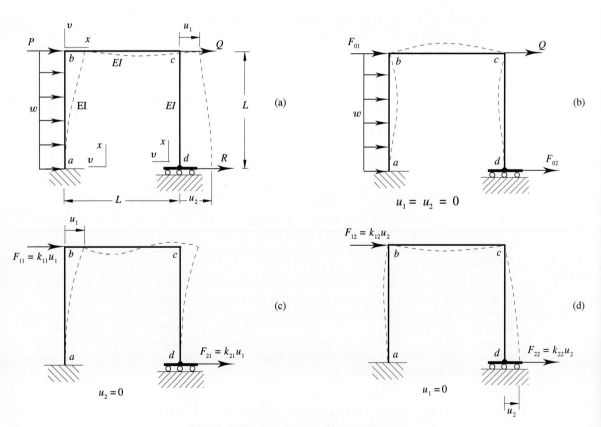

FIGURE 10.23 Two sidesway degrees of freedom.

Case 2 $u_1 = 1, u_2 = 0$, and no prescribed loads.

$$\theta_{(1)b} = -\frac{3}{5L}, \quad \theta_{(1)c} = -\frac{3}{5L}, \quad m_{(1)ab} = \frac{4.8EI}{L^2}, \quad m_{(1)ba} = \frac{3.6EI}{L^2}$$

$$m_{(1)bc} = -\frac{3.6EI}{L^2}, \quad m_{(1)cb} = -\frac{3.6EI}{L^2}, \quad m_{(1)cd} = \frac{3.6EI}{L^2}, \quad m_{(1)dc} = \frac{4.8EI}{L^2}$$

$$k_{11} = \frac{16.8EI}{L^3}, \quad k_{21} = -\frac{8.4EI}{L^3}$$

Case 3 $u_1 = 0, u_2 = 1$, and no prescribed loads.

$$\theta_{(2)b} = -\frac{1}{5L}, \quad \theta_{(2)c} = \frac{4}{5L}, \quad m_{(2)ab} = -\frac{0.4EI}{L^2}, \quad m_{(2)ba} = -\frac{0.8EI}{L^2}$$

$$m_{(2)bc} = \frac{0.8EI}{L^2}, \quad m_{(2)cb} = \frac{2.8EI}{L^2}, \quad m_{(2)cd} = -\frac{2.8EI}{L^2}, \quad m_{(2)dc} = -\frac{4.4EI}{L^2}$$

$$k_{12} = -\frac{8.4EI}{L^3}, \quad k_{22} = \frac{7.2EI}{L^3}$$

It is to be noted that in agreement with the reciprocal theorem, $k_{12} = k_{21}$. With the aid of the foregoing results the superposition equations reduce to

$$\left. \begin{array}{l} P = -Q - \dfrac{9wL}{20} + 16.8\dfrac{EIu_1}{L^3} - 8.4\dfrac{EIu_2}{L^3} \\[2mm] R = \dfrac{wL}{60} - 8.4\dfrac{EIu_1}{L^3} + 7.2\dfrac{EIu_2}{L^3} \end{array} \right\} \Rightarrow \begin{array}{l} u_1 = \dfrac{L^3}{7EI}\left(\dfrac{7}{6}R + P + Q + \dfrac{31}{72}wL\right) \\[2mm] u_2 = \dfrac{L^3}{6EI}\left(2R + P + Q + \dfrac{5}{12}wL\right) \end{array}$$

The end-moment are also obtained by superposition. For $R = P = Q = 0$ the end-moments are

$$M_{ab} = M_{(0)ab} + m_{(1)ab}u_1 + m_{(2)ab}u_2 = \frac{23.5}{63}wL^2$$

$$M_{ba} = M_{(0)ba} + m_{(1)ba}u_1 + m_{(2)ba}u_2 = \frac{8}{63}wL^2 = -M_{bc}$$

$$M_{cb} = M_{(0)cb} + m_{(1)cb}u_1 + m_{(2)cb}u_2 = -\frac{1}{63}wL^2 = -M_{cd}$$

$$M_{dc} = M_{(0)dc} + m_{(1)dc}u_1 + m_{(2)dc}u_2 = -\frac{1}{63}wL^2$$

10.11 MOMENT DISTRIBUTION

Moment distribution[1] is an iterative method that employs superposition to solve the (joint) moment equations of equilibrium of the slope-deflection method for the joint

---

[1] Moment distribution is an invention of Hardy Cross (see Timoshenko). Its introduction in an era when digital computers were nonexistent represented a giant step forward in numerical methods in engineering.

340 Chapter 10 Displacement Methods of Analysis

rotations with automatic back substitution into the slope-deflection equations. Practical application of the method is limited to problems in which joint displacements are prescribed. Two important advantages of the method are: (1) by its use, the problem of formulating the slope-deflection equations is completely circumvented, (2) convergence is guaranteed for stable linear elastic structures that undergo small deformations and displacements. The requirement that the joint displacements be prescribed does not exclude application of moment distribution to problems in which sidesway occurs. As has already been shown, sidesway problems can be solved with the aid of the principle of superposition and data generated by solving a sequence of problems in which displacements are prescribed.

The basic concepts and terminology of the method are introduced by way of a problem that initially appears to be purely of academic interest. Consider a frame for which there are no joint displacement and the loading consists only of couples applied to the joints. A portion of such a frame is shown in Figure 10.24.

Clearly, the portion consists of: (1) a central rigid joint, joint A, (2) all members that connect at the joint, members AB, AC, AD, AE, etc., and (3) adjacent joints, joints B, C, D, E, etc. The end of a member that connects into joint A is referred to as the *near end* of the member, and the opposite end of the member is referred to as the *far end*. Let $M_{(c)A}$ be a prescribed external couple applied to joint A. Let $M_{(r)B}, M_{(r)C}, M_{(r)D}$, etc., be external couples required at joints B, C, D, etc., to prevent rotations of joints B, C, D, etc. All external couples are taken to be positive in the counterclockwise sense. The problem is: Given $M_{(c)A}$, calculate the end-moments in all members, and calculate the required external couples $M_{(r)B}, M_{(r)C}, M_{(r)D}$, etc. The slope-deflection method is used to solve the problem. From the given conditions and the slope-deflection equations, it is clear that the end-moments and the external couples are zero for all members and joints not shown in Figure 10.24. In addition, for the portion of the structure

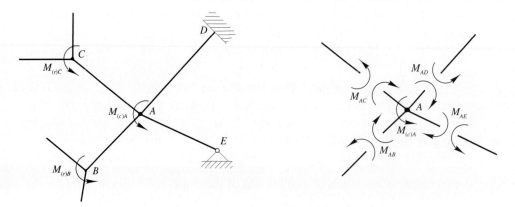

FIGURE 10.24 Part of a rigid frame.

shown in Figure 10.24, only joints A and E undergo non-zero rotations. The slope-deflection equations may be written in the form

$$M_{AB} = 4\left(\frac{EI}{L}\right)_{AB}\theta_A = K_{AB}\theta_A, \qquad M_{BA} = 2\left(\frac{EI}{L}\right)_{AB}\theta_A = \frac{1}{2}K_{AB}\theta_A$$

$$M_{AC} = 4\left(\frac{EI}{L}\right)_{AC}\theta_A = K_{AC}\theta_A, \qquad M_{CA} = 2\left(\frac{EI}{L}\right)_{AC}\theta_A = \frac{1}{2}K_{AC}\theta_A \qquad \text{(a)}$$

$$M_{AD} = 4\left(\frac{EI}{L}\right)_{AD}\theta_A = K_{AD}\theta_A, \qquad M_{DA} = 2\left(\frac{EI}{L}\right)_{AD}\theta_A = \frac{1}{2}K_{AD}\theta_A$$

$$M_{AE} = 3\left(\frac{EI}{L}\right)_{AE}\theta_A = K_{AE}\theta_A, \qquad M_{EA} = 0 \qquad\qquad\qquad = 0K_{AE}\theta_A$$

where θ_A is the rotation of joint A, taken positive in the counterclockwise sense, and

$$K_{A.} = 4\left(\frac{EI}{L}\right)_{A.}$$

where A identifies the near end of the member and the dot represents the far end of the member, is understood to be the stiffness of the near end of the member with respect to rotation when the far end of the member is fixed or when rotation of the far end is prevented. Similarly,

$$K_{A.} = 3\left(\frac{EI}{L}\right)_{A.}$$

is understood to be the stiffness of the near end of the member with respect to rotation when the far end of the member is free to rotate. The stiffness with respect to rotation is a joint property of the member and how the member is framed into the joints.

The partial free-body diagram of joint A and the members coming into the joint is representative of such a free-body diagram for all joints in the structure. The members are shown to emphasize the fact that by Newton's third law, the moment exerted on the joint by the end of a member is equal and opposite to the end-moment acting on the member. For moment equilibrium of joint A

$$M_{(c)A} = \Sigma M_{A.} = (\Sigma M_{A.})\theta_A = K_A \theta_A \Rightarrow \theta_A = \frac{M_{(c)A}}{K_A} \qquad \text{(b)}$$

In the foregoing relationship, K_A is understood to be the stiffness of the joint with respect to rotation. As in the case of member stiffness, the joint stiffness is a property of the structure. Moment equilibrium of the joint at the far end of any member requires that the external resisting moment be equal to the end-moment acting on the member, i.e.,

$$M_{(r).} = M_{.A} \qquad \text{(c)}$$

The required end-moments are obtained by substituting θ_A as given by Equation (b), into slope-deflection equations (a), and the required resisting moments are obtained from Equation (c). The results may be expressed in the form

$$M_{AB} = \frac{K_{AB}}{K_A} M_{(c)A} = DF_{AB} M_{(c)A}, \quad M_{BA} = \frac{1}{2} M_{AB} = COF_{AB} M_{AB}, \quad M_{(r)B} = M_{BA}$$

$$M_{AC} = \frac{K_{AC}}{K_A} M_{(c)A} = DF_{AC} M_{(c)A}, \quad M_{CA} = \frac{1}{2} M_{AC} = COF_{AC} M_{AC}, \quad M_{(r)C} = M_{CA} \quad \text{(d)}$$

$$M_{AD} = \frac{K_{AD}}{K_A} M_{(c)A} = DF_{AD} M_{(c)A}, \quad M_{DA} = \frac{1}{2} M_{AD} = COF_{AD} M_{AD}, \quad M_{(r)D} = M_{DA}$$

$$M_{AE} = \frac{K_{AE}}{K_A} M_{(c)A} = DF_{AE} M_{(c)A}, \quad M_{EA} = 0 M_{AE} = COF_{AE} M_{AE}, \quad M_{(r)E} = M_{EA}$$

The quantities $DF_{AB}, DF_{AC}, DF_{AD},$ and DF_{AE} are called *distribution factors*. The distribution factors are properties of the structure. From the equilibrium condition

$$M_{(c)A} = \Sigma M_{A.} = (\Sigma DF_{A.}) M_{(c)A}$$

it follows that the sum of the distribution factors is unity, i.e.,

$$\Sigma DF_{A.} = 1$$

Motivation for the terminology regarding distribution factors follows from the fact that DF_{AB} is the proportion of $M_{(c)A}$ that is induced at or distributed to the A-end of member AB, DF_{AC} is the proportion of $M_{(c)A}$ that is induced at or distributed to the A end of member AC, etc.

The quantities $COF_{AB}, COF_{AC}, COF_{AD}$, etc., are called *carryover factors*. The carryover factor COF_{AB} is the fraction of the moment that is distributed to the A (or near)-end of member AB that is in turn induced or carried over to the B (or far)- end of the member. Clearly, for the uniform members under consideration, the carryover factor is one-half if the connection at the far end of the member is rigid, and the carryover factor is zero if the connection at the far end is a pinned.

The results expressed in Equations (d), together with an understanding of how the distribution and carryover factors are defined and evaluated, point to a shortcut technique for solving the original problem without writing the slope-deflection equations or the equations of equilibrium. The procedure consists in calculating the distribution and carryover factors for the members that form joint A and then carrying out the *moment distribution* and *carryover* processes represented analytically by Equations (d). Preliminary steps in the solution of any problem by moment-distribution are:

1. Calculation of the distribution factors for each joint at which the moment distribution will be performed.
2. Identification of the carryover factors.
3. Calculation of the fixed-end moments due to prescribed loads between the ends of a member and/or due to prescribed joint displacements when all joint rotations are zero.
4. Specification of the value of the prescribed external moment $M_{(p)}$ at each joint at which moment distribution will be performed.

10.12 ONE INDEPENDENT JOINT ROTATION

The development to this point is sufficient to solve a simple problem using superposition and moment distribution. Consider the beam shown in Figure 10.25. $M_{(p)2}$ is a prescribed externally applied couple. It is desired to determine the end-moments for $M_{(p)2} = 0$ and for $M_{(p)2} = -20$ k-ft. In this problem, there is only one joint at which the process is to be applied; namely, joint 2. The member stiffnesses, joint stiffness, distribution factors, and carryover factors for joint 2 are given in the table below.

The fixed end-moments are

$$M_{F12} = 62.5 \text{ k-ft}, \quad M_{F21} = -62.5 \text{ k-ft}, \quad M_{F23} = 48 \text{ k-ft}, \quad M_{F32} = -72 \text{ k-ft}$$

TABLE 10.1 Joint 2: Stiffness, Distribution Factors, and Carryover Factors

(1)	(2)	(3)	(4)
Member	K	DF	COF
21	0.16EI	0.5	0.5
23	0.16EI	0.5	0.5
	0.32EI		

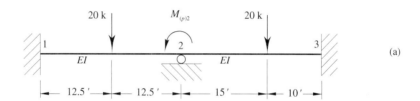

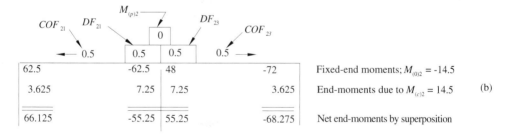

FIGURE 10.25 Moment distribution, one joint rotates.

Now, observe that the required equilibrium state of the given beam can be obtained as the superposition of:

1. the equilibrium state that exists when the beam is subjected to all applied loads between the ends of the members and whatever external couple $M_{(0)2}$ must be applied at joint 2 to enforce the condition of no rotation of joint 2, i.e., $\theta_2 = 0$
2. the equilibrium state that exists when the only load acting is an applied couple $M_{(c)2}$ at joint 2.

For the first state, the required external couple is readily obtained as the sum of the fixed end-moments, i.e.,

$$M_{(0)2} = M_{F21} + M_{F23} = -62.5 + 48 = -14.5 \text{ k-ft}$$

For the second state, the couple $M_{(c)2}$ is chosen so that after superposition the net external couple at joint 2 is equal to the prescribed value, i.e., $M_{(c)2}$ is chosen so that

$$M_{(p)2} = M_{(0)2} + M_{(c)2} = -14.5 + M_{(c)2}$$

The end-moments corresponding to $M_{(c)2}$ are found by moment-distribution. For $M_{(p)2} = 0$, $M_{(c)2} = 14.5$ k-ft, and for $M_{(p)2} = -20$ k-ft, $M_{(c)2} = -5.5$ k-ft. Details of the moment distribution and superposition processes are set out in tabular form as indicated in Figure 10.25b for $M_{(p)2} = 0$ and in Figure 10.25c for $M_{(p)2} = -20$ k-ft. That the final values for the end-moments are correct can be verified by direct application of the slope-deflection method.

10.13 TWO INDEPENDENT JOINT ROTATIONS

Consider the determination of the end-moments for the beam shown in Figure 10.26 by the combined use of superposition and moment-distribution. Moment distribution is required only at joints 2 and 3. The distribution and carryover factor data for joints 2 and 3 are given in Table 10.2.

The non-zero fixed-end moments are $M_{F12} = 24$ k-ft and $M_{F21} = -24$ k-ft, and the prescribed external moments are $M_{(p)2} = 0$, $M_{(p)3} = -10$ k-ft, $M_{(p)4} = 0$.

As in the first problem, the equilibrium state of the given beam can be obtained as the superposition of:

1. the equilibrium state that exists when the beam is subjected to the applied loads between the ends of the members and whatever external couples $M_{(0)2}$ and $M_{(0)3}$ are required at the joints so that $\theta_2 = \theta_3 = 0$;
2. the equilibrium state that exists when joint 2 is permitted to rotate under the action of applied couple $M_{(c)2}$ with no rotation of joint 3;
3. the equilibrium state that exists when joint 3 is permitted to rotate under the action of applied couple $M_{(c)3}$ with no rotation of joint 2.

There is no need to consider joint 4 because the end of the member is free to rotate and the end-moment M_{43} is zero in all cases. The end-moments and external applied couples required for each of the states are displayed in Figures 10.26b, c, and d. The external couples $M_{(0)2}$ and $M_{(0)3}$ (corresponding to $\theta_2 = \theta_3 = 0$ in Figure 10.26b) are found from the requirement of moment equilibrium of the joints. As indicated in

Section 10.13 Two Independent Joint Rotations 345

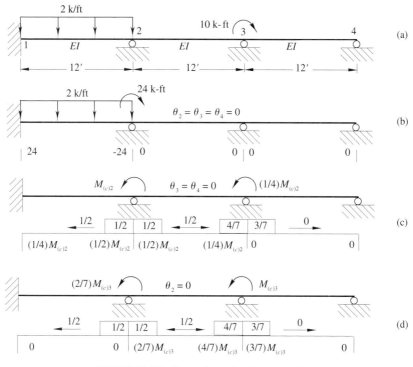

FIGURE 10.26 Two independent joint rotations.

TABLE 10.2 Stiffnesses, Distribution Factors, and Carryover Factors

	Joint 2				Joint 3		
Member	K/EI	DF	COF	Member	K/EI	DF	COF
21	1/3	0.5	0.5	32	1/3	4/7	0.5
23	1/3	0.5	0.5	34	1/4	3/7	0
	2/3				7/12		

Figure 10.26c, the end-moments induced by $M_{(c)2}$ with no rotation of joint 3 are determined by moment distribution, and $M_{(r)3}$ is found from the requirement of moment equilibrium of joint 3. The results displayed in Figure 10.26d also follow from moment distribution and joint equilibrium. By superposition, the net external moment $M_{(e)2}$ and $M_{(e)3}$ at joints 2 and 3 are

$$M_{(e)2} = M_{(0)2} + M_{(c)2} + \frac{2}{7} M_{(c)3}$$

$$M_{(e)3} = M_{(0)3} + \frac{1}{4} M_{(c)2} + M_{(c)3}$$

(e)

For the given beam, $M_{(0)2} = -24$ k-ft, $M_{(0)3} = 0$, the required values for the external moments are $M_{(e)2} = M_{(p)2} = 0$, and $M_{(e)3} = M_{(p)3} = -10$ k-ft. Substituting these values into Equations (e) yields

$$M_{(c)2} + \frac{2}{7} M_{(c)3} = 24$$

$$\frac{1}{4} M_{(c)2} + M_{(c)3} = -10 \tag{f}$$

the solution values of which are

$$M_{(c)2} = 28.92 \text{ k-ft}, \quad M_{(c)3} = -17.23 \text{ k-ft}$$

By superposition, the values for the end-moments are

$$M_{12} = 24 + \frac{1}{4} M_{(c)2} = 31.23 \text{ k-ft}, \quad M_{21} = -24 + \frac{1}{2} M_{(c)2} = -9.54 \text{ k-ft}$$

$$M_{23} = \frac{1}{2} M_{(c)2} + \frac{2}{7} M_{(c)3} = 9.54 \text{ k-ft}, \quad M_{32} = \frac{1}{4} M_{(c)2} + \frac{4}{7} M_{(c)3} = -2.62 \text{ k-ft},$$

$$M_{34} = \frac{3}{7} M_{(c)3} = -7.38 \text{ k-ft}$$

10.14 SEVERAL INDEPENDENT JOINT ROTATIONS

The foregoing procedure can be implemented to determine the end-moments for any number of independent unknown joint rotations. However, as it stands, the approach is clumsy and requires the formulation and solution of simultaneous equations (the superposition equations corresponding to Equations (e) above) equal in number to the number of independent joint rotations. Moment distribution and iteration can be combined to determine the end-moments and solve the equations while bypassing the operation of formulating the equations. To bring out the underlying concepts of the technique, consider the following table of successive approximations to $M_{(c)2}$ and $M_{(c)3}$ when Equations (f) are solved by the by the Gauss-Seidel method.

It is obvious from the entries in columns (2) and (3) that the iterates are converging to the solution of the equations and that the entries in Line 6 are sufficiently accurate approximations to $M_{(c)2}$ and $M_{(c)3}$. The end-moments due to $M_{(c)2}$ and $M_{(c)3}$ (as

TABLE 10.3 Gauss-Seidel Iteration

(1)	(2)	(3)	(4)	(5)
Step	$M_{(c)2}$	$M_{(c)3}$	$\Delta M_{(c)2}$	$\Delta M_{(c)3}$
1	24	0	24	0
2	24	−16	0	−16
3	28.571	−16	4.571	0
4	28.571	−17.143	0	−1.143
5	28.898	−17.143	0.327	0
6	28.898	−17.224	0	−0.081

given in Line 6) are found by moment distribution as shown in Figure 10.27c and Figure 10.27d, respectively. Superposition of the loadings and end-moments represented by Figure 10.27b, 10.27c, and 10.27d yields the loading and end-moments displayed in Figure 10.27e. Because the differences in the loadings shown in Figure 10.27a and 10.27e are negligibly small, the end-moments shown in Figure 10.27e represent the solution of the problem.

Clearly, the final values for $M_{(c)2}$ and $M_{(c)3}$ are nothing more than the sums (or the superposition) of their respective incremental values as shown in columns 5 and 6 of the Gauss-Seidel iteration table. It follows that the final values for the end-moments are the sums (or the superposition) of the incremental end-moments induced by the successive increments in $M_{(c)2}$ and $M_{(c)3}$. Because increments in the end-moments due to any increment $\Delta M_{(c)2}$ or $\Delta M_{(c)3}$ can be found by moment distribution, the key to coming up with the final values of the end-moments (without writing any equations) is

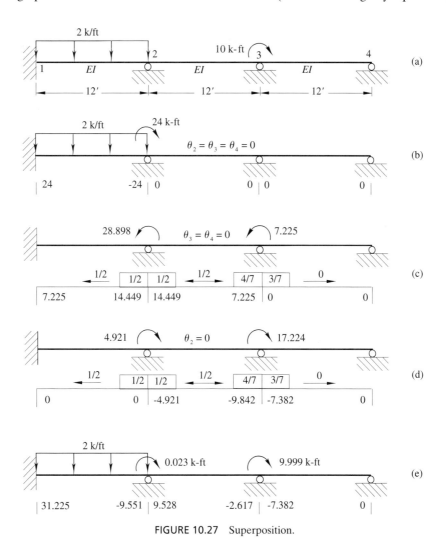

FIGURE 10.27 Superposition.

the rule for determining the successive incremental values in $M_{(c)2}$ and $M_{(c)3}$. The rule is that the increment at any joint be chosen so that the sum of: the end-moments when all joint rotations are zero, the end-moments that have been induced by all incremental external moments that have already been applied at all joints, and the increment to be applied, add up to the prescribed value for the external moment at the joint under consideration. Incremental external moments or corrective external moments determined in this way are understood to be chosen so as to "balance" the moments at the joint. In practice, the procedure is repeated until the corrective moments are negligibly small. The foregoing technique of determining corrective external moments combined with moment distribution and superposition is the Hardy Cross moment distribution method.

Application of the Hardy Cross moment distribution method to the determination of the end-moments for the beam in Figure 10.26 is illustrated in Figure 10.28. The distribution and carryover factors and the prescribed values for the external moments at joints 2 and 3 are displayed as headers to the lines that constitute the moment distribution table. The sequence of the operations displayed in the figure corresponds to the sequence employed in generating the results by the Gauss-Seidel method of iteration discussed previously. However, because convergence is guaranteed and the solution is unique, the end results are independent of the order in which the operations are carried out. The entries in Line (1) of the table are nothing more than the fixed-end moments. Because the sum of the fixed end-moments at joint 2 is -24 k-ft, while the prescribed value is $M_{(p)2} = 0$, the initial external corrective moment at joint 2 is $+24$ k-ft. The end-moments due to the corrective moment are determined by moment distribution and have the values displayed in Line 2 of the table. The line drawn beneath the end-moments induced in the members at joint 2 as a consequence of the corrective external moment is used to indicate that the sum of the end-moments above the line

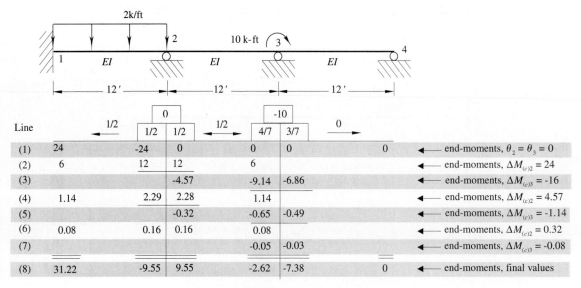

FIGURE 10.28 Moment distribution.

Section 10.14 Several Independent Joint Rotations 349

satisfy the condition that the sum of the end-moments of all members coming into a joint must be equal to the prescribed external moment. As indicated under joint 3 in Line (2) of the table, the initial correction $\Delta M_{(c)2} = 24$ k-ft induces an end-moment of 6 k-ft at the 3-end of member 23. By superposition of the end-moments shown in lines (1) and (2), the sum of the end-moments for the members forming joint 3 is 6 k-ft. However, the prescribed value for the external moment at joint 3 is -10 k-ft. Accordingly, the initial corrective external moment at joint 3 is -16 k-ft as indicated in the table. The corresponding induced end-moments are determined by the moment distribution process and are as indicated in Line (3). Again, the line drawn under the entries in Line (3) beneath joint 3 is used to indicate that the sum of the end-moments (obtained by superposition) of the members coming into the joint is equal to the

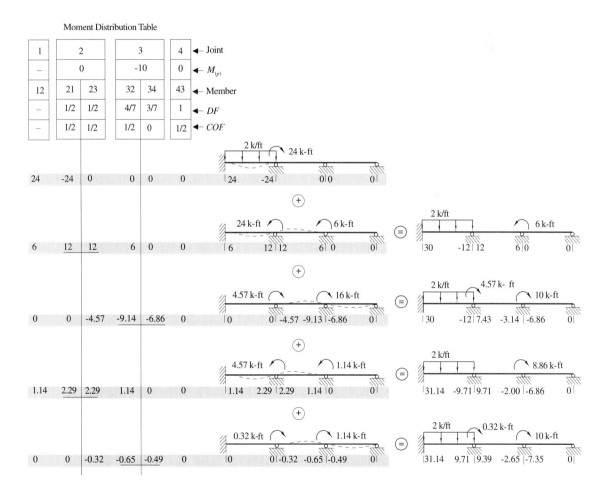

FIGURE 10.29 Physical content.

prescribed value for the external moment. The moment carried over to the 2-end of member 23 as a consequence of the corrective moment of -16 k-ft at joint 3 is -4.57 k-ft. Because joint 2 was balanced before the correction was applied at joint 3, the additional correction required at joint 2 is $+4.57$ k-ft. As indicated by the entries in Line 4, this correction at joint 2 requires an additional correction at joint 3 in the amount of -1.14 k-ft. Successive corrections are applied at joints 2 and 3 until the corrections are sufficiently small. The final values of the end moments are obtained by superposition as indicated in Line 8.

Basic ideas regarding the physical content of the moment distribution-superposition procedure are illustrated in Figure 10.29. Figures in the middle column represent the individual load cases which are to be superimposed. The order of superposition is from top to bottom. Figures in the right column represent the successive stages of the beam under the superimposed loads.

An alternate format for a moment distribution table is also shown in Figure 10.29. The alternate format eliminates the need for a sketch of the structure in recording results from each stage of moment distribution. As will be shown by example, the alternate format is particularly useful in the analysis of rigid frame structures.

Example 10.15

Construct the shear and bending moment diagrams and calculate the deflection at the point of application of the 20-k load for the continuous beam shown in Figure 10.30. Because the beam is statically indeterminate, the solution process begins with the determination of the end-moments for the individual members; here, by moment distribution. Data essential to implementation of the moment distribution procedure are shown below the sketch of the beam. The moment distribution procedure must be carried out at joints 1 and 4 as well as at joints 2 and 3.

Because rotation of joint 2 is prevented when external moment $M_{(c)1}$ is applied, $K_{12} = 4(2EI/12)$ and $COF_{12} = 1/2$. However, because the "1" end of member 12 is free to rotate when external couple $M_{(c)2}$ applied, $K_{21} = 3(2EI/12)$ and $COF_{21} = 0$. The reasoning used to determine the stiffnesses and carryover factors for member 12 also applies to member 34. The net values of the fixed end-moments for member 12 are determined by superposition as indicated in Figure 10.30. The prescribed value for the external moment at joints 1, 2, and 3 is zero. The effect of the 20-k load on the 6' overhanging member 45 is to induce a prescribed external moment of -120 k-ft at joint 4. The order in which the joints are selected for application of the moment distribution process does not affect convergence to the solution values of the end-moments; however, it does affect the number of steps required to obtain a sufficiently accurate solution. Because $COF_{21} = 0$ ($COF_{34} = 0$), moment distribution at joint 2 (Joint 3) will have no effect on M_{12} (M_{43}). It follows that only one distribution need be carried out at joint 1 (Joint 4). Good practice is to perform the distribution process at joints 1 and 4 before going on to joints 2 and 3. The results of the moment distribution process are displayed in the moment distribution table shown in Figure 10.30. The sequence in which the joints were "balanced" is identified by means of a number in a triangle.

With the sign conventions for shear and bending moment as illustrated in Figure 10.31, and with the end-moments as given in the moment distribution table contained in Figure 10.30, the shear and bending moment diagrams are as shown in Figure 10.31.

Virtual work or the moment-area method can be employed to calculate the deflection at any point in any span. Regardless of which method is employed, the calculations are simplified

Section 10.14 Several Independent Joint Rotations

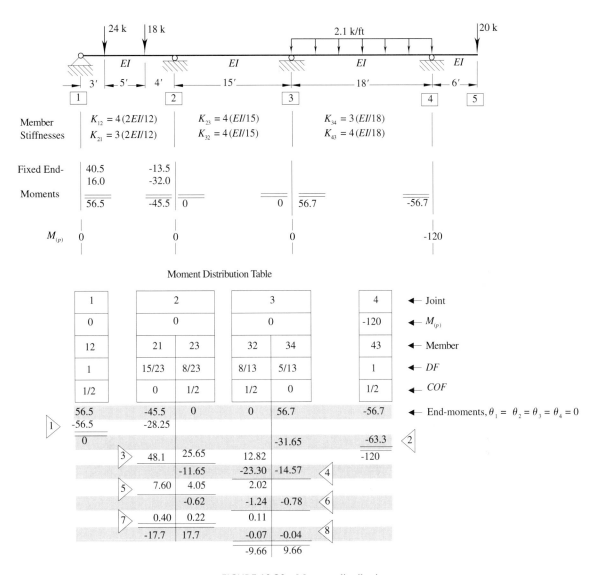

FIGURE 10.30 Moment distribution.

when the rotation of the tangent at a support point is known. In the section on the slope-deflection equations, formulas (Equations B) for the rotations at the ends of a member in terms of quantities that are determined as part of the moment distribution process were derived. The formulas are

$$\theta_{ab} = \frac{L}{6EI}[(2M_{ab} - M_{ba}) - (2M_{Fab} - M_{Fba})] + \frac{v_{ba} - v_{ab}}{L}$$

$$\theta_{ba} = \frac{L}{6EI}[(2M_{ba} - M_{ab}) - (2M_{Fba} - M_{Fab})] + \frac{v_{ba} - v_{ab}}{L}$$

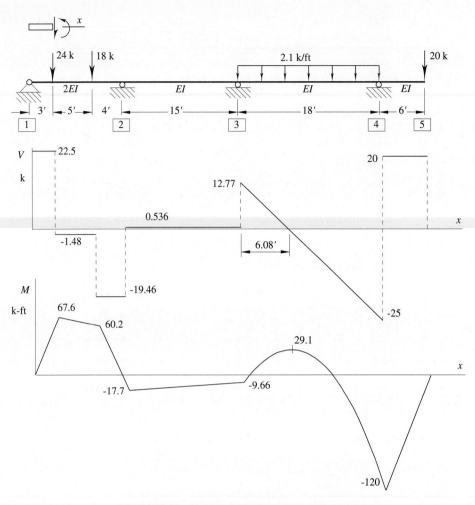

FIGURE 10.31 Shear and bending moment diagrams.

Application of the second equation to span 34 and the condition of compatibility at joint 4 gives

$$\theta_{45} = \theta_4 = \theta_{43} = \frac{3}{EI}[(-2 \times 120 - 9.66) - (-2 \times 56.7 - 56.7)] = -\frac{238.7}{EI}\text{ k-ft}^2$$

Because the deflection at joint 4 is zero, by geometry and the second moment-area theorem,

$$v_5 = 6\theta_4 + d[5t4] = -6\left(\frac{238.7}{EI}\right) - 4\left(\frac{3 \times 120}{EI}\right) = -\frac{2872}{EI}\text{ k-ft}^3$$

i.e., the deflection is 2872 k-ft³/EI downward or in the direction of the $-v$-axis.

Example 10.16

The members of the frame shown in Figure 10.32a are axially rigid. Calculate the deflection at the point of application of the 12-k load. A force analysis of the frame is required to obtain the information needed to calculate the required deflection. As a consequence of construction of the

Section 10.14 Several Independent Joint Rotations 353

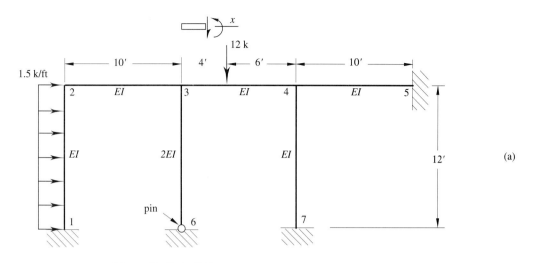

Moment Distribution Table

2	
0		← $M_{(p)}$
21	23	← Member
5/11	6/11	← DF
1/2	1/2	← COF
-18	0
8.18	9.82
	-3.42
1.56	1.86
	-0.55
0.25	0.30
	-0.04
0.02	0.02
-7.99	7.99

3		
0		
32	34	36
4/13	4/13	5/13
1/2	1/2	0
0	17.28	0
4.91		
-6.83	-6.83	-8.53
2.64		
0.93		
-1.10	-1.10	-1.37
	0.10	
0.15		
-0.08	-0.08	-0.09
-2.02	12.01	-9.99

4		
0		
43	45	47
6/17	6/17	5/17
1/2	1/2	1/2
-11.52	0	0
-3.42		
5.27	5.27	4.40
-0.55		
0.19	0.19	0.17
-0.04		
0.02	0.02	0
-10.05	5.48	4.57

← Joint
← $M_{(p)}$
← Member
← DF
← COF
← End-moments, $\theta_2 = \theta_3 = \theta_4 = 0$

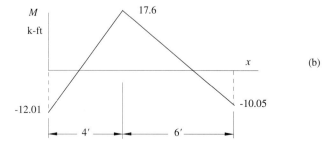

FIGURE 10.32 Statically indeterminate frame, $d = 8$.

354 Chapter 10 Displacement Methods of Analysis

frame and the condition of inextensional deformation, there is no joint translation. Because of the pin at support 6, the frame may be treated as being kinematically indeterminate to the third degree; it is statically indeterminate to the eighth degree. Accordingly, a displacement method of analysis is indicated, and moment distribution is the logical choice. With the exception of member 36, all members stiffnesses are determined from the generic relationship $K = 4(EI/L)$ and all carryover factors are $1/2$. For member 36, because of the pin connection at the support, $K = 3(EI/L)_{36}$ and $COF_{36} = 0$. The end-moment for the ends of the members that define joints 2, 3, and 4 are obtained by moment distribution as indicated in the table shown in Figure 10.32.

Because moment distribution is not required at joints 1, 5, 6, or 7, the end-moment in the member at any one of these joints is simply the end moment that exists when $\theta_2 = \theta_3 = \theta_4 = 0$ plus the moment induced as a consequence of moment distribution at the joint at the opposite end of the member. Thus,

$M_{12} = 18 + 0.5(8.18 + 1.56 + 0.25 + 0.020) = 23.01$ k-ft,

$M_{54} = 0 + 0.5(5.27 + 0.19 + 0.020) = 2.74$ k-ft

$M_{63} = 0 + 0(-8.53 - 1.37 - 0.09) = 0, \quad M_{74} = 0 + 0.5(4.40 + 0.17 + 0) = 2.29$ k-ft

With coordinates and sign convention as indicated in Figure 10.32a, the bending moment diagram for member 34 is as shown in Figure 10.32b.

With the aid of the formulas for the end-rotation of a member in terms of displacements, end-moments, and fixed-end moments,

$$\theta_{34} = \frac{10}{6EI}[(2 \times 12.01 + 10.05) - (2 \times 17.28 + 11.52)] = -\frac{20}{EI} \text{ k-ft}^2$$

(Note: because $\theta_{34} = \theta_{32} = \theta_{36}$, either of the last two rotations could have been used to obtain θ_{34}, and with slightly less computational effort). Denote by p the point of application of the 12-k load. Then, since $v_3 = 0$ it follows by geometry and the second moment-area theorem that

$$EIv_p = 4\theta_{34} + d[pt3] \Rightarrow EIv_p = -4 \times 20 + \left(-12.01 \times 2 \times \frac{8}{3} + 17.6 \times 2 \times \frac{4}{3}\right)$$

$$= -97.1 \text{ k-ft}^3$$

10.15 NON-ZERO PRESCRIBED DISPLACEMENTS AND MOMENT DISTRIBUTION

Structures for which the prescribed joint displacements have non-zero values may be analyzed by moment distribution in much the same way as a structure which is subjected to applied loads and the prescribed displacements are zero. The point to be kept in mind is that, in applying the moment distribution process, the mechanical state of a structure is expressed as the superposition of an initial state and states in which an external couple $M_{(c)j}$ is applied to joint j and external couples $M_{(r)j}$ are applied to joints k, where $k \neq j$. The characteristic feature of the initial state is that the rotation at each joint at which rotation *may* take place as a consequence of the loading is assigned the value zero. Satisfaction of the conditions on the displacements and the condition of equilibrium for each joint requires the application of an external couple of moment $M_{(0)j}$ at each location where joint rotation may take place. The

Section 10.15   Non-Zero Prescribed Displacements and Moment Distribution

end-moments for the initial conditions are found by direct substitution into the slope-deflection equations, and $M_{(0)j}$ is obtained from the condition of moment equilibrium. The characteristic features of a state for which the external couples $M_{(c)j}$ and $M_{(r)k}$ act are: (1) all joints are in equilibrium with respect to moments, and (2) the joint rotation θ_j is that which is induced by $M_{(c)j}$ while $\theta_k = 0$ for all $k \neq j$, except when k identifies the end of a member at a pin connection. It follows that the only distinction between structural analysis by moment distribution when non-zero displacements are prescribed and when loads are prescribed is the source of the initial values of the end-moments. In both situations, the initial values are determined directly from the slope deflection equations.

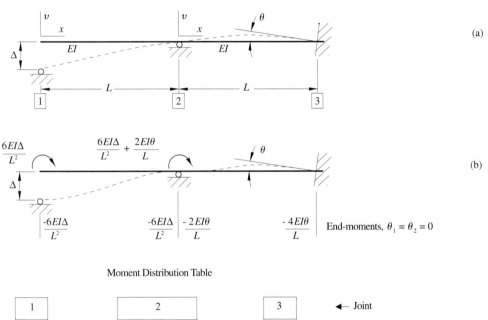

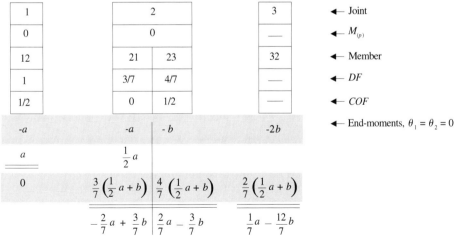

FIGURE 10.33   Support settlements.

To reinforce the basic ideas, consider the beam shown in Figure 10.33a. The non-zero prescribed displacements are $v_1 = -\Delta$, and $\theta_3 = -\theta$. Joints 1 and 2 are free to rotate. The end-moments and the external couples $M_{(0)1}$ and $M_{(0)2}$ are displayed in Figure 10.33b as they act physically. The mechanical state of the beam as defined by the conditions shown in Figure 10.33a is generated by applying corrective external couples $M_{(c)2}$ and $M_{(c)3}$. The procedures are those followed in previous applications of moment distribution. The results are displayed in the moment distribution table shown in Figure 10.33, in which $a = 6EI\Delta/L^2$ and $b = 2EI\theta/L$. From the moment distribution table and the definitions of a and b

$$M_{32} = \frac{1}{7}a - \frac{12}{7}b = \frac{1}{7}\left(\frac{6EI\Delta}{L^2}\right) - \frac{12}{7}\left(\frac{2EI\theta}{L}\right) = \frac{6EI\Delta}{7L^2} - \frac{24EI\theta}{7L}$$

$$M_{21} = -\frac{2}{7}a + \frac{3}{7}b = -\frac{2}{7}\left(\frac{6EI\Delta}{L^2}\right) + \frac{3}{7}\left(\frac{2EI\theta}{L}\right) = -\frac{12EI\Delta}{7L^2} + \frac{6EI\theta}{7L}$$

The results agree with those obtained by direct application of the slope-deflection equations as well as those obtained by force methods of analysis.

10.16 SIDESWAY ANALYSIS AND MOMENT DISTRIBUTION

The moment distribution procedure is limited to the calculation of end-moments when joint translational displacements are known; however, the joint displacements are unknown quantities in the sidesway type problem. In view of this, an analysis of sidesway with the aid of moment distribution must be arranged so that displacements are known whenever moment distribution is to be applied. The necessary conditions for application of the moment distribution procedure are satisfied when the sidesway problem is set up for solution by superposition. A detailed description and examples that illustrate how superposition may be employed in sidesway analysis is given in this chapter in the section, "**Sidesway Problems and Superposition**." Accordingly, the only aspect of the sidesway problem that is taken up here is the calculation of the required end-moments by moment distribution.

The frame shown in Figure 10.34a is the frame discussed in Example 10.13 of the section "**Sidesway Problems and Superposition**." Sidesway analysis of the frame by superposition requires the determination of: (1) the end-moments $M_{(0)}$ and the force F_0 when the applied loads act and there is no joint translation as shown in Figure 10.34b, and (2) the end-moments m, and the force k due to a unit sidesway displacement as shown in Figure 10.34c.

The end-moments for no sidesway are obtained by moment-distribution. Details of the calculation are displayed in the moment distribution table below Figure 10.34b. The results agree (as they must) with those obtained previously by the slope-deflection method.

The first step in the determination of the end-moments for the situation shown in Figure 10.34c by moment distribution is to evaluate the end-moments when $\theta_2 = \theta_3 = 0$. These end-moments are found by direct substitution into the slope-deflection

Section 10.16   Sidesway Analysis and Moment Distribution   357

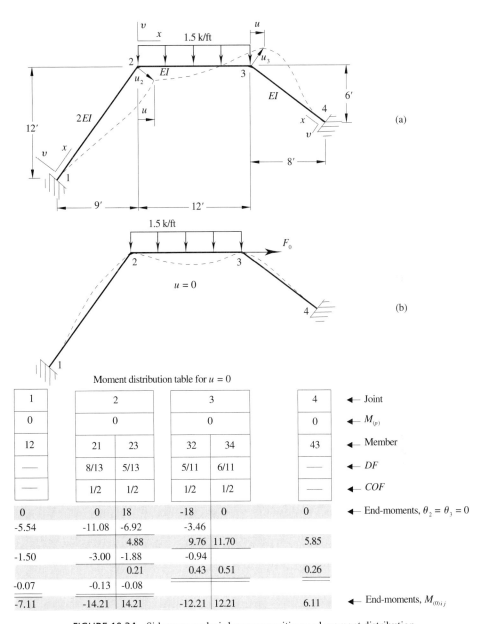

FIGURE 10.34 Sidesway analysis by superposition and moment distribution.

equations. With local coordinates as shown in Figure 10.34a, the transverse components of the member end-displacements in terms of the horizontal component of displacement u are

$$v_{12} = 0, \quad v_{21} = -\frac{5}{4}u, \quad v_{23} = -\frac{3}{4}u, \quad v_{32} = \frac{4}{3}u, \quad v_{43} = 0, \quad v_{34} = -\frac{5}{3}u$$

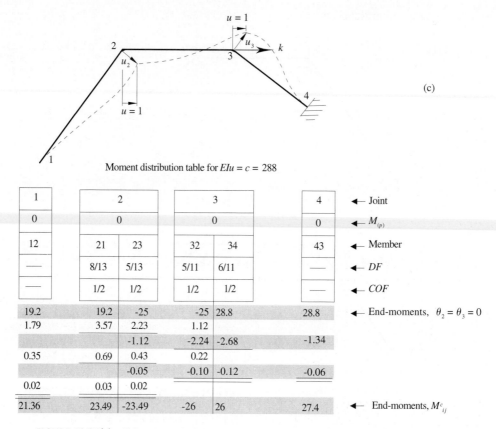

FIGURE 10.34(c) Sidesway analysis by superposition and moment distribution (continued).

For any value of u, the end-moments with $\theta_2 = \theta_3 = 0$ are

$$M_{12} = M_{21} = \frac{2EI}{15}\left[\frac{6}{15}\left(\frac{5u}{4}\right)\right] = \frac{EIu}{15}$$

$$M_{23} = M_{32} = \frac{EI}{12}\left[-\frac{6}{12}\left(\frac{3u}{4}\right) - \frac{6}{12}\left(\frac{4u}{3}\right)\right] = -\frac{25EIu}{288}$$

$$M_{34} = M_{43} = \frac{EI}{10}\left[\frac{6}{10}\left(\frac{5u}{3}\right)\right] = \frac{EIu}{10}$$

Assignment of a unit value to the displacement u leads to an awkward situation in dealing with numbers in the moment distribution process. The situation is sidestepped by carrying out the calculations for an assigned value of u such that $EIu = c$, where c is any convenient number. Let M^c be an end-moment when $EIu = c$. Then, by the principle of superposition, the corresponding end-moment m due to $u = 1$ is $m = EIM^c/c$. Details of the calculation of the end-moments for $c = 288$ are

Section 10.17   Equations of Equilibrium by Virtual Work Matrix Methods   359

displayed in the moment distribution table shown in Figure 10.34c. Therefore,

$$m_{12} = 21.36 \times \frac{EI}{288} = 0.0742EI, \quad m_{21} = 23.49 \times \frac{EI}{288} = 0.0816EI = -m_{23}$$

$$m_{43} = 27.40 \times \frac{EI}{288} = 0.0951EI, \quad m_{34} = 26.00 \times \frac{EI}{288} = 0.0903EI = -m_{32}$$

Again, the results agree with those obtained by direct application of the slope-deflection method.

10.17 EQUATIONS OF EQUILIBRIUM BY VIRTUAL WORK MATRIX METHODS

Perhaps the most challenging element in the displacement method of analysis of a framed structure that may undergo joint translation as well as joint rotation is formulation of the governing equations of equilibrium corresponding to the independent position variables. With the aid of the principle of virtual work, the task is simplified and reduced to a systematic procedural application of basic operations from matrix algebra. The information needed in a practical implementation of the formulation consists of only that information which is essential to the displacement method of analysis without regard to a particular methodology. The formulation presented herein is limited to structures that undergo small (infinitesimal) deformations and displacements. In this case, the equations of equilibrium are formulated on the basis of the undeformed geometry of the structure, and displacements due to any cause always qualify as virtual displacements.

The external forces that act on a portion of a structure are shown in Figure 10.35a. $\mathbf{F}$ is the prescribed force at a representative joint and M^e is the moment of the prescribed couple at a representative joint. $w(x)$ and $p(x)$ are the prescribed distributed transverse and axial loads on a representative member.[1] Deformations and displacements are small and the structure is assumed to be in equilibrium. Accordingly, as noted above, the equations of equilibrium are formulated on the basis of the undeformed geometry of the structure. Representative virtual displacements are shown in Figure 10.35b to a grossly exaggerated scale. The derivation employs the differential equations of equilibrium, the differential deformation-displacement relations, and associated boundary conditions. The notation used to identify the end-forces exerted on the ends of a representative member by the joints at its ends is displayed in Figure 10.35c. The corresponding notation with respect to virtual displacements is illustrated in Figure 10.35d.

The differential equations of equilibrium and the boundary conditions on force are

$$V' = -w(x), \quad V(0^+) = V_{ab}, \quad V(L^-) = -V_{ba}$$
$$M' = V(x), \quad M(0^+) = -M_{ab}, \quad M(L^-) = M_{ba}$$
$$N' = -p(x), \quad N(0^+) = -N_{ab}, \quad N(L^-) = N_{ba}$$

in which primes are used to indicate derivatives with respect to x.

---

[1] Concentrated forces and or couples that act between the ends of a member do not alter the results of the derivation. Accordingly, such loads have been omitted to achieve compactness in the presentation.

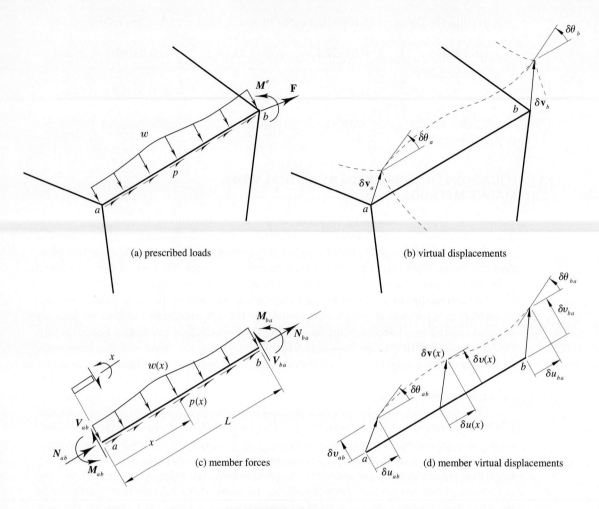

FIGURE 10.35 Notation.

The deformation-displacement relations and the displacement boundary conditions are

$$\delta v'' = \delta\kappa(x), \quad \delta v'(0^+) = \delta\theta_{ab}, \quad \delta v'(L^-) = \delta\theta_{ba}$$
$$\delta v' = \delta\theta(x), \quad \delta v(0^+) = \delta v_{ab}, \quad \delta v(L^-) = \delta v_{ba}$$
$$\delta u' = \delta\varepsilon(x), \quad \delta u(0^+) = \delta u_{ab}, \quad \delta u(L^-) = \delta u_{ba}$$

Because the structure is in equilibrium the virtual work of all the forces on arbitrary virtual displacements must be zero, i.e.,

$$\delta W = \delta W_e - \delta W_\sigma = \left(\sum_j (\mathbf{F} \cdot \delta \mathbf{v} + M^e \, \delta\theta) + \sum_m \int_0^L -w \, \delta v \, dx + \sum_m \int_0^L p \, \delta u \, dx \right)$$
$$- \left(\sum_m \int_0^L (M \, \delta\kappa + N \, \delta\varepsilon) \, dx \right) = 0$$

Section 10.17 Equations of Equilibrium by Virtual Work Matrix Methods

in which m (j) indicates that the summation is taken over all members (joints). The integrals involving δv and δu can be eliminated from the formulation by making use of the differential equations of equilibrium, the deformation-displacement relations, the boundary conditions, and integration by parts. Thus

$$-\int_0^L w\,\delta v\,dx = \int_0^L V'\,\delta v\,dx = [V\,\delta v]_0^L - \int_0^L M'\,\delta v'\,dx$$

$$= [V\,\delta v]_0^L - [M\,\delta v']_0^L + \int_0^L M\,\delta v''\,dx$$

$$= -[V_{ab}\,\delta v_{ab} + V_{ba}\,\delta v_{ba}] - [M_{ab}\,\delta \theta_{ab} + M_{ba}\,\delta \theta_{ba}] + \int_0^L M\,\delta\kappa\,dx$$

$$\int_0^L p\,\delta u\,dx = -\int_0^L N'\,\delta u\,dx = -[N\,\delta u]_0^L + \int_0^L N\,\delta u'\,dx$$

$$= -[N_{ab}\,\delta u_{ab} + N_{ba}\,\delta u_{ba}] + \int_0^L N\,\delta\varepsilon\,dx$$

With the aid of the foregoing results, the work equation becomes

$$\sum_j (\mathbf{F}\cdot\delta\mathbf{v} + M^e\,\delta\theta)$$

$$-\sum_m (M_{ab}\,\delta\theta_{ab} + M_{ba}\,\delta\theta_{ba} + V_{ab}\,\delta v_{ab} + V_{ba}\,\delta v_{ba} + N_{ab}\,\delta u_{ab} + N_{ba}\,\delta u_{ba}) = 0$$

For the great majority of rigid frames, the members may be treated as being axially rigid. In this case, the work equation takes on a slightly different form. By the condition of inextensional deformation, $\delta u(x)$ is constant over the length of the member and $\delta u(x) = \delta u_{ab} = \delta u_{ba} = \delta p_m$. Also, from the equations of equilibrium and the force boundary conditions

$$\int_0^L p\,dx = -\int_0^L N'\,dx = -(N_{ba} + N_{ab}) \equiv P_m$$

Therefore, for the axial loads on an axially rigid member

$$\int_0^L p\,\delta u\,dx = P_m\,\delta p_m$$

and the work equation for a structure in which all members are treated as being axially rigid is reduced to

$$\sum_j (\mathbf{F}\cdot\delta\mathbf{v} + M^e\,\delta\theta) + \sum_m P_m\,\delta p_m$$

$$-\sum_m (M_{ab}\,\delta\theta_{ab} + M_{ba}\,\delta\theta_{ba} + V_{ab}\,\delta v_{ab} + V_{ba}\,\delta v_{ba}) = 0$$

which must be satisfied for arbitrary virtual displacements. To put the work equation in a more useable form, the prescribed force at a joint, the displacement of a joint, and the virtual displacement at a joint are referenced to rectangular Cartesian global coordinates. Let X and Y identify the global coordinate axes, let F_X and F_Y be the rectangular components of force, let u_X and u_Y be the rectangular components of displacement, and let δu_X and δu_Y be the rectangular components of virtual displacement. Then the expression for the virtual work of the prescribed forces at a joint becomes

$$\sum_j (\mathbf{F} \cdot \delta \mathbf{v} + M^e \delta\theta) = \sum_j (F_X \delta u_X + F_Y \delta u_Y + M^e \delta\theta)$$

and, in the case of inextensional deformation, the work equation is expressed as

$$\sum_j (F_X \delta u_X + F_Y \delta u_Y + M^e \delta\theta) + \sum_m P_m \delta p_m$$

$$- \sum_m (M_{ab} \delta\theta_{ab} + M_{ba} \delta\theta_{ba} + V_{ab} \delta v_{ab} + V_{ba} \delta v_{ba}) = 0$$

The work expressions are examples of bilinear forms which can be written in a more compact and useful way in terms of matrices. Define column matrices (column vectors) $Q_j, P_m, R_j, \delta q_j, \delta p_m, \delta r_m$ by their row vector equivalents (transposes) as follows:

$$Q_j^T = (F_X, F_Y, M^e)_j, \qquad P_m^T = (P_m)_m, \qquad R_m^T = (M_{ab}, M_{ba}, V_{ab}, V_{ba})_m$$

$$\delta q_j^T = (\delta u_X, \delta u_Y, \delta\theta)_j, \qquad \delta p_m^T = (\delta p_m)_m, \qquad \delta r_m^T = (\delta\theta_{ab}, \delta\theta_{ba}, \delta v_{ab}, \delta v_{ba})_m$$

Then, written in matrix notation, the work equation for inextensional deformation is

$$\delta W = \sum_j \delta q_j^T Q_j + \sum_m \delta p_m^T P_m - \sum_m \delta r_m^T R_m = 0 \qquad (a)$$

Each vector (in the sense of matrix algebra) appearing in the work equation is given a name appropriate to the elements that constitute the vector. Q_j is the applied *load vector* at a joint, P_m is the *axial load vector* for a member, and R_m is the *end-force vector* for a member. δq_j is the *virtual displacement vector* at a joint, δp_m is the *axial virtual displacement vector* for a member, and δr_m is the virtual *end-displacement vector* for a member. Because virtual displacements are infinitesimally small changes in displacements, the virtual displacement vectors $\delta q_j, \delta p_m$, and δr_m are the changes in associated displacement vectors defined by

$$q_j^T = (u_X, u_Y, \theta)_j, \qquad p_m^T = (p_m)_m, \qquad r_m^T = (\theta_{ab}, \theta_{ba}, v_{ab}, v_{ba})_m$$

By definition, the statement that a framed structure is an n-degree-freedom mechanical system means that the configuration of the structure is fully defined by n independent position or displacement variables. For the class of structures under consideration, the configuration of any member is fully defined by the displacements (which include rotation of the tangent as well as translation) at its ends. It follows that any set of n independent variables that define the positions or displacements of the joints also define the configuration of the structure. Because the displacements and rotations (and the virtual displacements and rotations) are infinitesimally small, they are linear functions of the independent position or displacement variables. The practical meaning of this is that each of the displacement vectors that has been introduced can be expressed as the product of a matrix whose elements are constants and a column vector the elements of which are the independent position or displacement variables. Let $z_1, z_2, \ldots, z_n$ be the independent displacement variables, and let z be the column vector whose elements are $z_1, z_2, \ldots, z_n$, so that $z^T = (z_1, z_2, \ldots, z_n)$. Because z

defines the displaced position of the structure, z is called the *displacement vector*. In view of the relationship between displacements and virtual displacements, the column vector defined by $\delta z^T = (\delta z_1, \delta z_2, \ldots, \delta z_n)$ is understood to be the *virtual displacement vector*. In terms of the displacement z, the joint and member displacement vectors, q_j, p_m, and r_m are

$$q_j = A_j z, \quad p_m = B_m z, \quad r_m = C_m z$$

in which A_j has 3 rows and n columns, B_m has one row and n columns, and C_m has 4 rows and n columns. In view of the relationship between displacements and virtual displacements, the virtual joint and member displacement vectors $\delta q_j, \delta p_m$, and δr_m are given by

$$\delta q_j = A_j \delta z, \quad \delta p_m = B_m \delta z, \quad \delta r_m = C_m \delta z$$

where δz is the virtual displacement vector. In all practical applications, there always exists at least one choice for the displacement variables $z_1, z_2, \ldots, z_n$ such that the matrices A_j, B_m, and C_m can be determined from small angle geometry. In most applications, the characteristics of the structure suggest one or more reasonably obvious and suitable choices for the displacement variables. The numerical value for most of the elements in A_j, B_m, and C_m is zero. The non-zero elements are essential in the displacement method of analysis and must be determined whether or not matrix methods are employed. The matrices B_m and C_m characterize the geometry of the structure and are understood to be properties of the structure.

By replacing $\delta q_j, \delta p_m$, and δr_m in the work equation, Equation (a), by their equivalents in terms of the virtual displacement δz, the work equation becomes

$$\delta W = \sum_j \delta z^T A_j^T Q_j + \sum_m \delta z^T B_m^T P_m - \sum_m \delta z^T C_m^T R_m = 0$$

or, since δz^T is a common factor,

$$\delta W = \delta z^T \left(\sum_j A_j^T Q_j + \sum_m B_m^T P_m - \sum_m C_m^T R_m \right) = \delta z^T f = 0 \tag{b}$$

Because A_j^T has n rows and 3 columns and Q_j has 3 rows and 1 column, the product $A_j^T Q_j$ is a column vector of n rows, and so are $B_m^T P_m$ and $C_m^T R_m$, and the sum of the terms in parentheses, i.e., f in Equation (b). The statement that δz is arbitrary means that the elements in δz can be assigned any values. Accordingly, taking $\delta z_j = 0$ for $j \neq k$ leads to the relation $\delta z_k f_k = 0$ for $\delta z_k \neq 0$ from which it follows that $f_k = 0$ for $k = 1, 2, \ldots, n$. Therefore, Equation (b) is satisfied for arbitrary δz if and only if

$$f = \sum_j A_j^T Q_j + \sum_m B_m^T P_m - \sum_m C_m^T R_m = 0 \tag{c}$$

Equations (c) are independent equations of equilibrium corresponding to the displacement variables $z_1, z_2, \ldots, z_n$. The equations are valid subject only to the condition that deformations and displacements are small; the equations are valid without regard to the mechanical properties of the material.

To illustrate the ideas that have been presented, consider the concrete problem of deriving the equations of equilibrium for the frame shown in Figure 10.36. This is the frame of Example 10.9. The members are axially rigid. Due to the axial rigidity, the

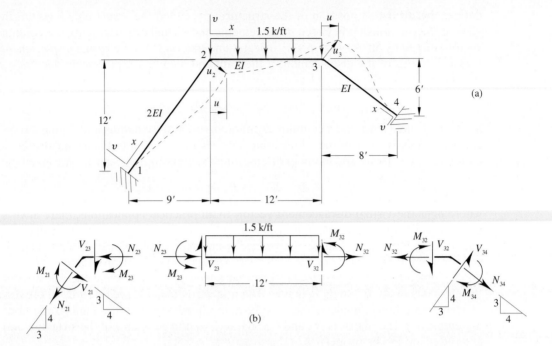

FIGURE 10.36 Matrix formulation, example frame.

frame has three degrees of freedom. As in the solution of Example 10.9, the joint rotations θ_2 and θ_3 and the lateral displacement u are taken as the independent displacement variables.

With $z_1 = \theta_2, z_2 = \theta_3$, and $z_3 = u$, the displacement vector z is

$$z^T = (\theta_2, \theta_3, u)$$

From the displacement boundary conditions, conditions of compatibility, and the geometric analysis of displacements in Sidesway Problem 2, the member end-displacements in terms of θ_2, θ_3, and u were determined to be given by

for member 12 ($m = 1$): $\theta_{12} = 0,$ $\theta_{21} = \theta_2,$ $v_{12} = 0,$ $v_{21} = -5u/4$
for member 23 ($m = 2$): $\theta_{23} = \theta_2,$ $\theta_{32} = \theta_3,$ $v_{23} = -3u/4,$ $v_{32} = 4u/3$
for member 43 ($m = 3$): $\theta_{43} = 0,$ $\theta_{34} = \theta_3,$ $v_{43} = 0,$ $v_{34} = -5u/3$

Each of the foregoing sets of relations can be cast in the form of a matrix equation. Thus, for member 12 ($m = 1$)

$$\begin{Bmatrix} \theta_{12} \\ \theta_{21} \\ v_{12} \\ v_{21} \end{Bmatrix} = \begin{bmatrix} 0 & 0 & 0 \\ 1 & 0 & 0 \\ 0 & 0 & 0 \\ 0 & 0 & -5/4 \end{bmatrix} \begin{Bmatrix} \theta_2 \\ \theta_3 \\ u \end{Bmatrix} \quad \text{or} \quad r_1 = C_1 z \Rightarrow C_1 = \begin{bmatrix} 0 & 0 & 0 \\ 1 & 0 & 0 \\ 0 & 0 & 0 \\ 0 & 0 & -5/4 \end{bmatrix}$$

Section 10.17 Equations of Equilibrium by Virtual Work Matrix Methods 365

for member 23 ($m = 2$)

$$\begin{Bmatrix} \theta_{23} \\ \theta_{32} \\ v_{23} \\ v_{32} \end{Bmatrix} = \begin{bmatrix} 1 & 0 & 0 \\ 0 & 1 & 0 \\ 0 & 0 & -3/4 \\ 0 & 0 & 4/3 \end{bmatrix} \begin{Bmatrix} \theta_2 \\ \theta_3 \\ u \end{Bmatrix} \quad \text{or} \quad r_2 = C_2 z \Rightarrow C_2 = \begin{bmatrix} 1 & 0 & 0 \\ 0 & 1 & 0 \\ 0 & 0 & -3/4 \\ 0 & 0 & 4/3 \end{bmatrix}$$

and for member 43 ($m = 3$)

$$\begin{Bmatrix} \theta_{43} \\ \theta_{34} \\ v_{43} \\ v_{34} \end{Bmatrix} = \begin{bmatrix} 0 & 0 & 0 \\ 0 & 1 & 0 \\ 0 & 0 & 0 \\ 0 & 0 & -5/3 \end{bmatrix} \begin{Bmatrix} \theta_2 \\ \theta_3 \\ u \end{Bmatrix} \quad \text{or} \quad r_3 = C_3 z \Rightarrow C_3 = \begin{bmatrix} 0 & 0 & 0 \\ 0 & 1 & 0 \\ 0 & 0 & 0 \\ 0 & 0 & -5/3 \end{bmatrix}$$

There are no prescribed loads at the joints and there are no prescribed axial forces between the ends of the members. Accordingly, $Q_j = 0$ and $P_m = 0$; therefore, there is no need to evaluate the matrices A_j and B_m. The member end-force vectors are

$$R_1^T = (M_{12}, M_{21}, V_{12}, V_{21}), \quad R_2^T = (M_{23}, M_{32}, V_{23}, V_{32}), \quad R_3^T = (M_{43}, M_{34}, V_{43}, V_{34})$$

Therefore,

$$C_1^T R_1 = \begin{bmatrix} 0 & 1 & 0 & 0 \\ 0 & 0 & 0 & 0 \\ 0 & 0 & 0 & -5/4 \end{bmatrix} \begin{Bmatrix} M_{12} \\ M_{21} \\ V_{12} \\ V_{21} \end{Bmatrix} = \begin{Bmatrix} M_{21} \\ 0 \\ -\frac{5}{4} V_{21} \end{Bmatrix},$$

$$C_2^T R_2 = \begin{bmatrix} 1 & 0 & 0 & 0 \\ 0 & 1 & 0 & 0 \\ 0 & 0 & -3/4 & 4/3 \end{bmatrix} \begin{Bmatrix} M_{23} \\ M_{32} \\ V_{23} \\ V_{32} \end{Bmatrix} = \begin{Bmatrix} M_{23} \\ M_{32} \\ -\frac{3}{4} V_{23} + \frac{4}{3} V_{32} \end{Bmatrix}$$

$$C_3^T R_3 = \begin{bmatrix} 0 & 0 & 0 & 0 \\ 0 & 1 & 0 & 0 \\ 0 & 0 & 0 & -5/3 \end{bmatrix} \begin{Bmatrix} M_{43} \\ M_{34} \\ V_{43} \\ V_{34} \end{Bmatrix} = \begin{Bmatrix} 0 \\ M_{34} \\ -\frac{5}{3} V_{34} \end{Bmatrix}$$

Because $Q_j = 0$ and $P_m = 0$, the equations of equilibrium, Equation (c), becomes

$$-f = C_1^T R_1 + C_2^T R_2 + C_3^T R_3 = \begin{Bmatrix} M_{21} \\ 0 \\ -\frac{5}{4} V_{21} \end{Bmatrix} + \begin{Bmatrix} M_{23} \\ M_{32} \\ -\frac{3}{4} V_{23} + \frac{4}{3} V_{32} \end{Bmatrix} + \begin{Bmatrix} 0 \\ M_{34} \\ -\frac{5}{3} V_{34} \end{Bmatrix} = 0$$

or, in expanded form

$$0 = M_{21} + M_{23}$$
$$0 = M_{32} + M_{34} \tag{i}$$
$$0 = -\frac{5}{4} V_{21} - \frac{3}{4} V_{23} + \frac{4}{3} V_{32} - \frac{5}{3} V_{34}$$

The equations of equilibrium derived in the solution of Example 10.9 are

$$0 = M_{21} + M_{23}$$
$$0 = M_{32} + M_{34} \qquad \text{(ii)}$$
$$0 = -5.25 + \frac{5}{4}V_{21} + \frac{5}{3}V_{34} - \frac{25}{144}(M_{21} + M_{34})$$

That the third equation in Equations (i) and (ii) are equivalent follows from equilibrium of the free-body diagrams shown in Figure 10.36b. For member 23

$$0 = \Sigma M_3 \Rightarrow V_{23} = 9 + \frac{(M_{23} + M_{32})}{12}, \quad 0 = \Sigma M_2 \Rightarrow V_{32} = 9 - \frac{(M_{23} + M_{32})}{12}$$

Moment equilibrium of joints 2 and 3 requires that $M_{21} = -M_{23}$ and $M_{32} = -M_{34}$. With the aid of the foregoing relations it is easy to show that

$$\frac{3}{4}V_{23} - \frac{4}{3}V_{32} = -5.25 - \frac{25}{144}(M_{21} + M_{34})$$

Substituting this result into the last of Equations (i) followed by multiplication by -1 reduces Equations (i) to Equations (ii).

10.18 HOOKE'S LAW

The advantages of the using the method of virtual work to derive the equations of equilibrium becomes much clearer when the members obey Hooke's law and the slope-deflection equations are used to express end-forces in terms of end displacements. The slope-deflection equations, written as a matrix equation, are

$$\begin{Bmatrix} M_{ab} \\ M_{ba} \\ V_{ab} \\ V_{ba} \end{Bmatrix} = \frac{EI}{L} \begin{bmatrix} 4 & 2 & 6/L & -6/L \\ 2 & 4 & 6/L & -6/L \\ 6/L & 6/L & 12/L^2 & -12/L^2 \\ -6/L & -6/L & -12/L^2 & 12/L^2 \end{bmatrix} \begin{Bmatrix} \theta_{ab} \\ \theta_{ba} \\ v_{ab} \\ v_{ba} \end{Bmatrix} + \begin{Bmatrix} M_{Fab} \\ M_{Fba} \\ V_{Fab} \\ V_{Fba} \end{Bmatrix}$$

which can be written compactly as

$$R_m = K_m r_m + R_{Fm} \qquad \text{(d)}$$

where R_m is the member end-force vector and r_m is the member end-displacement vector, both of which have been defined previously. R_{Fm} is the member *fixed end-force* vector

$$R_{Fm}^T = (M_{Fab}, M_{Fba}, V_{Fab}, V_{Fba})_m$$

and the symmetric matrix K_m is the *member stiffness matrix*

$$K_m = K_m^T = \left\{\frac{EI}{L}\right\}_m \begin{bmatrix} 4 & 2 & 6/L & -6/L \\ 2 & 4 & 6/L & -6/L \\ 6/L & 6/L & 12/L^2 & -12/L^2 \\ -6/L & -6/L & -12/L^2 & 12/L^2 \end{bmatrix}_m$$

To get the member end-force vector in terms of the displacement vector z, insert $r_m = C_m z$ into Equation (d). Thus

$$R_m = K_m C_m z + R_{Fm} \qquad (e)$$

Substitution of R_m from Equation (e) into the equilibrium Equation (c) followed by a rearrangement of terms gives

$$\sum_m (C_m^T K_m^T C_m) z = \left(\sum_m C_m^T K_m C_m \right) z = \sum_j A_j^T Q_j + \sum_m B_m^T P_m - \sum_m C_m^T R_{Fm}$$

The foregoing equation of equilibrium can be written in the more compact form

$$Kz = \sum_j A_j^T Q_j + \sum_m B_m^T P_m - \sum_m C_m^T R_{Fm} = L_z \qquad (f)$$

in which the symmetric matrix

$$K = \sum_m C_m^T K_m C_m$$

is the *structure stiffness matrix* and L_z is the *effective applied load vector* corresponding to the displacement z. Equation (f) shows that when the material obeys Hooke's law and the deformations and displacements are small, and the principle of virtual work is used to derive the equations of equilibrium, the coefficient matrix of the unknown displacements must be symmetric.

For the structure shown in Figure 10.36, the member stiffness matrices K_m, the geometric matrices C_m, and the member fixed-end force vectors R_{Fm} are:

$$\text{member 12 } (m = 1) \quad K_1 = \frac{2EI}{15} \begin{bmatrix} 4 & 2 & 2/5 & -2/5 \\ 2 & 4 & 2/5 & -2/5 \\ 2/5 & 2/5 & 4/75 & -4/75 \\ -2/5 & -2/5 & -4/75 & 4/75 \end{bmatrix},$$

$$C_1 = \begin{bmatrix} 0 & 0 & 0 \\ 1 & 0 & 0 \\ 0 & 0 & 0 \\ 0 & 0 & -5/4 \end{bmatrix}, \quad R_{F1} = 0$$

$$\text{member 23 } (m = 2) \quad K_2 = \frac{EI}{12} \begin{bmatrix} 4 & 2 & 1/2 & -1/2 \\ 2 & 4 & 1/2 & -1/2 \\ 1/2 & 1/2 & 1/12 & -1/12 \\ -1/2 & -1/2 & -1/12 & 1/12 \end{bmatrix},$$

$$C_2 = \begin{bmatrix} 1 & 0 & 0 \\ 0 & 1 & 0 \\ 0 & 0 & -3/4 \\ 0 & 0 & 4/3 \end{bmatrix}, \quad R_{F2} = \begin{Bmatrix} 18 \\ -18 \\ 9 \\ 9 \end{Bmatrix}$$

member 43 ($m = 3$) $\quad K_3 = \dfrac{EI}{10}\begin{bmatrix} 4 & 2 & 3/5 & -3/5 \\ 2 & 4 & 3/5 & -3/5 \\ 3/5 & 3/5 & 3/25 & -3/25 \\ -3/5 & -3/5 & -3/25 & 3/25 \end{bmatrix}$,

$$C_3 = \begin{bmatrix} 0 & 0 & 0 \\ 0 & 1 & 0 \\ 0 & 0 & 0 \\ 0 & 0 & -5/3 \end{bmatrix}, \quad R_{F3} = 0$$

The matrices $K_m C_m$ and $C_m^T K_m C_m$, and the vectors $C_m^T R_{Fm}$ are:

member 12 ($m = 1$)

$$K_1 C_1 = \dfrac{2EI}{15}\begin{bmatrix} 2 & 0 & 1/2 \\ 4 & 0 & 1/2 \\ 2/5 & 0 & 1/15 \\ -2/5 & 0 & -1/15 \end{bmatrix}, \quad C_1^T K_1 C_1 = EI\begin{bmatrix} 8/15 & 0 & 1/15 \\ 0 & 0 & 0 \\ 1/15 & 0 & 1/90 \end{bmatrix}, \quad C_1^T R_{F1} = 0$$

member 23 ($m = 2$) $\quad K_2 C_2 = \dfrac{EI}{12}\begin{bmatrix} 4 & 2 & -25/24 \\ 2 & 4 & -25/24 \\ 1/2 & 1/2 & -25/144 \\ -1/2 & -1/2 & 25/144 \end{bmatrix}$,

$$C_2^T K_2 C = EI\begin{bmatrix} 1/3 & 1/6 & -25/288 \\ 1/6 & 1/3 & -25/288 \\ -25/288 & -25/288 & 625/20736 \end{bmatrix}, \quad C_2^T R_{F2} = \begin{Bmatrix} 18 \\ -18 \\ 21/4 \end{Bmatrix}$$

member 43 ($m = 3$)

$$K_3 C_3 = \dfrac{EI}{10}\begin{bmatrix} 0 & 2 & 1 \\ 0 & 4 & 1 \\ 0 & 3/5 & 1/5 \\ 0 & -3/5 & -1/5 \end{bmatrix}, \quad C_3^T K_3 C_3 = EI\begin{bmatrix} 0 & 0 & 0 \\ 0 & 2/5 & 1/10 \\ 0 & 1/10 & 1/30 \end{bmatrix}, \quad C_3^T R_{F3} = 0$$

Therefore, the structure stiffness matrix is

$$K = \Sigma C_m^T K_m C_m = EI\begin{bmatrix} 13/15 & 1/6 & -2.9/144 \\ 1/6 & 11/15 & 1.9/144 \\ -2.9/144 & 1.9/144 & 1546.6/20736 \end{bmatrix}$$

There are no forces applied to the joints and there are no axial loads so $Q_j = 0$ and $P_m = 0$, and the effective load vector is

$$L_z = \sum_j A_j^T Q_j + \sum_m B_m^T P_m - \sum_m C_m^T R_{Fm} = -\begin{Bmatrix} 18 \\ -18 \\ 21/4 \end{Bmatrix}$$

The equations of equilibrium $Kz = L_z$ are

$$EI\begin{bmatrix} 13/15 & 1/6 & -2.9/144 \\ 1/6 & 11/15 & 1.9/144 \\ -2.9/144 & 1.9/144 & 1546.6/20736 \end{bmatrix} \begin{Bmatrix} \theta_2 \\ \theta_3 \\ u \end{Bmatrix} = \begin{Bmatrix} -18 \\ 18 \\ -21/4 \end{Bmatrix}$$

In matrix notation, the solution of the equations is

$$EIz^T = EI(\theta_2, \theta_3, u) = (-29.0, 32.65, -84.0)$$

The member end-forces are found by direct substitution into Equation (f). Thus:

member 12 ($m = 1$)

$$R_1 = K_1 C_1 z + R_{F1} = \frac{2}{15}\begin{bmatrix} 2 & 0 & 1/2 \\ 4 & 0 & 1/2 \\ 2/5 & 0 & 1/15 \\ -2/5 & 0 & -1/15 \end{bmatrix}\begin{Bmatrix} -29.0 \\ 32.65 \\ -84 \end{Bmatrix} = \begin{Bmatrix} -13.33 \\ -21.1 \\ -2.29 \\ 2.29 \end{Bmatrix} = \begin{Bmatrix} M_{12} \\ M_{21} \\ V_{12} \\ V_{21} \end{Bmatrix}$$

member 23 ($m = 2$)

$$R_2 = K_2 C_2 z + R_{F2} = \frac{1}{12}\begin{bmatrix} 4 & 2 & -25/24 \\ 2 & 4 & -25/24 \\ 1/2 & 1/2 & -25/144 \\ -1/2 & -1/2 & 25/144 \end{bmatrix}\begin{Bmatrix} -29.0 \\ 32.65 \\ -84.0 \end{Bmatrix} + \begin{Bmatrix} 18 \\ -18 \\ 9 \\ 9 \end{Bmatrix}$$

$$= \begin{Bmatrix} 21.1 \\ -4.66 \\ 10.37 \\ 7.63 \end{Bmatrix} = \begin{Bmatrix} M_{23} \\ M_{32} \\ V_{23} \\ V_{32} \end{Bmatrix}$$

member 43 ($m = 3$)

$$R_2 = K_3 C_3 z + R_{F3} = \frac{1}{10}\begin{bmatrix} 0 & 2 & 1 \\ 0 & 4 & 1 \\ 0 & 3/5 & 1/5 \\ 0 & -3/5 & -1/5 \end{bmatrix}\begin{Bmatrix} -29.0 \\ 32.65 \\ -84 \end{Bmatrix} = \begin{Bmatrix} -1.87 \\ 4.66 \\ 0.279 \\ -0.279 \end{Bmatrix} = \begin{Bmatrix} M_{43} \\ M_{34} \\ V_{43} \\ V_{34} \end{Bmatrix}$$

10.19 REACTIONS AND VIRTUAL WORK

Forces on a structure are classified as prescribed or reactive. Reactive forces are developed at points where the displacements are prescribed. Let $S_1, S_2, \ldots, S_k$ denote the reactive forces. The objective of analysis is to determine the internal forces, the displacements where the forces are prescribed, and the reaction forces where the displacements are prescribed so that the equilibrium conditions, compatibility conditions, deformation-displacement relations and force-deformation relations are satisfied everywhere. In the displacement method of structural analysis, as exemplified by the slope-deflection method or moment distribution, for all practical purposes the procedures yield all of the required results directly except for the forces of reaction (In the case of inextensional deformation, internal axial forces are treated as reactive forces;

however, these forces are not treated explicitly herein). The reactive forces are defined in terms of the internal forces and, when the internal forces are known, the conditions of equilibrium are available for the calculation of the forces of reaction. In many situations, the principle of virtual work is especially useful in accomplishing this task. The technique is relatively simple, and parallels the procedure used to accomplish the same task in the case of statically determinate structures (in which case the members may be treated as being rigid with no loss of flexibility in application). The mechanical devices that are perceived to produce the prescribed displacements are removed, and are replaced by the forces of reaction $S_1, S_2, \ldots, S_k$ that they generate. In this way, the degrees of freedom of the deformable structure is increased by one for each constraint that is removed. As a consequence of removing the constraints, additional independent displacement variables, one corresponding to each constraint, must be introduced so that the original displacement variables $z_1, z_2, \ldots, z_n$ together with the additional variables still define the configuration of the structure. Let z_1^* be the position variable that corresponds to the removal of the first constraint which develops reaction force S_1, let z_2^* be the position variable that corresponds to the removal of the second constraint which develops reaction force S_2, etc., up through z_k^* which corresponds to the last of the constraints and develops reaction force S_k, and let z^* be the displacement vector defined by $z^{*T} = (z_1^*, z_2^*, \ldots, z_k^*)$. Then the two independent displacement vectors, z as previously defined and z^*, define all possible joint displacements and joint rotations for the structure, and they also define all possible virtual joint displacements and rotations of the joints. Included in the many ways in which the z_i^* may be chosen is that for which z_i^* is displacement measured in the direction of reaction force S_i. In the discussion that follows the z_i^* are assumed to be chosen in this way. The displacement vector z^* defines joint displacement vectors q_j^*, member axial displacement vectors p_m^*, and member end-displacement vectors r_m^*. For the intended application, the relationship of the virtual displacement vectors $\delta q_j^*, \delta p_m^*$, and δr_m^* to the virtual displacement δz^* is needed. The relationships between the vectors are linear and are determined by an analysis involving only small angle geometry. The relationships are expressed in the form

$$\delta q_j^* = A_j^* \, \delta z^*, \quad \delta p_m^* = B_m^* \, \delta z^*, \quad \delta r_m^* = C_m^* \, \delta z^*, \tag{g}$$

Let S be the column vector whose elements are the reactive forces $S_1, S_2, \ldots, S_k$, i.e., let S be such that $S^T = (S_1, S_2, \ldots, S_k)$. S is called the *reaction force vector*. The virtual work on arbitrary virtual displacements defined by δz and δz^* is the sum of the virtual works due to each of the virtual displacements. By construction, δz cannot induce displacements such that virtual work is done by the reaction forces. Also by construction of δz^* and the reaction force vector, the virtual work of the reactions forces is given by $\delta z^{*T} S$. Therefore, the total virtual work is given by

$$\delta W = \delta z^T \left(\sum_j A_j^T Q_j + \sum_m B_m^T P_m - \sum_m C_m^T R_m \right)$$
$$+ \delta z^{*T} \left(S + \sum_j A_j^{*T} Q_j + \sum_m B_m^{*T} P_m - \sum_m C_m^{*T} R_m \right)$$

which, by the principle of virtual work, must be zero for arbitrary δz and arbitrary δz^*. It has been assumed that the displacement analysis has been carried out. Because the equations to be solved in the displacement analysis are summed up in the relation (see Equation (c))

$$\sum_j A_j^T Q_j + \sum_m B_m^T P_m - \sum_m C_m^T R_m = 0$$

it follows that the first term in the foregoing work expression is automatically zero for all δz. Thus, in application of the principle of virtual work to the calculation of reactions there is nothing to be gained by assigning the virtual displacement δz any value other than zero. The work equation for the determination of the reactions by the principle of virtual work is now reduced to

$$\delta W = \delta z^{*T}\left(S + \sum_j A_j^{*T} Q_j + \sum_m B_m^{*T} P_m - \sum_m C_m^{*T} R_m\right) \tag{h}$$

The only unknown term in the parentheses is the reaction force vector S. To satisfy the requirement that the virtual work be zero for arbitrary δz^*, S must be such that the sum of the terms in braces is zero. Therefore, the reaction force vector, S, is

$$S = -\sum_j A_j^{*T} Q_j - \sum_m B_m^{*T} P_m + \sum_m C_m^{*T} R_m \tag{i}$$

The foregoing formulation was set out under the assumption that all of the forces of reaction are to be determined. However, the formulation is easily adapted to the calculation of any combination of the reaction forces. This is accomplished by replacing the virtual displacement vector δz^* in the formulation by a reduced virtual displacement vector $\delta z_{(r)}^*$ that contains only those virtual displacements that correspond to the desired reactions. For example, assume that in all there are 10 reaction forces but only S_3 and S_7 are required. Then $\delta z_{(r)}^{*T} = (\delta z_3^*, \delta z_7^*)$, and the matrices A_j^*, B_m^*, and B_m^* will contain only the two columns corresponding to δz_3^*, and δz_7^* and Equation (i) will yield the reduced reaction force vector $S_{(r)}$ whose elements are S_3 and S_7.

To illustrate the ideas and to develop confidence in the procedures, consider the frame shown in Figure 10.37a. The members are axially rigid. The problem is to find the force F_0 to prevent lateral displacement of the frame and to determine the horizontal components of reaction H_1 and H_4. Because of the built-in supports at 1 and 4, the independent prescribed displacements referenced to the global coordinates are

$$u_{1X} = u_{1Y} = \theta_1 = 0, \quad u_{4X} = u_{4Y} = \theta_4 = 0, \quad u_{3X} = 0$$

An analysis using either the slope-deflection equations or moment distribution yields the member end-force vectors

$$R_1 = \begin{Bmatrix} M_{12} \\ M_{21} \\ V_{12} \\ V_{21} \end{Bmatrix} = \begin{Bmatrix} -7.11 \\ -14.22 \\ -1.42 \\ 1.42 \end{Bmatrix}, \quad R_2 = \begin{Bmatrix} M_{23} \\ M_{32} \\ V_{23} \\ V_{32} \end{Bmatrix} = \begin{Bmatrix} 14.22 \\ -12.24 \\ 9.17 \\ 8.84 \end{Bmatrix}$$

$$R_3 = \begin{Bmatrix} M_{43} \\ M_{34} \\ V_{43} \\ V_{34} \end{Bmatrix} = \begin{Bmatrix} 6.12 \\ 12.24 \\ 1.84 \\ -1.84 \end{Bmatrix}$$

where the end-moment have units of k-ft and the end-shears have units of kips.

372 Chapter 10 Displacement Methods of Analysis

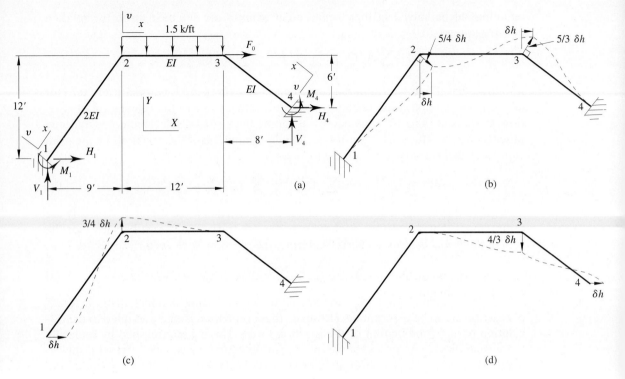

FIGURE 10.37 Reactions by virtual work.

The virtual displacements for the calculation of F_0 are shown in Figure 10.37b. Because there are no prescribed forces at the joints and there are no prescribed axial loads, only the member virtual end-displacement vectors need be determined. From the geometry shown in Figure 10.37b, these vectors are

$$\delta r_1^* = \delta h\, C_1^* = \begin{Bmatrix} \delta\theta_{12} \\ \delta\theta_{21} \\ \delta v_{12} \\ \delta v_{21} \end{Bmatrix} = \delta h \begin{Bmatrix} 0 \\ 0 \\ 0 \\ -5/4 \end{Bmatrix}, \quad \delta r_2^* = \delta h\, C_2^* = \begin{Bmatrix} \delta\theta_{23} \\ \delta\theta_{32} \\ \delta v_{23} \\ \delta v_{32} \end{Bmatrix} = \delta h \begin{Bmatrix} 0 \\ 0 \\ -3/4 \\ 4/3 \end{Bmatrix},$$

$$\delta r_3^* = \delta h\, C_3^* = \begin{Bmatrix} \delta\theta_{43} \\ \delta\theta_{34} \\ \delta v_{43} \\ \delta v_{34} \end{Bmatrix} = \delta h \begin{Bmatrix} 0 \\ 0 \\ 0 \\ -5/3 \end{Bmatrix}$$

By Equation (i)

$$F_0 = \sum_m C_m^{*T} R_m = C_1^{*T} R_1 + C_2^{*T} R_2 + C_1^{*T} R_3 = -1.78 + 4.91 + 3.07 = 6.20\ \text{k}$$

which, to within a round-off error, agrees with the result obtained previously by the conventional vector method using free-body diagrams.

The virtual displacements for the calculation of H_1 are shown in Figure 10.37c. From the geometry of small angles, the member virtual end-displacement vectors are

$$\delta r_1^* = \delta h\, C_1^* = \begin{Bmatrix} \delta\theta_{12} \\ \delta\theta_{21} \\ \delta v_{12} \\ \delta v_{21} \end{Bmatrix} = \delta h \begin{Bmatrix} 0 \\ 0 \\ -4/5 \\ 9/20 \end{Bmatrix}, \quad \delta r_2^* = \delta h\, C_2^* = \begin{Bmatrix} \delta\theta_{23} \\ \delta\theta_{32} \\ \delta v_{23} \\ \delta v_{32} \end{Bmatrix} = \delta h \begin{Bmatrix} 0 \\ 0 \\ 3/4 \\ 0 \end{Bmatrix},$$

$$\delta r_3^* = \delta h\, C_3^* = \begin{Bmatrix} \delta\theta_{43} \\ \delta\theta_{34} \\ \delta v_{43} \\ \delta v_{34} \end{Bmatrix} = \delta h \begin{Bmatrix} 0 \\ 0 \\ 0 \\ 0 \end{Bmatrix}$$

By Equation (i)

$$H_1 = \sum_m C_m^{*T} R_m = C_1^{*T} R_1 + C_2^{*T} R_2 + C_3^{*T} R_3 = -1.78 + 6.88 + 0 = 8.66\ \text{k}$$

The virtual displacements for the calculation of H_4 are shown in Figure 10.37d. Again, by small angle geometry

$$\delta r_1^* = \delta h\, C_1^* = \begin{Bmatrix} \delta\theta_{12} \\ \delta\theta_{21} \\ \delta v_{12} \\ \delta v_{21} \end{Bmatrix} = \delta h \begin{Bmatrix} 0 \\ 0 \\ 0 \\ 0 \end{Bmatrix}, \quad \delta r_2^* = \delta h\, C_2^* = \begin{Bmatrix} \delta\theta_{23} \\ \delta\theta_{32} \\ \delta v_{23} \\ \delta v_{32} \end{Bmatrix} = \delta h \begin{Bmatrix} 0 \\ 0 \\ 0 \\ -4/3 \end{Bmatrix},$$

$$\delta r_3^* = \delta h\, C_3^* = \begin{Bmatrix} \delta\theta_{43} \\ \delta\theta_{34} \\ \delta v_{43} \\ \delta v_{34} \end{Bmatrix} = \delta h \begin{Bmatrix} 0 \\ 0 \\ -3/5 \\ 16/15 \end{Bmatrix}$$

Again, by Equation (i)

$$H_4 = \sum_m C_m^{*T} R_m = C_1^{*T} R_1 + C_2^{*T} R_2 + C_3^{*T} R_3 = 0 - 11.77 - 3.0 = -14.84\ \text{k}$$

Force equilibrium in the horizontal direction requires that $F_0 + H_1 + H_4 = 0$. The calculated values satisfy the requirement to within an error due to round-off.

10.20 STRAIN ENERGY OF BENDING

When axial deformation is negligible, the total strain energy of a framed structure is just the sum of the strain energies of bending of its members. Of particular interest in the displacement method of analysis of a frame is the strain energy of bending of a beam under a prescribed load between its ends for all possible end-displacements as shown in Figure 10.38.

For a beam that obeys Hooke's law the strain energy of bending is given by

$$U = \int_0^L \frac{EI v''^2}{2}\, dx$$

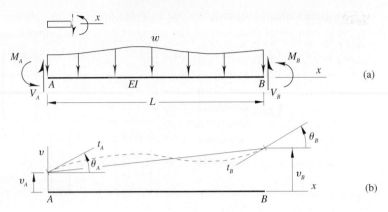

FIGURE 10.38 Beam.

For a uniform beam, which is considered herein, an expression for v'' can be formulated directly in terms of loads and end-displacements. Such a formulation is, however, clumsy. A more convenient approach is to formulate the expression in terms of the load and the end-moments M_A and M_B, evaluate the strain energy in terms of the load and M_A and M_B, and then use the slope-deflection equations to eliminate the end-moments to obtain the strain energy in the desired form. From statics the bending moment at any section is

$$M(x) = -M_A\left(1 - \frac{x}{L}\right) + M_B\left(\frac{x}{L}\right) + M_0(x) = EIv''$$

where $M_0(x)$ is the bending moment due to the load between the ends of the beam (the beam being treated as simply supported). Substitution of the expression for $M(x)$ into the energy integral followed by integration gives

$$U = \frac{1}{2EI}\left(\frac{L}{3}M_A^2 - \frac{L}{3}M_AM_B + \frac{L}{3}M_B^2\right) - M_A\int_0^L\left(1 - \frac{x}{L}\right)\frac{M_0(x)}{EI}dx$$
$$+ M_B\int_0^L\left(\frac{x}{L}\right)\frac{M_0(x)}{EI}dx + \int_0^L\frac{[M_0(x)]^2}{2EI}dx$$

By the moment-area theorems or by virtual work

$$\int_0^L\left(1 - \frac{x}{L}\right)\frac{M_0(x)}{EI}dx = \theta_{A0}, \quad \int_0^L\left(\frac{x}{L}\right)\frac{M_0(x)}{EI}dx = \theta_{B0}$$

where θ_{A0} and θ_{B0} are the rotations of the tangents at A and B due to the load between the ends of the beam, measured positive in the counterclockwise direction. Only changes in energy play a role in structural analysis so the integral involving only $M_0(x)$ is an irrelevant constant in applications. Therefore, the strain energy can be taken to be given by

$$U = \frac{L}{6EI}(M_A^2 - M_AM_B + M_B^2) + M_A\theta_{A0} + M_B\theta_{B0}$$

From the derivation of the slope-deflection equations

$$M_A = \frac{EI}{L}\left(4\theta_A + 2\theta_B + \frac{6}{L}v_A - \frac{6}{L}v_B\right) + M_{FA},$$

$$M_B = \frac{EI}{L}\left(2\theta_A + 4\theta_B + \frac{6}{L}v_A - \frac{6}{L}v_B\right) + M_{FB}$$

$$\theta_{A0} = \frac{L}{6EI}(-2M_{FA} + M_{FB}), \qquad \theta_{B0} = \frac{L}{6EI}(M_{FA} - 2M_{FB})$$

Substitution of the foregoing expressions into the energy equation yields the desired result. Matrix methods are used to simplify the derivation. Define matrices as follows:

$$m^T = (M_A, M_B), \qquad m_F^T = (M_{FA}, M_{FB}), \qquad \theta_0^T = (\theta_{A0}, \theta_{B0})$$
$$R^T = (M_A, M_B, V_A, V_B), \qquad R_F^T = (M_{FA}, M_{FB}, V_{FA}, V_{FB}), \qquad r^T = (\theta_A, \theta_B, v_A, v_B)$$

$$A = \begin{bmatrix} 1 & -1/2 \\ -1/2 & 1 \end{bmatrix}, \qquad C = \frac{EI}{L}\begin{bmatrix} 4 & 2 & 6/L & -6/L \\ 2 & 4 & 6/L & -6/L \end{bmatrix},$$

$$K = \frac{EI}{L}\begin{bmatrix} 4 & 2 & 6/L & -6/L \\ 2 & 4 & 6/L & -6/L \\ 6/L & 6/L & 12/L^2 & -12/L^2 \\ -6/L & -6/L & -12/L^2 & 12/L^2 \end{bmatrix}$$

Then

$$m = Cr + m_F, \qquad \theta_0 = -\frac{L}{3EI}Am_F, \qquad R = Kr + R_F$$

and the formulation for the strain energy becomes

$$U = \frac{L}{6EI}m^T Am + \theta_0^T m$$

Now,

$$\frac{L}{6EI}m^T Am = \frac{L}{6EI}(r^T C^T + m_F^T)(ACr + Am_F)$$

$$= \frac{L}{6EI}r^T C^T ACr + \frac{L}{3EI}m_F^T ACr + \frac{L}{6EI}m_F^T Am_F$$

and

$$\theta_0^T m = -\frac{L}{3EI}m_F^T A^T(Cr + m_F)$$

With the aid of the last two results and the symmetry of matrix A, the expression for the strain energy reduces to

$$U = \frac{1}{2}r^T\left(\frac{L}{3EI}C^T AC\right)r - \frac{L}{6EI}m_F^T Am_F$$

The second term in the above expression is an irrelevant constant and the term in parentheses is the stiffness matrix $K = K^T$. Thus, to within an irrelevant constant the strain energy of the beam is given by

$$U = \frac{1}{2} r^T K r$$

and the total strain energy of the structure is

$$U = \frac{1}{2} \sum_m r_m^T K_m r_m$$

10.21 POTENTIAL ENERGY OF PRESCRIBED LOADS

The prescribed loads on a representative portion of a structure are shown in Figure 10.39. Concentrated forces and/or couples that may act between the ends of a member do not alter the results of the derivation. Accordingly, such loads have been omitted to achieve compactness of the development.

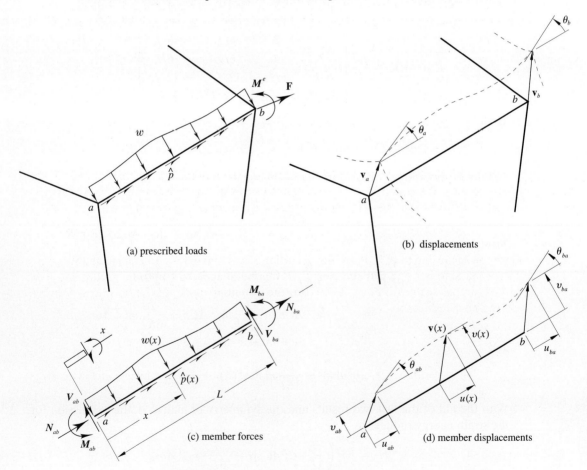

FIGURE 10.39 Potential energy of applied loads, notation.

Section 10.21 Potential Energy of Prescribed Loads

The prescribed loads are constant in magnitude and direction. Accordingly, the potential energy of the loads is

$$\Omega = -\sum_j (\mathbf{F}\cdot\mathbf{v} + M^e\theta) + \sum_m \int_0^L wv\,dx - \sum_m \int_0^L \hat{p}u\,dx$$

in which the index j (m) indicates summation over all joints (members). The integrals can be cast in more useful forms. Inextensional deformation is assumed. Therefore, the axial displacement u is constant over the length of a member. Let p be the constant value of $u(x)$. Then

$$\int_0^L \hat{p}u\,dx = p\int_0^L \hat{p}\,dx = pP$$

where P is simply the resultant of the axial forces acting between the ends of the member.

The integral involving the transverse load w and the transverse displacement v is simplified by making use of the differential equations of equilibrium, the displacement-deformation equations, and the boundary conditions. Two integrations by parts gives

$$\int_0^L wv\,dx = M_{ab}\theta_{ab} + M_{ba}\theta_{ba} + V_{ab}v_{ab} + V_{ba}v_{ba} - \int_0^L Mv''\,dx$$

The material obeys Hooke's law so $M = EIv''$ and

$$\int_0^L Mv''\,dx = \int_0^L EIv''^2\,dx = 2U$$

where U is the strain energy of deformation of the beam. The potential energy of the transverse loads is cast in a more compact form with the aid of definitions and results from the discussion of strain energy wherein it was, in effect, shown that

$$M_{ab}\theta_{ab} + M_{ba}\theta_{ba} + V_{ab}v_{ab} + V_{ba}v_{ba} = r^T R = r^T(Kr + R_F) = 2U + r^T R_F$$

where, according to definitions given earlier, r is the member end-displacement vector, R is the member end-force vector, R_F is the member fixed end-force vector, and K is the member stiffness matrix. Therefore, the potential energy of the transverse load on a member reduces to

$$\int_0^L wv\,dx = r^T R_F$$

The potential energy of the prescribed loads applied to a joint expressed in terms of matrices is

$$-\sum_j (\mathbf{F}\cdot\mathbf{v} + M^e\theta) = -q^T Q$$

in which

$$q^T = (u_X, u_Y, \theta), \quad Q^T = (F_X, F_Y, M^e)$$

where u_X and u_Y, and F_X and F_Y are rectangular components of displacement and force, respectively, referred to global Cartesian coordinates.

With the aid of the foregoing results, the formulation for the total potential energy of the prescribed loads is transformed to

$$\Omega = -\sum_j q_j^T Q_j - \sum_m p_m^T P_m + \sum_m r_m^T R_{Fm}$$

where

$$q_j^T = (u_X, u_Y, \theta)_j, \qquad Q_j^T = (F_X, F_Y, M^e)_j$$
$$p_m^T = (p_m)_m, \qquad P_m^T = (P_m)_m$$
$$r_m^T = (\theta_{ab}, \theta_{ba}, v_{ab}, v_{ba})_m, \qquad R_{Fm}^T = (M_{Fab}, M_{Fba}, V_{Fab}, V_{Fba})_m$$

10.22 PRINCIPLE OF STATIONARY POTENTIAL ENERGY: EQUATIONS OF EQUILIBRIUM

The total potential energy, V, of a structure is the sum of the potential energy of the prescribed forces and the strain energy. Thus,

$$V = \Omega + U = -\sum_j q_j^T Q_j - \sum_m p_m^T P_m + \sum_m r_m^T R_{Fm} + \frac{1}{2} \sum_m r_m^T K_m r_m = V(q_j, p_m, r_m)$$

The elements in the member joint-displacement vectors, q_j, the axial displacements, p_m, and the elements in the member end-displacement vectors, r_m, are not independent quantities. They must satisfy the equations that reflect the condition of inextensional deformation and the compatibility of displacements at the joints. For a structure that undergoes small deformations and displacements, the elements in q_j, p_m, and r_m are linear functions of n independent variables $z_1, z_2, \ldots, z_n$ that define the displaced configuration of the structure. The displacement variables can always be chosen so that small angle geometry can be employed to express the elements in q_j, p_m, and r_m, in terms of a structure displacement vector z whose elements are $z_1, z_2, \ldots, z_n$. Thus, in terms of the displacement vector z

$$q_j = A_j z, \quad p_m = B_m z, \quad r_m = C_m z$$

where A_j has 3 rows and n columns, B_m has one row and n columns, and C_m has 4 rows and n columns. Substitution of the foregoing relationships into the potential energy function yields

$$V(z) = z^T \left(-\sum_j A_j^T Q_j - \sum_m B_m^T P_m + \sum_m C_m^T R_{Fm} \right) + \frac{1}{2} z^T \left(\sum_m C_m^T K_m C_m \right) z$$
$$\equiv -z^T L_z + \frac{1}{2} z^T K z$$

where, by previous definition, L_z is the effective applied load vector and K is the structure stiffness matrix.

The principle of stationary potential energy asserts that if z defines an equilibrium position, then the first order change in V due to an arbitrary virtual displacement δz from that position must be zero. The potential energy of the structure in the position defined by displacement $z + \delta z$ is

$$V(z + \delta z) = -(z^T + \delta z^T)L_z + \frac{1}{2}(z^T + \delta z^T)K(z + \delta z)$$

$$= -z^T L_z + \frac{1}{2} z^T K z + \delta z^T(-L_z + K z) + \frac{1}{2}\delta z^T K \,\delta z$$

$$= V(z) + \delta V + \frac{1}{2}\delta^2 V$$

and the first order change in V due to the virtual displacement δz is the term

$$\delta V = \delta z^T(-L_z + K z)$$

The condition that $\delta V = 0$ for arbitrary δz can be satisfied only if $-L_z + K z = 0$ or, only if

$$-\sum_j A_j^T Q_j - \sum_m B_m^T P_m + \sum_m C_m^T R_{Fm} + K z = 0 \tag{j}$$

Equations (j) are the equilibrium conditions expressed in terms of displacements as derived from the principle of stationary potential energy; they are exactly the same as Equations (f), which are the equilibrium conditions derived from the principle of virtual work. That the two equations should be the same is an expected result.

10.23 MATRIX FORMULATIONS—SUMMARY

Definitions of the basic quantities that enter in the matrix formulation of the governing equations in the displacement analysis of a framed structure for which the members are treated as being axially rigid and the material obeys Hooke's law are summarized below.

X, Y	global rectangular coordinates
x, y	local (member) coordinates; origin of these coordinates is at the a-end of the member
u_X, u_Y	components of displacement referenced to the global coordinates, positive in the senses of the global coordinate axes
u, v	components of displacement referenced to local coordinates, positive in the senses of the local coordinates axes
θ	rotation, positive in the counterclockwise sense
z	vector of independent displacement variables, $z^T = (z_1, z_2, \ldots, z_n)$
q_j	vector of joint displacement components, $q_j^T = (u_X, u_Y, \theta)_j$; by geometry the relationship between q_j and z is determined to be given in the form $q_j = A_j z$, where A_j is a $3 \times n$ matrix of constants
p_m	common value of the axial component of displacement along the length of axially rigid member m; p_m is a scalar which can be thought of as a column or a row vector with one element; by geometry the relationship between p_m and z is determined to be given in the form $p_m = B_m z$, where B_m is a $1 \times n$ matrix of constants
r_m	member end-displacement vector, $r_m^T = (\theta_{ab}, \theta_{ba}, v_{ab}, v_{ba})_m$; by geometry the relationship between r_m and z is determined to be given in the form $r_m = C_m z$, where C_m is a $4 \times n$ matrix of constants

Q_j load vector for joint j; $Q_j^T = (F_X, F_Y, M^e)_j$; F_X and F_Y are positive in the senses of the global coordinate axes and M^e is positive in the counterclockwise sense

P_m axial component of the resultant of the loads between the ends of member m; P_m is a scalar which can be thought of as a column vector with one element

R_m end-force vector for member m, $R_m^T = (M_{ab}, M_{ba}, V_{ab}, V_{ba})_m$; M_{ab} and M_{ba} are end-moments, positive in the counterclockwise sense; V_{ab} and V_{ba} are end-shears, positive in the sense of the y-axis of the local coordinates

R_{Fm} fixed end-force vector for member m, $R_{Fm}^T = (M_{Fab}, M_{Fba}, V_{Fab}, V_{Fba})_m$; M_{Fab} and M_{Fba} are fixed end-moments and V_{Fab} and V_{Fba} are fixed end-shears

K_m member flexural stiffness matrix for member m

$$K_m = K_m^T = \left(\frac{EI}{L}\right)_m \begin{bmatrix} 4 & 2 & 6/L & -6/L \\ 2 & 4 & 6/L & -6/L \\ 6/L & 6/L & 12/L^2 & -12/L^2 \\ -6/L & -6/L & -12/L^2 & 12/L^2 \end{bmatrix}_m$$

The member end-force vector for member m is given by

$$R_m = K_m C_m z + R_{Fm}$$

and the vector z of the independent position variables is the solution of the equations

$$\left(\sum_m C_m^T K_m C_m\right) z = Kz = \sum_j A_j^T Q_j + \sum_m B_m^T P_m - \sum_m C_m^T R_{Fm} = L_z$$

in which the symmetric matrix K is the structure stiffness matrix and L_z is the effective applied load vector corresponding to the displacement variables $z_1, z_2, \ldots, z_n$.

10.24 APPLICATION PROCEDURE

Basic application of the matrix method focusses on the determination of the displacement vector z and the member end-force vectors R_m of plane-framed structures. With z and R_m known, all other quantities of interest may be obtained by application of statics and element of mechanics of materials. The following eight-step procedure is used in solving the example problems.

1. Define global coordinates for the problem.
2. Define local coordinates for each member.
3. Determine the degrees of freedom and select independent displacement variables $z_1, z_2, \ldots, z_n$.
4. Use geometry to construct the matrices A_j, B_m, and C_m.
5. Construct the load vectors Q_j, P_m, and R_{Fm}.
6. Construct the member stiffness matrices K_m.
7. Solve the equations

$$\left(\sum_m C_m^T K_m C_m\right) z = Kz = \sum_j A_j^T Q_j + \sum_m B_m^T P_m - \sum_m C_m^T R_{Fm} = L_z$$

for the displacement vector z.

8. Evaluate the member end-force vectors by substitution into

$$R_m = K_m C_m z + R_{Fm}$$

Example 10.17

Use the matrix method of analysis to determine the member end-forces for the frame shown in Figure 10.40.

The global and local coordinates for the problem are chosen as shown in Figure 10.40. Because the members are axially rigid, the structure has three degrees of freedom, two with respect to joint rotation and one with respect to sidesway. (Joint and chord line displacements consistent with the condition of inextensional deformation are shown with dashed lines in Figure 10.40). The independent displacement variables are taken to be the joint rotations θ_2 and θ_3 and the horizontal displacement, u, of joint 3 as shown in Figure 10.40. With this choice $z^T = (\theta_2, \theta_3, u)$. From the axial rigidity of member 23 ($m = 2$) it follows that the horizontal component of displacement of joint 2 is u. From the axial rigidity of member 12 ($m = 1$) it follows that the displacement of joint 2 must be perpendicular to the member. Accordingly, the displacement of joint 2 must be as shown in Figure 10.40. Joint 1 is fixed so $A_1 = 0$. For joints 2 and 3,

$$\begin{Bmatrix} u_X \\ u_Y \\ \theta \end{Bmatrix}_2 = \begin{bmatrix} 0 & 0 & 1 \\ 0 & 0 & -3/4 \\ 1 & 0 & 0 \end{bmatrix} \begin{Bmatrix} \theta_2 \\ \theta_3 \\ u \end{Bmatrix}, \quad \begin{Bmatrix} u_X \\ u_Y \\ \theta \end{Bmatrix}_3 = \begin{bmatrix} 0 & 0 & 1 \\ 0 & 0 & 0 \\ 0 & 1 & 0 \end{bmatrix} \begin{Bmatrix} \theta_2 \\ \theta_3 \\ u \end{Bmatrix}$$

from which it follows that the matrices A_2 and A_3 are

$$A_2 = \begin{bmatrix} 0 & 0 & 1 \\ 0 & 0 & -3/4 \\ 1 & 0 & 0 \end{bmatrix}, \quad A_3 = \begin{bmatrix} 0 & 0 & 1 \\ 0 & 0 & 0 \\ 0 & 1 & 0 \end{bmatrix}$$

The axial component of displacement, p_1, is zero for member 12 ($m = 1$) and the axial component of displacement, p_2, for member 23 ($m = 2$) is equal to u. Thus, the (transposes of the)

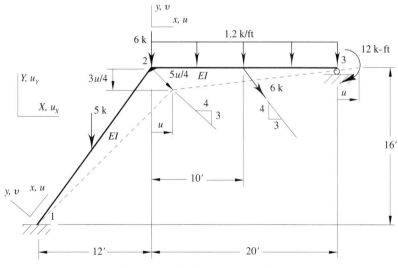

FIGURE 10.40 Frame example 1.

matrices B_1 and B_2 are

$$B_1 = 0, \quad B_2 = (0, 0, 1)$$

The member end-displacement vectors are

$$\begin{Bmatrix} \theta_{12} \\ \theta_{21} \\ v_{12} \\ v_{21} \end{Bmatrix} = \begin{bmatrix} 0 & 0 & 0 \\ 1 & 0 & 0 \\ 0 & 0 & 0 \\ 0 & 0 & -5/4 \end{bmatrix} \begin{Bmatrix} \theta_2 \\ \theta_3 \\ u \end{Bmatrix}, \quad \begin{Bmatrix} \theta_{23} \\ \theta_{32} \\ v_{23} \\ v_{32} \end{Bmatrix} = \begin{bmatrix} 1 & 0 & 0 \\ 0 & 1 & 0 \\ 0 & 0 & -3/4 \\ 0 & 0 & 0 \end{bmatrix} \begin{Bmatrix} \theta_2 \\ \theta_3 \\ u \end{Bmatrix}$$

Accordingly, the matrices C_1 and C_2 are

$$C_1 = \begin{bmatrix} 0 & 0 & 0 \\ 1 & 0 & 0 \\ 0 & 0 & 0 \\ 0 & 0 & -5/4 \end{bmatrix}, \quad C_2 = \begin{bmatrix} 1 & 0 & 0 \\ 0 & 1 & 0 \\ 0 & 0 & -3/4 \\ 0 & 0 & 0 \end{bmatrix}$$

By inspection, the (transposes of the) joint load vectors are

$$Q_2^T = (0, -6, 0), \quad Q_3^T = (0, 0, -12)$$

The axial component of the resultant force between the ends of member 12 and member 23 are $P_1 = -4$ and $P_2 = 3.6$, respectively, and the (transposes of the) member fixed-end force vectors are

$$R_{F1}^T = (7.5, -7.5, 1.5, 1.5), \quad R_{F2}^T = (52, -52, 14.4, 14.4)$$

The member stiffness matrices are

$$K_1 = K_2 = \frac{EI}{20} \begin{bmatrix} 4 & 2 & 0.3 & -0.3 \\ 2 & 4 & 0.3 & -0.3 \\ 0.3 & 0.3 & 0.3 & -0.03 \\ -0.3 & -0.3 & -0.03 & 0.03 \end{bmatrix}$$

Substitution of the foregoing data into the defining relations yields

$$A_2^T Q_2 = \begin{Bmatrix} 0 \\ 0 \\ 4.5 \end{Bmatrix}, \quad A_3^T Q_3 = \begin{Bmatrix} 0 \\ -12 \\ 0 \end{Bmatrix}, \quad B_1^T P_1 = \begin{Bmatrix} 0 \\ 0 \\ 0 \end{Bmatrix}, \quad B_2^T P_2 = \begin{Bmatrix} 0 \\ 0 \\ 3.6 \end{Bmatrix},$$

$$C_1^T R_{F1} = \begin{Bmatrix} -7.5 \\ 0 \\ -1.875 \end{Bmatrix}, \quad C_2^T R_{F2} = \begin{Bmatrix} 52 \\ -52 \\ -10.8 \end{Bmatrix},$$

and

$$L_z = \sum_j A_j^T Q_j + \sum_m B_m^T P_m - \sum_m C_m^T R_{Fm} = \begin{Bmatrix} -44.5 \\ 40 \\ 20.775 \end{Bmatrix}$$

For the member stiffness matrices,

$$K_1 C_1 = \frac{EI}{20000} \begin{bmatrix} 2000 & 0 & 375 \\ 4000 & 0 & 375 \\ 300 & 0 & 37.5 \\ -300 & 0 & -37.5 \end{bmatrix}, \quad C_1^T K_1 C_1 = \frac{EI}{20000} \begin{bmatrix} 4000 & 0 & 375 \\ 0 & 0 & 0 \\ 375 & 0 & 46.875 \end{bmatrix}$$

$$K_2 C_2 = \frac{EI}{20000} \begin{bmatrix} 4000 & 2000 & -225 \\ 2000 & 4000 & -225 \\ 300 & 300 & -22.5 \\ -300 & -300 & 22.5 \end{bmatrix}, \quad C_2^T K_2 C_2 = \frac{EI}{20000} \begin{bmatrix} 4000 & 2000 & -225 \\ 2000 & 4000 & -225 \\ -225 & -225 & 16.875 \end{bmatrix}$$

and the structure stiffness matrix is

$$K = \sum_m C_m^T K_m C_m = \frac{EI}{20000} \begin{bmatrix} 8000 & 2000 & 150 \\ 2000 & 4000 & -225 \\ 150 & -225 & 63.75 \end{bmatrix}$$

The solution values of the equations $Kz = L_z$ are

$$z_1 = \theta_2 = -646.9 \text{ k-ft}^2/EI, \quad z_2 = \theta_3 = 1217.4 \text{ k-ft}^2/EI, \quad z_3 = u = 12336 \text{ k-ft}^3/EI$$

and the end-force vectors are

$$R_1 = K_1 C_1 z + R_{F1} = \begin{Bmatrix} M_{12} \\ M_{21} \\ V_{12} \\ V_{21} \end{Bmatrix} = \begin{Bmatrix} 174.1 \text{ k-ft} \\ 94.4 \text{ k-ft} \\ 14.93 \text{ k} \\ -11.93 \text{ k} \end{Bmatrix},$$

$$R_2 = K_2 C_2 z + R_{F2} = \begin{Bmatrix} M_{23} \\ M_{32} \\ V_{23} \\ V_{32} \end{Bmatrix} = \begin{Bmatrix} -94.4 \text{ k-ft} \\ -12 \text{ k-ft} \\ 9.08 \text{ k} \\ 19.72 \text{ k} \end{Bmatrix}$$

Example 10.18

The members of the rigid frame shown in Figure 10.41 are axially rigid. Use the matrix method of analysis to determine the member end-forces. Take global and local coordinates as shown in Figure 10.41. The members are identified by number in Figure 10.41b.

The frame has five degrees of freedom, three with respect to joint rotation and two with respect to sidesway. (Joint and chord line displacements consistent with the condition of inextensional deformation are shown in Figure 10.41b). Selected for the independent displacement variables are the joint rotations $\theta_2, \theta_3, \theta_4$, and the joint translations u_2 and u_4. With this choice, $z^T = (\theta_2, \theta_3, \theta_4, u_2, u_4)$. No loads applied to the joints; accordingly, for the joint load vectors, $Q_j = 0, j = 2, 3, 4$. Only member 2 carries a load between its ends and, therefore, the fixed end-force vectors R_{Fm} and the axial loads P_m are zero for members 1, 3, and 4, i.e., $R_{Fm} = 0$ and $P_m = 0$ for $m = 1, 3, 4$.

Member end-displacements in the directions of the local coordinates in terms of u_2 and u_4 are needed for the construction of B_2 and for the construction of the matrices C_m. The relationships could be obtained directly from the geometry shown in Figure 10.41b; however, a better procedure is to determine the displacements due to u_2 and u_4 separately and then make use of

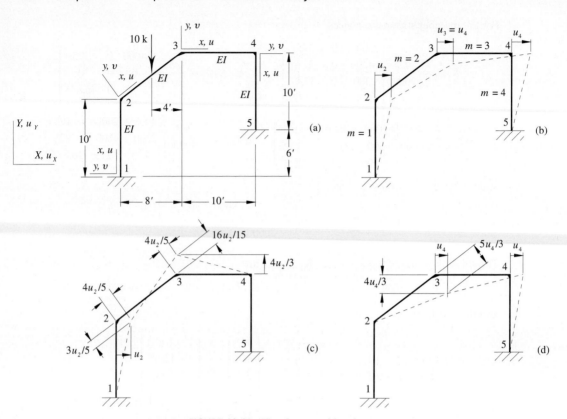

FIGURE 10.41 Five degrees of freedom.

the principle of superpositon. On the basis of the information shown in Figure 10.41c and d, it follows that the essential components of displacement are

$$v_{12} = 0, \qquad v_{21} = -u_2$$
$$v_{23} = -3u_2/5, \qquad v_{32} = 16u_2/15 - 5u_4/3, \qquad p_2 = 4u_2/5$$
$$v_{34} = 4u_2/3 - 4u_4/3, \qquad v_{43} = 0$$
$$v_{45} = u_4, \qquad v_{54} = 0$$

From the definitions $p_m = B_m z$ and $r_m = C_m z$ and the above relationships,

$$B_2 = (0, 0, 0, 4/5, 0)$$

$$C_1 = \begin{bmatrix} 0 & 0 & 0 & 0 & 0 \\ 1 & 0 & 0 & 0 & 0 \\ 0 & 0 & 0 & 0 & 0 \\ 0 & 0 & 0 & -1 & 0 \end{bmatrix}, \qquad C_2 = \begin{bmatrix} 1 & 0 & 0 & 0 & 0 \\ 0 & 1 & 0 & 0 & 0 \\ 0 & 0 & 0 & -3/5 & 0 \\ 0 & 0 & 0 & 16/15 & -5/3 \end{bmatrix}$$

$$C_3 = \begin{bmatrix} 0 & 1 & 0 & 0 & 0 \\ 0 & 0 & 1 & 0 & 0 \\ 0 & 0 & 0 & 4/3 & -4/3 \\ 0 & 0 & 0 & 0 & 0 \end{bmatrix}, \qquad C_4 = \begin{bmatrix} 0 & 0 & 1 & 0 & 0 \\ 0 & 0 & 0 & 0 & 0 \\ 0 & 0 & 0 & 0 & 1 \\ 0 & 0 & 0 & 0 & 0 \end{bmatrix}$$

The axial component of the resultant force between the ends of member 23 is $P_2 = -6$ k and the fixed end-force vector is

$$R_{F2}{}^T = (10, -10, 4, 4)$$

All members have the same member stiffness matrix given by

$$K = \frac{EI}{30}\begin{bmatrix} 12 & 6 & 1.8 & -1.8 \\ 6 & 12 & 1.8 & -1.8 \\ 1.8 & 1.8 & 0.36 & -0.36 \\ -1.8 & -1.8 & -0.36 & 0.36 \end{bmatrix}$$

The remainder of the analysis involves elementary substitutions into defining relations. Thus

$$B_2^T P_2 = \begin{Bmatrix} 0 \\ 0 \\ 0 \\ -4.8 \\ 0 \end{Bmatrix}, \quad C_2^T R_{F2} = \begin{Bmatrix} 10 \\ -10 \\ 0 \\ 28/15 \\ -20/3 \end{Bmatrix},$$

$$L_z = \sum_j A_j^T Q_j + \sum_m B_m^T P_m - \sum_m C_m^T R_{Fm} = \begin{Bmatrix} -10 \\ 10 \\ 0 \\ -20/3 \\ 20/3 \end{Bmatrix}$$

$$K_1 C_1 = \frac{EI}{30}\begin{bmatrix} 6 & 0 & 0 & 1.8 & 0 \\ 12 & 0 & 0 & 1.8 & 0 \\ 1.8 & 0 & 0 & 0.36 & 0 \\ -1.8 & 0 & 0 & -0.36 & 0 \end{bmatrix}, \quad C_1^T K_1 C_1 = \frac{EI}{30}\begin{bmatrix} 12 & 0 & 0 & 1.8 & 0 \\ 0 & 0 & 0 & 0 & 0 \\ 0 & 0 & 0 & 0 & 0 \\ 1.8 & 0 & 0 & 0.36 & 0 \\ 0 & 0 & 0 & 0 & 0 \end{bmatrix}$$

$$K_2 C_2 = \frac{EI}{30}\begin{bmatrix} 12 & 6 & 0 & -3 & 3 \\ 6 & 12 & 0 & -3 & 3 \\ 1.8 & 1.8 & 0 & -0.6 & 0.6 \\ -1.8 & -1.8 & 0 & 0.6 & -0.6 \end{bmatrix}, \quad C_2^T K_2 C_2 = \frac{EI}{30}\begin{bmatrix} 12 & 6 & 0 & -3 & 3 \\ 6 & 12 & 0 & -3 & 3 \\ 0 & 0 & 0 & 0 & 0 \\ -3 & -3 & 0 & 1 & -1 \\ 3 & 3 & 0 & -1 & 1 \end{bmatrix}$$

$$K_3 C_3 = \frac{EI}{30}\begin{bmatrix} 0 & 12 & 6 & 2.4 & -2.4 \\ 0 & 6 & 12 & 2.4 & -2.4 \\ 0 & 1.8 & 1.8 & 0.48 & -0.48 \\ 0 & -1.8 & -1.8 & -0.48 & 0.48 \end{bmatrix},$$

$$C_3^T K_3 C_3 = \frac{EI}{30}\begin{bmatrix} 0 & 0 & 0 & 0 & 0 \\ 0 & 12 & 6 & 2.4 & -2.4 \\ 0 & 6 & 12 & 2.4 & -2.4 \\ 0 & 2.4 & 2.4 & 0.64 & -0.64 \\ 0 & -2.4 & -2.4 & -0.64 & 0.64 \end{bmatrix}$$

$$K_4 C_4 = \frac{EI}{30}\begin{bmatrix} 0 & 0 & 12 & 0 & 1.8 \\ 0 & 0 & 6 & 0 & 1.8 \\ 0 & 0 & 1.8 & 0 & 0.36 \\ 0 & 0 & -1.8 & 0 & -0.36 \end{bmatrix}, \quad C_4^T K_4 C_4 = \frac{EI}{30}\begin{bmatrix} 0 & 0 & 0 & 0 & 0 \\ 0 & 0 & 0 & 0 & 0 \\ 0 & 0 & 12 & 0 & 1.8 \\ 0 & 0 & 0 & 0 & 0 \\ 0 & 0 & 1.8 & 0 & 0.36 \end{bmatrix}$$

$$K = \sum C_m^T K_m C_m = \frac{EI}{30} \begin{bmatrix} 24 & 6 & 0 & -1.2 & 3 \\ 6 & 24 & 6 & -0.6 & 0.6 \\ 0 & 6 & 24 & 2.4 & -0.6 \\ -1.2 & -0.6 & 2.4 & 2 & -1.64 \\ 3 & 0.6 & -0.6 & -1.64 & 2 \end{bmatrix}$$

The solution values of the equations $Kz = L_z$ are

$\theta_2 = -37.6 \dfrac{\text{k-ft}^2}{EI}$, $\theta_3 = 19.17 \dfrac{\text{k-ft}^2}{EI}$, $\theta_4 = -3.49 \dfrac{\text{k-ft}^2}{EI}$, $u_2 = 30.71 \dfrac{\text{k-ft}^3}{EI}$, $u_4 = 174.8 \dfrac{\text{k-ft}^3}{EI}$

The end-force vectors are

$$R_1 = K_1 C_1 z = \begin{Bmatrix} M_{12} \\ M_{21} \\ V_{12} \\ V_{21} \end{Bmatrix} = \begin{Bmatrix} -5.68 \text{ k-ft} \\ -13.2 \text{ k-ft} \\ -1.888 \text{ k} \\ 1.888 \text{ k} \end{Bmatrix}, \quad R_2 = K_2 C_2 z + R_{F2} = \begin{Bmatrix} M_{23} \\ M_{32} \\ V_{23} \\ V_{32} \end{Bmatrix} = \begin{Bmatrix} 13.2 \text{ k-ft} \\ 4.56 \text{ k-ft} \\ 5.78 \text{ k} \\ 2.22 \text{ k} \end{Bmatrix}$$

$$R_3 = K_3 C_3 z = \begin{Bmatrix} M_{34} \\ M_{43} \\ V_{34} \\ V_{43} \end{Bmatrix} = \begin{Bmatrix} -4.56 \text{ k-ft} \\ -9.09 \text{ k-ft} \\ -1.365 \text{ k} \\ 1.365 \text{ k} \end{Bmatrix}, \quad R_4 = K_4 C_4 z = \begin{Bmatrix} M_{45} \\ M_{54} \\ V_{45} \\ V_{54} \end{Bmatrix} = \begin{Bmatrix} 9.09 \text{ k-ft} \\ 9.79 \text{ k-ft} \\ 1.888 \text{ k} \\ -1.888 \text{ k} \end{Bmatrix}$$

KEY POINTS

1. In the analysis of a structure using manual computational techniques, the displacement method of analysis should be used when the degree of statical indeterminacy is greater than the degree of kinematical indeterminacy.
2. Formulation of a complete independent set of equations of equilibrium corresponding to the independent position variables is central to successful application of the displacement method.
3. Where the necessary conditions are satisfied, the modified slope deflection equations should be employed.
4. Moment distribution (which is a displacement method of analysis) is an efficient, practical procedure for calculating end-moments when joint displacements are known.
5. Problems involving sidesway can always be solved by making combined use of moment distribution and superposition.
6. The principle of virtual work can always be used to derive the independent equations of equilibrium (See Point 2 above). Only that information which is essential to any technique of structural analysis by displacement methods is needed to apply the principle.
7. If the material obeys Hooke's law, application of either the principle of virtual work or the principle of stationary potential energy leads naturally to equations of equilibrium in which the coefficients of the unknown displacements form a symmetric matrix called the structure stiffness matrix.
8. Matrix algebra is a useful tool in structural analysis; it provides the basis for systematic problem formulation, and the associated numerical work is well suited for execution by a programmed digital computer.

PROBLEMS

10.1 a) Determine the degree of kinematical indeterminacy and the degree of statical indeterminacy, b) determine the end-moments using the basic slope-deflection equations, c) construct the shear and bending moment diagrams, and d) calculate the deflection at point A.

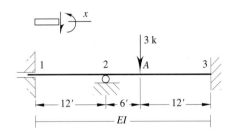

PROBLEM 10.1

10.2 a) Determine the degree of kinematical indeterminacy and the degree of statical indeterminacy, b) determine the end-moments using the basic slope-deflection equations, c) construct the shear and bending moment diagrams, and d) calculate the deflection at point A.

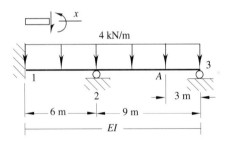

PROBLEM 10.2

10.3 a) Determine the degree of kinematical indeterminacy and the degree of statical indeterminacy, b) determine the end-moments using the basic slope-deflection equations, c) construct the shear and bending moment diagrams, and d) calculate the deflection at point A.

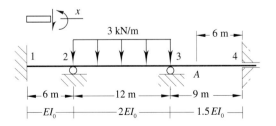

PROBLEM 10.3

10.4 a) Determine the degree of kinematical indeterminacy and the degree of statical indeterminacy, b) determine the end-moments using the basic slope-deflection equations, c) construct the shear and bending moment diagrams.

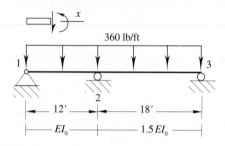

PROBLEM 10.4

10.5 a) Determine the degree of kinematical indeterminacy and the degree of statical indeterminacy, b) determine the end-moments using the basic slope-deflection equations, c) construct the shear and bending moment diagrams.

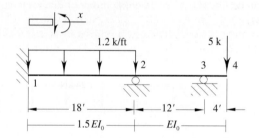

PROBLEM 10.5

10.6 a) Determine the degree of kinematical indeterminacy and the degree of statical indeterminacy, b) determine the end-moments using the basic slope-deflection equations, c) construct the shear and bending moment diagrams.

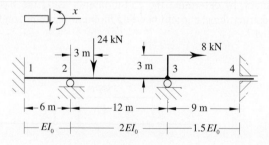

PROBLEM 10.6

10.7 The members are axially rigid: a) determine the degree of kinematical indeterminacy and the degree of statical indeterminacy, b) select local coordinates for each member, c) determine the end-moments using the basic slope-deflection equations, and d) construct the shear and bending moment diagrams for member 23.

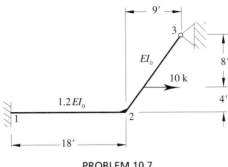

PROBLEM 10.7

10.8 The members are axially rigid: a) determine the degree of kinematical indeterminacy and the degree of statical indeterminacy, b) select local coordinates for each member, c) determine the end-moments using the basic slope-deflection equations, and d) construct the shear and bending moment diagrams for member 23.

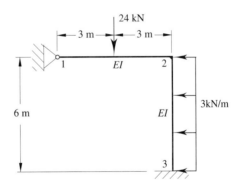

PROBLEM 10.8

10.9 The members are axially rigid: a) determine the degree of kinematical indeterminacy and the degree of statical indeterminacy, b) select local coordinates for each member, c) determine the end-moments using the basic slope-deflection equations, and d) construct the shear and bending moment diagrams for member 34.

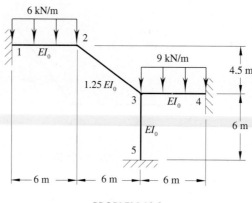

PROBLEM 10.9

10.10 The members are axially rigid: a) determine the degree of kinematical indeterminacy and the degree of statical indeterminacy, b) select local coordinates for each member, c) determine the force F_0 to prevent sidesway of the frame. Use the basic slope-deflection equations.

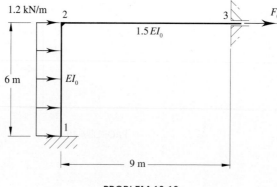

PROBLEM 10.10

10.11 The members are axially rigid: a) determine the degree of kinematical indeterminacy and the degree of statical indeterminacy, b) select local coordinates for each member, c) determine the force F_0 to prevent sidesway of the frame. Use the basic slope-deflection equations.

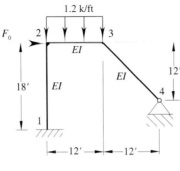

PROBLEM 10.11

10.12 The members are axially rigid: a) determine the degree of kinematical indeterminacy and the degree of statical indeterminacy, b) select local coordinates for each member, c) determine the forces F_{02} and F_{03} to prevent sidesway of the frame. Use the basic slope-deflection equations.

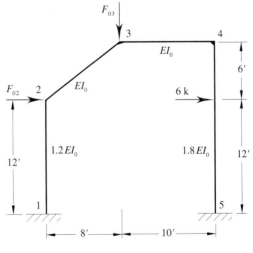

PROBLEM 10.12

10.13 The members are axially rigid. Express the axial and transverse components of the member end-displacements in terms of the joint displacement u_2. Use local coordinates as shown.

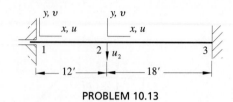

PROBLEM 10.13

10.14 The members are axially rigid. Express the axial and transverse components of the member end-displacements in terms of the joint displacement u_2. Use local coordinates as shown.

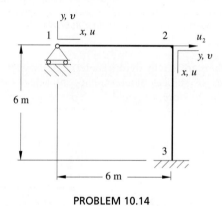

PROBLEM 10.14

10.15 The members are axially rigid. Express the axial and transverse components of the member end-displacements in terms of the joint displacement u_3. Use local coordinates as shown.

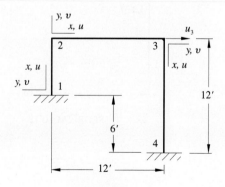

PROBLEM 10.15

10.16 The members are axially rigid. Express the axial and transverse components of the member end-displacements in terms of the joint displacement u_3. Use local coordinates as shown.

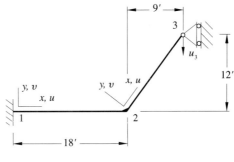

PROBLEM 10.16

10.17 The members are axially rigid. Express the axial and transverse components of the member end-displacements in terms of the joint displacement u_1. Use local coordinates as shown.

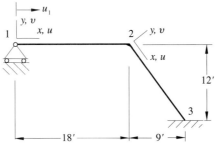

PROBLEM 10.17

10.18 The members are axially rigid. Express the axial and transverse components of the member end-displacements in terms of the joint displacement u_3. Use local coordinates as shown.

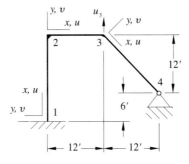

PROBLEM 10.18

10.19 The members are axially rigid. Express the axial and transverse components of the member end-displacements in terms of the joint displacement u_3. Use local coordinates as shown.

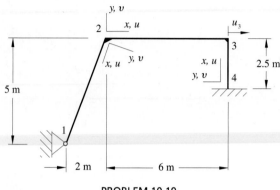

PROBLEM 10.19

10.20 The members are axially rigid. Express the axial and transverse components of the member end-displacements in terms of the joint displacement u_1. Use local coordinates as shown.

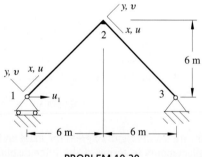

PROBLEM 10.20

10.21 The members are axially rigid. Express the axial and transverse components of the member end-displacements in terms of the joint displacement u_2. Use local coordinates as shown.

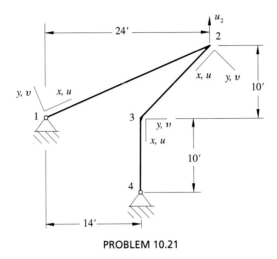

PROBLEM 10.21

10.22 Determine the end-moments using the modified slope-deflection equations where appropriate and then calculate the reactions.

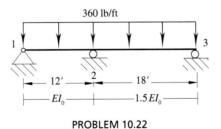

PROBLEM 10.22

10.23 Determine the end-moments using the modified slope-deflection equations where appropriate and then calculate the reactions.

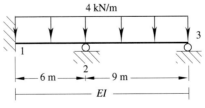

PROBLEM 10.23

10.24 Determine the end-moments using the modified slope-deflection equations where appropriate and then calculate the reactions.

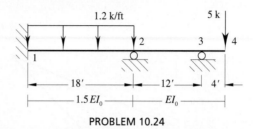

PROBLEM 10.24

10.25 Determine the end-moments using the modified slope-deflection equations where appropriate and then calculate the reactions. The members are axially rigid.

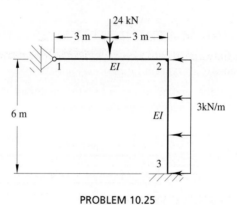

PROBLEM 10.25

10.26 Determine the end-moments using the modified slope-deflection equations where appropriate and then calculate the reactions. The members are axially rigid.

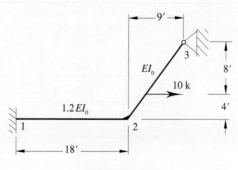

PROBLEM 10.26

10.27 Determine the end-moments using the modified slope-deflection equations where appropriate and then calculate the reactions. The members are axially rigid.

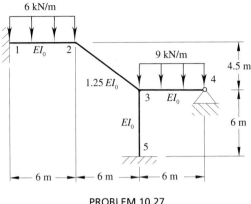

PROBLEM 10.27

10.28 Use the slope-deflection equations to determine the sidesway displacement u_2 and the end-moments. Use local coordinates as shown.

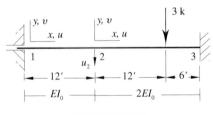

PROBLEM 10.28

10.29 Use the slope-deflection equations to determine the sidesway displacement u_3 and the end-moments. Use local coordinates as shown.

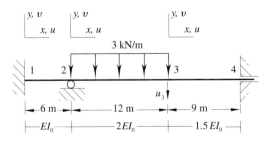

PROBLEM 10.29

10.30 Identify local coordinates for each member. Then use the slope-deflection equations to determine the sidesway displacement u_2 and the end-moments.

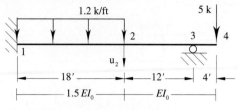

PROBLEM 10.30

10.31 The members are axially rigid. Use the slope-deflection equations to determine the sidesway displacement u_2 and the end-moments, and then determine the reactions. Use local coordinates as shown.

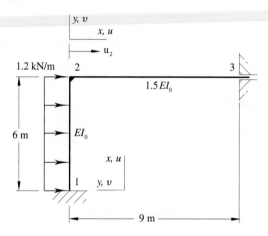

PROBLEM 10.31

10.32 The members are axially rigid. Use the slope-deflection equations to determine the sidesway displacement u_3 and the end-moments, and then determine the reactions. Use local coordinates as shown.

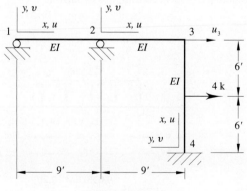

PROBLEM 10.32

10.33 The members are axially rigid. Use the slope-deflection equations to determine the side-sway displacement u_3 and the end-moments, and then determine the reactions. Use local coordinates as shown.

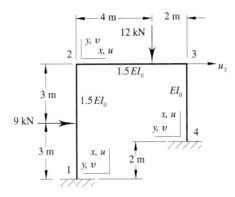

PROBLEM 10.33

10.34 The members are axially rigid. Use the slope-deflection equations to determine the side-sway displacement u_3 and the end-moments, and then determine the reactions. Use local coordinates as shown.

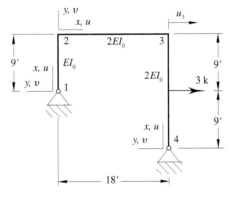

PROBLEM 10.34

10.35 The members are axially rigid. Select local coordinates for each member. Use the slope-deflection equations to determine the sideway displacement u_3 and the end-moments, and then determine the reactions.

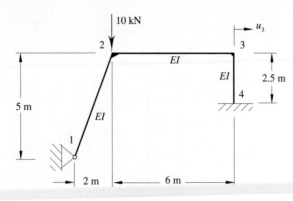

PROBLEM 10.35

10.36 The members are axially rigid. Select local coordinates for each member. Use the slope-deflection equations to determine the sideway displacement u_1 and the end-moments, and then determine the reactions.

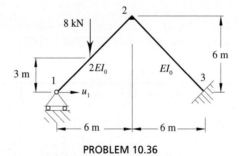

PROBLEM 10.36

10.37 The members are axially rigid. Select local coordinates for each member. Use the slope-deflection equations to determine the sideway displacement u_2 and the end-moments, and then determine the reactions.

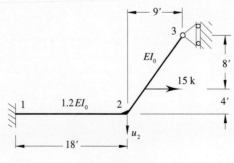

PROBLEM 10.37

10.38 The members are axially rigid. Select local coordinates for each member. Use the slope-deflection equations to determine the sidesway displacements u_2 and u_3 and the end-moments, and then determine the reactions.

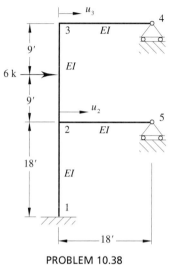

PROBLEM 10.38

10.39 Use the slope-deflection equations to determine the end-moments and the reactions due to a downward settlement of 0.36 inches of support A. $EI = 360,000$ k-ft².

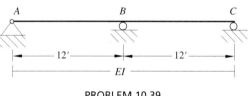

PROBLEM 10.39

10.40 Use the slope-deflection equations to determine the end-moments and the reactions due to a downward settlement of 12 mm of support A and a heaving (upward displacement) of 10 mm of support B. $EI = 166$ MN · m².

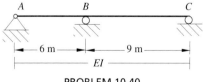

PROBLEM 10.40

10.41 Use the slope-deflection equations to determine the end-moments and the reactions due to a downward settlement of 0.36 inches of support C. $EI = 240{,}000$ k-ft^2.

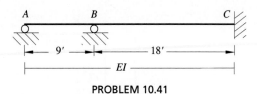

PROBLEM 10.41

10.42 Use the slope-deflection equations to determine the end-moments and the reactions due to a downward settlement of 0.36 inches and a horizontal settlement to the right of 0.24 inches of support C. $EI = 240{,}000$ k-ft^2.

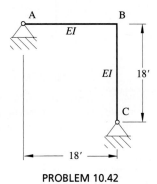

PROBLEM 10.42

10.43 Use the slope-deflection equations to determine the sidesway displacement u_3, the end-moments, and the reactions due to a downward settlement of 0.24 in. of support 1. $EI = 240{,}000$ k·ft^2.

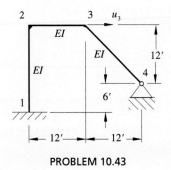

PROBLEM 10.43

10.44 Use moment distribution to calculate the end-moments and then calculate the deflection at A.

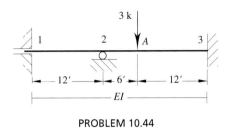

PROBLEM 10.44

10.45 Use moment distribution to calculate the end-moments and then calculate the rotation of joint 2.

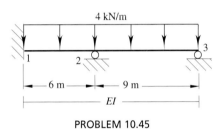

PROBLEM 10.45

10.46 Use moment distribution to calculate the end-moments and then calculate the deflection of A.

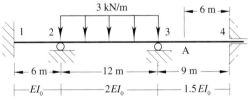

PROBLEM 10.46

10.47 Use moment distribution to calculate the end-moments and then calculate the rotation of joint 2.

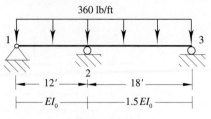

PROBLEM 10.47

10.48 Use moment distribution to calculate the end-moments and then calculate the deflection of point 4.

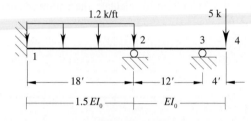

PROBLEM 10.48

10.49 Use moment distribution to calculate the end-moments.

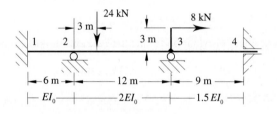

PROBLEM 10.49

10.50 The members are axially rigid. Use moment distribution to calculate the end-moments.

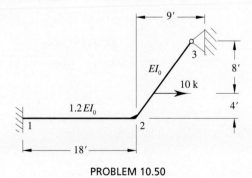

PROBLEM 10.50

10.51 The members are axially rigid. Use moment distribution to calculate the end-moments.

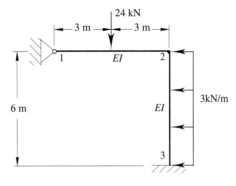

PROBLEM 10.51

10.52 The members are axially rigid. Use moment distribution to calculate the end-moments.

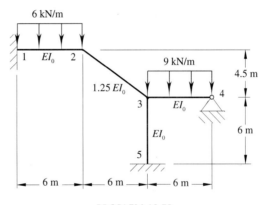

PROBLEM 10.52

10.53 The members are axially rigid. Use moment distribution to calculate the end-moments.

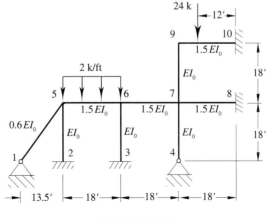

PROBLEM 10.53

10.54 The members are axially rigid. Use moment distribution and superposition to determine the sidesway displacement and the end-moments.

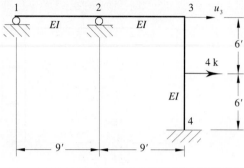

PROBLEM 10.54

10.55 The members are axially rigid. Use moment distribution and superposition to determine the sidesway displacement and the end-moments.

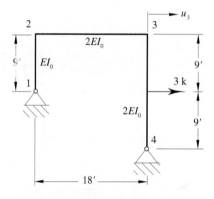

PROBLEM 10.55

10.56 The members are axially rigid. Use moment distribution and superposition to determine the sidesway displacements and the end-moments.

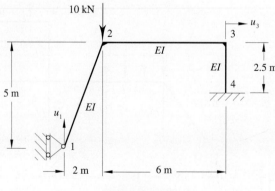

PROBLEM 10.56

10.57 The members are axially rigid. Use moment distribution and superposition to determine the sidesway displacement and the end-moments.

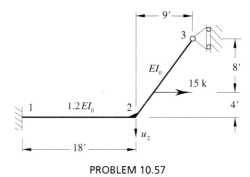

PROBLEM 10.57

10.58 The members are axially rigid. Use moment distribution and superposition to determine the sidesway displacements and the end-moments.

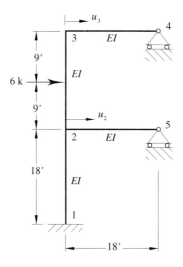

PROBLEM 10.58

10.59 The members are axially rigid. Employ moment distribution to calculate the end-moments.

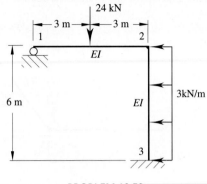

PROBLEM 10.59

10.60 The members are axially rigid. Employ moment distribution to calculate the end-moments.

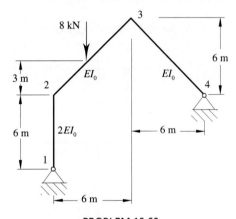

PROBLEM 10.60

10.61 The members are axially rigid. Employ moment distribution to calculate the end-moments.

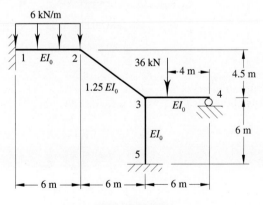

PROBLEM 10.61

10.62–10.78 General instructions: The problems are to be formulated and solved using matrix methods. Review the sections **Matrix Formulations—Summary** and **Application Procedure** before moving on to the solution process. Where they are included as part of the problem figure, use global and/or local coordinates as shown. When included as part of the problem statement, take the independent position variables to be the assigned quantities. In all cases the members are axially rigid.

10.62 Determine the end-force vectors and calculate the reaction at the roller support.

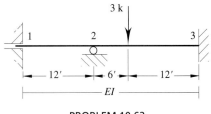

PROBLEM 10.62

10.63 Determine the end-force vectors and calculate the reactions at the roller supports.

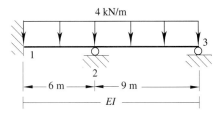

PROBLEM 10.63

10.64 Determine the end-force vectors and calculate the reactions.

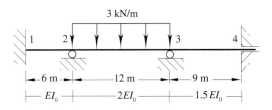

PROBLEM 10.64

10.65 Determine the end-force vectors and calculate the reactions. Suggestion: Consider the structure to be made up of members 12 and 23 and that the external loading at joint 3 consists of the force and couple exerted on the joint by member 34.

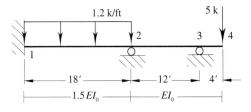

PROBLEM 10.65

10.66 Determine the end-force vectors and calculate the reactions. Suggestion: Consider the structure to be made up of members 12, 23, and 34 and that the external loading at joint 3 consists of the force and couple exerted on the joint by the vertical member at joint 3.

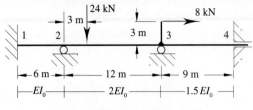

PROBLEM 10.66

10.67 Determine the end-force vectors and calculate the reactions. Take local coordinates as shown.

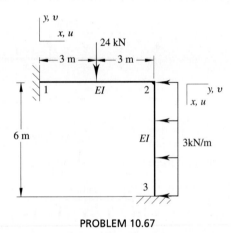

PROBLEM 10.67

10.68 Determine the end-force vectors and calculate the reactions. Take local coordinates as shown.

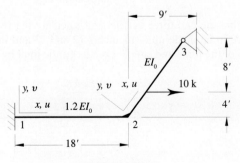

PROBLEM 10.68

10.69 Determine the end-force vectors and calculate the reactions. Take global and local coordinates as shown.

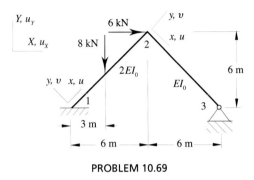

PROBLEM 10.69

10.70 Determine the end-force vectors and calculate the reactions. Take local coordinates as shown.

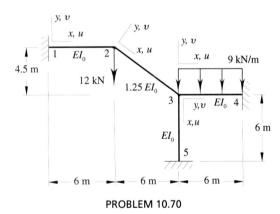

PROBLEM 10.70

10.71 The frame has one degree of freedom with respect to sidesway. Determine the end-force vectors.

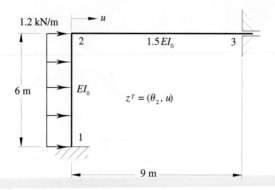

PROBLEM 10.71

10.72 The frame has one degree of freedom with respect to sidesway. Determine the end-force vectors.

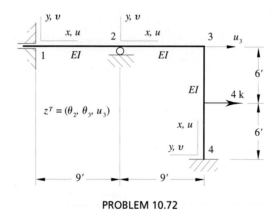

PROBLEM 10.72

10.73 The frame has one degree of freedom with respect to sidesway. Determine the end-force vectors.

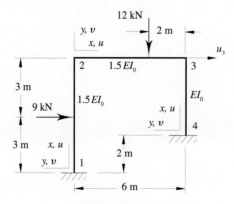

PROBLEM 10.73

10.74 The frame has one degree of freedom with respect to sidesway. Determine the end-force vectors.

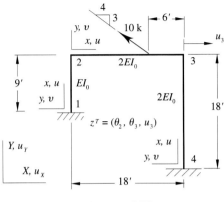

PROBLEM 10.74

10.75 The frame has one degree of freedom with respect to sidesway. Determine the end-force vectors.

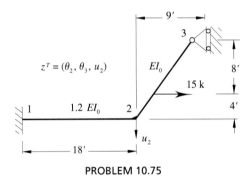

PROBLEM 10.75

10.76 The frame has one degree of freedom with respect to sidesway. Determine the end-force vectors.

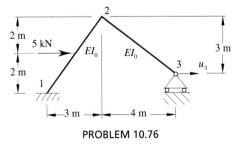

PROBLEM 10.76

10.77 The frame has one degree of freedom with respect to sidesway. Determine the end-force vectors.

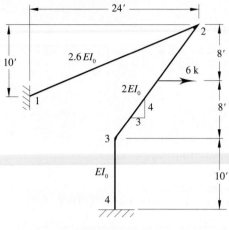

PROBLEM 10.77

10.78 The frame has two degree of freedom with respect to sidesway. Determine the end-force vectors.

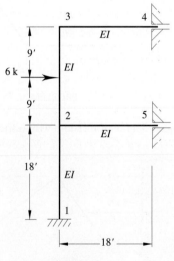

PROBLEM 10.78

CHAPTER 11

Influence Lines

11.1 INTRODUCTION

Practical application of the concept of an influence function or influence line is based on the assumption that the principle of superposition is valid. For a statically indeterminate structure, the principle of superposition holds only when the deformations and displacements are small and the material obeys Hooke's law. Direct application of the definition can be used as the basis for carrying out the calculations to construct an influence line. However, a substantial reduction in the effort required to construct an influence line for a force in a statically indeterminate structure is obtained by applying the Müller-Breslau principle.

11.2 MÜLLER-BRESLAU PRINCIPLE

The Müller-Breslau principle asserts that the geometry of an influence line for force at a given location in a structure that undergoes small deformation and displacement and obeys Hooke's law is proportional to the deflections induced in an associated cut-back structure subjected to an appropriate prescribed loading at the location under consideration. The cut-back structure corresponding to a reactive force at a support (a pair of interactive internal forces at a section) is derived by removing the mechanical device (material) that develops the force (force pair) of interest. The prescribed loading consists of a force (force pair) that is characteristic of the reactive force (interactive internal forces) developed by the mechanical device (material) that has been removed. Consider the statically indeterminate structure shown in Figure 11.1a. The location of the unit load for the construction of influence lines is defined by the position variable s as shown. The reaction forces and the internal forces at any specific location in the structure are functions of the position s of the unit load as indicated by the notation in Figure 11.1a.

Assume it is required to construct the influence line for the reactive moment $M_B(s)$ at support B. The cut-back structure to be considered for this purpose is derived by removing that part of the support that develops the reactive moment. Accordingly, the cut-back structure has a pin support at B. The cut-back structure and the prescribed loading for which the Müller-Breslau principle applies are shown in Figure 11.1b. By the principle, $M_B(s)$ is proportional to $v(s)$, where $v(s)$ is the deflection induced at location s in the cut-back structure by the moment M^B, as shown in Figure 11.1b. To be more specific, the principle asserts that $M_B(s) = v(s)/\theta^B$. Thus, with the aid of the Müller-Breslau principle, construction of the influence line is reduced to the problem

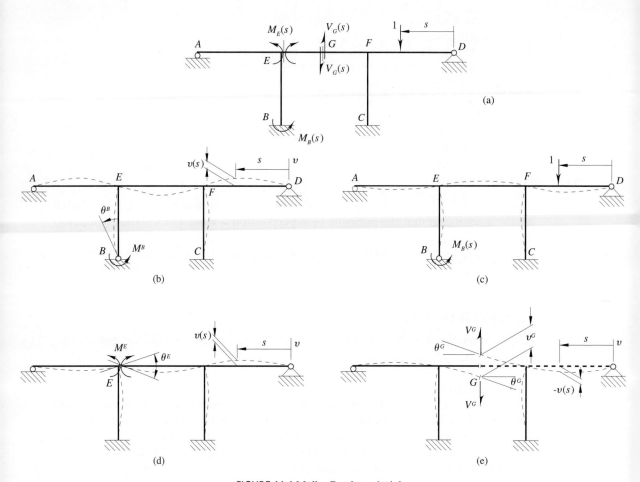

FIGURE 11.1 Müller-Breslau principle.

of analyzing the cut-back structure for one case of loading. It is to be noted that the degree of indeterminacy of the cut-back structure is one less than the degree of indeterminacy of the given structure.

The validity of the result obtained by application of the Müller-Breslau principle is established with the aid of Betti's theorem. Consider the cut-back structure under the loads shown in Figure 11.1c, in which $M_B(s)$ is set equal to the reactive moment developed at support B in the given structure. The mechanical state of the cut-back structure in Figure 11.1c coincides with the mechanical state of the given structure shown in Figure 11.1a. Accordingly, for the cut-back structure shown in Figure 11.1c the rotation of the tangent at B is zero. Consider the cut-back structure under "p" loads as shown in Figure 11.1b and "q" loads as shown in Figure 11.1c. By Betti's theorem

$$W_e^{qp} = M_B(s)\theta^B - 1v(s) = W_e^{pq} = 0$$

which, indeed, states that the influence function $M_B(s)$ is proportional to the deflection function $v(s)$, the proportionality factor being the reciprocal of θ^B.

The cut-back structure appropriate to the determination of the influence line for the bending moment M_E at the E-end of Member EF is derived by removing the material that develops the bending moment at that section while retaining the material that develops the axial and shear forces. The result is a cut-back structure with a pin connection at the E-end of Member EF as shown in Figure 11.1d. By the Müller-Breslau principle, the influence line for the bending moment at E is proportional to the deflection curve $v(s)$ induced by the couples M^E and the proportionality factor is the reciprocal of the discontinuity θ^E in the rotation of the tangent on either side of the pin as shown in Figure 11.1d.

The cut-back structure appropriate to the determination of the influence line for the shear V_G at section G is derived by removing the material that develops the shear while retaining the material that develops the bending moment and the axial force. The result is a cut-back structure with a dovetail connection at section G as shown in Figure 11.1e. By the Müller-Breslau principle, the influence line for the shear at G is proportional to the deflection curve $v(s)$ induced by the shear forces V^G and the proportionality factor is the reciprocal of the discontinuity v^G in deflection on either side of the dovetail connection as shown in Figure 11.1e.

11.3 INFLUENCE LINES FOR DEFLECTION

Efficient construction of an influence line for deflection at a point in a structure follows from a simple application of either Betti's theorem or Maxwell's reciprocal theorem. Consider the influence line for the deflection at point G of the structure shown in Figure 11.2a. By definition, the influence line is a plot of $v_G(s)$ against s, where $v_G(s)$ is the deflection at point G when the unit load is placed at position s. By the reciprocal theorem, the deflection at G due to a unit load at s is equal to the deflection at point s when the unit load is placed at point G as shown in Figure 11.2b. Accordingly, the deflection curve $v(s)$ shown in Figure 11.2b is the influence line for the deflection at G, i.e., $v_G(s) = v(s)$.

11.4 PRACTICAL APPLICATION

In general, the construction of an influence line for a statically indeterminate structure with the aid of the Müller-Breslau principle will require a force analysis of an associated cut-back structure that is also statically indeterminate (an exception occurs

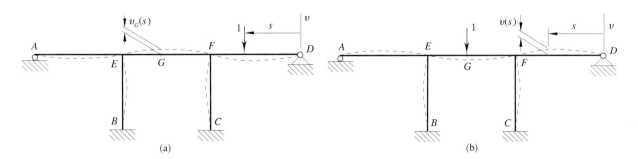

FIGURE 11.2 Influence line for deflection.

Chapter 11 Influence Lines

when the given structure is statically indeterminate to the first degree and an influence line for force is required). Ordinarily, moment distribution will prove to be the most convenient method for the force analysis. In all cases, deflections must also be calculated. The required deflections can always be obtained by numerical methods. However, the required deflection depends only on the end-displacements and the end-moments and are relatively easy to calculate. For a uniform member with end-displacements and end-moments (as shown in Figure 11.3), the equation of the deflection curve, which is readily determined by direct integration of the moment-curvature equation, is given by

$$v(x) = v_{ab} + \left(\frac{x}{L}\right)[v_{ba} - v_{ab}] + \left[2\left(\frac{x}{L}\right) - 3\left(\frac{x}{L}\right)^2 + \left(\frac{x}{L}\right)^3\right]\frac{L^2 M_{ab}}{6EI}$$
$$- \left[\left(\frac{x}{L}\right) - \left(\frac{x}{L}\right)^3\right]\frac{L^2 M_{ba}}{6EI} \quad \text{(a)}$$

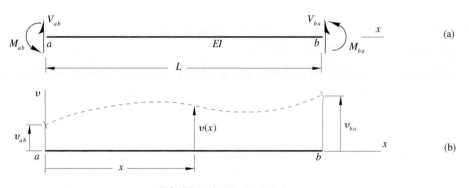

FIGURE 11.3 Deflection curve.

Example 11.1

For the beam shown in Figure 11.4a, construct the equations of the influence lines for the reaction at supports A and B and for the shear at the section just to the right of B. The beam is statically indeterminate to the second degree. As will be shown later, once the influence lines have been determined for any two of the required quantities, the influence line for the third quantity can be determined from statics. The cut-back structure and the associated loading for the calculation of the influence line for R_A are shown in Figure 11.4b. Analysis of the cut-back structure is carried out using moment distribution. The end-moments for $\theta_1 = \theta_2 = 0$ and R^A such that $6EIv^A/L^2 = 140$ are shown in the first line of the moment distribution table in Figure 11.4. One distribution at joint A followed by one distribution at joint B yields final values for the end-moments as shown beneath the double lines in the moment distribution table. By the principle of superposition, the end-moments for any value of v^A are

$$M_{AB} = 0, \quad M_{BA} = \frac{12EIv^A}{7L^2}, \quad M_{BC} = -\frac{12EIv^A}{7L^2}, \quad M_{CB} = -\frac{6EIv^A}{7L^2}$$

Substitution of the foregoing results into Equation (a) yields equations that define the deflection curve $v(s)$ shown in Figure 11.4b in terms of local coordinates appropriate to each span. By

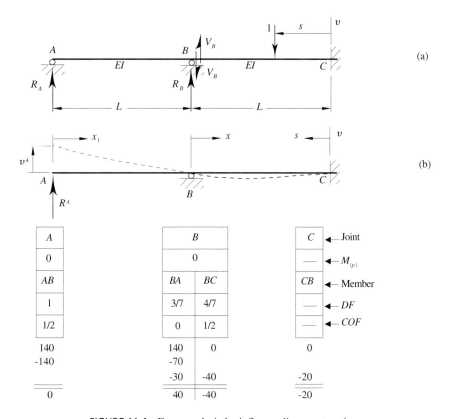

FIGURE 11.4 Force analysis for influence line construction.

Betti's theorem, the equation of the influence line for R_A is $v(s)/v^A$. Accordingly, the equation of the influence line is

$$0 < s < L, \quad s = L - x, \quad \frac{v}{v^A} = R_A = -\frac{3}{7}\left[\left(\frac{x}{L}\right) - 2\left(\frac{x}{L}\right)^2 + \left(\frac{x}{L}\right)^3\right]$$

$$L < s < 2L, \quad s = 2L - x_1, \quad \frac{v}{v^A} = R_A = 1 - \frac{1}{7}\left[9\left(\frac{x_1}{L}\right) - 2\left(\frac{x_1}{L}\right)^3\right]$$

The cut-back structure and the associated loading for the calculation of the influence line for R_B are shown in Figure 11.4 (continuation 1). The end-moments for $\theta_1 = \theta_2 = 0$ and R^B such that $6EIv^B/L^2 = 140$ are shown in the first line of the moment distribution table. From the results in the table and the principle of superposition, the end-moments for any value of v^B are

$$M_{AB} = 0, \quad M_{BA} = -\frac{30EIv^B}{7L^2}, \quad M_{BC} = \frac{30EIv^B}{7L^2}, \quad M_{CB} = \frac{36EIv^B}{7L^2}$$

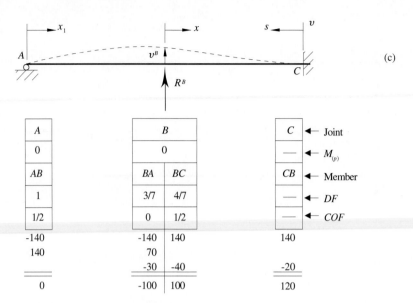

A	B		C	← Joint
			← $M_{(p)}$	
AB	BA	BC	CB	← Member
1	3/7	4/7	—	← DF
1/2	0	1/2	—	← COF
-140	-140	140	140	
140	70			
	-30	-40	-20	
0	-100	100	120	

FIGURE 11.4 (continuation 1) Force analysis for influence line construction.

The equation of the influence line for R_B is derived by following the procedures used to establish the influence line for R_A. Thus, the equation of the influence line for R_B is

$$0 < s < L, \quad s = L - x, \quad \frac{v}{v^B} = R_B = \frac{1}{7}\left[7 - 3\left(\frac{x}{L}\right) - 15\left(\frac{x}{L}\right)^2 + 11\left(\frac{x}{L}\right)^3\right]$$

$$L < s < 2L, \quad s = 2L - x_1, \quad \frac{v}{v^B} = R_B = \frac{1}{7}\left[12\left(\frac{x_1}{L}\right) - 5\left(\frac{x_1}{L}\right)^3\right]$$

The cut-back structure and the associated loading for the calculation of the influence line for V_B are shown in Figure 11.4 (continuation 2). The cut-back structure has a dovetail connection at the section where the shear V_B acts. The connection enforces the condition that the rotations of the tangents on either side of the connection be equal. The connection develops equal and opposite internal bending moments which need not be shown. The end-moments for $\theta_1 = \theta_2 = 0$ and V^B such that $6EI\Delta/L^2 = 140$ are shown in the first line of the moment distribution table below Figure 11.4d. From the results in the table and the principle of superposition, the end-moments for any value of Δ are

$$M_{AB} = 0, \quad M_{BA} = -\frac{18EI\Delta}{7L^2}, \quad M_{BC} = \frac{18EI\Delta}{7L^2}, \quad M_{CB} = \frac{30EI\Delta}{7L^2}$$

The previously established procedure is used to construct the equation of the influence line for V_B. The equation is

$$0 < s < L, \quad s = L - x, \quad \frac{v}{\Delta} = V_B = \frac{1}{7}\left[7 - 6\left(\frac{x}{L}\right) - 9\left(\frac{x}{L}\right)^2 + 8\left(\frac{x}{L}\right)^3\right]$$

$$L < s < 2L, \quad s = 2L - x_1, \quad \frac{v}{\Delta} = V_B = \frac{3}{7}\left[\left(\frac{x_1}{L}\right) - \left(\frac{x_1}{L}\right)^3\right]$$

The influence functions for R_A, R_B, and V_B are not independent. From the forces shown in Figure 11.4a and force equilibrium, it follows that for $0 < s < L$, $V_B(x) = R_A(x) + R_B(x)$, and for $L < s < 2L$, $V_B(x_1) = R_A(x_1) + R_B(x_1) - 1$. It can be verified that the relationships that have been constructed satisfy the foregoing requirements.

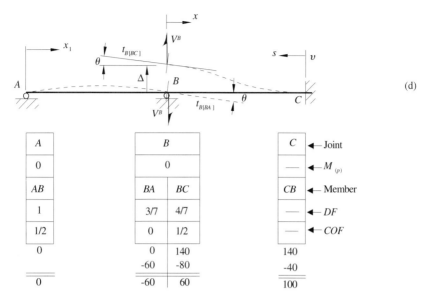

FIGURE 11.4 (continuation 2)   Force analysis for influence line construction.

KEY POINTS

1. Influence lines can always be constructed on the basis of definition.
2. The Müller-Breslau principle can be used to determine the general shape of an influence function for force in a statically indeterminate structure.
3. The Müller-Breslau principle can be used to reduce the effort to construct an influence line for force in a statically indeterminate structure.
4. The Müller-Breslau principle does not apply to statically determinate structures.

PROBLEMS

11.1 Construct the influence functions and plot the influence lines for R_A and M_B.

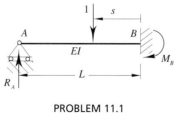

PROBLEM 11.1

11.2 Construct the influence function and plot the influence line for the reaction at B.

11.3 Construct the influence function and plot the influence line for the reaction at C.

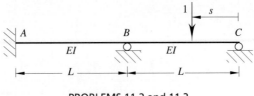

PROBLEMS 11.2 and 11.3

11.4 Construct the influence function and plot the influence line for the reaction at B.

11.5 Construct the influence function and plot the influence line for the bending moment over support C.

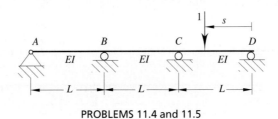

PROBLEMS 11.4 and 11.5

11.6 Construct the influence function and plot the influence line for the reaction at C.

11.7 Construct the influence function and plot the influence line for the moment of the reaction couple at D.

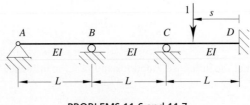

PROBLEMS 11.6 and 11.7

11.8 Construct the influence function and plot the influence line for the bending moment at the B-end of member AB.

11.9 Construct the influence function and plot the influence line for the shear at the B-end of member BC.

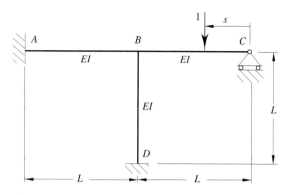

PROBLEMS 11.8 and 11.9

Appendix A

A.1 SUPPLEMENTAL NOTES ON THE PRINCIPLE OF VIRTUAL WORK

Consider a body for which the deformations due to some cause are known. Associated with the deformations are compatible displacements. If the deformations and displacements are infinitesimally small, they qualify as virtual deformations and displacements. As illustrated in Figure A.1, let the body be subjected to an assigned external load Q, and let $R_1, R_2, R_3, \ldots, R_r$ and $C_a, C_b, \ldots, C_c$ be any set of external forces and couples such that the body is in equilibrium. For a plane body, the force system must satisfy the equations of equilibrium

$$0 = \Sigma F_x, \quad 0 = \Sigma F_y, \quad 0 = \Sigma M_a$$

In general, there is an infinite number of internal force systems such that the differential equations of equilibrium are satisfied throughout the body. In the following formulations, it is assumed that one such system has been constructed. Accordingly, the internal and external force systems satisfy the conditions imposed on the forces in the formulation of the principle of virtual work. Thus, by the principle of virtual work

$$\delta W_e = Q u_q + \sum_r R u_r + \sum_c C \theta_c = \delta W_\sigma$$

which can be rearranged as

$$Q u_q = \delta W_\sigma - \sum_r R u_r - \sum_c C \theta_c$$

where u_q is the component of displacement at the point of application of Q, u_r is the component of displacement at the point of application of force R in the direction of R,

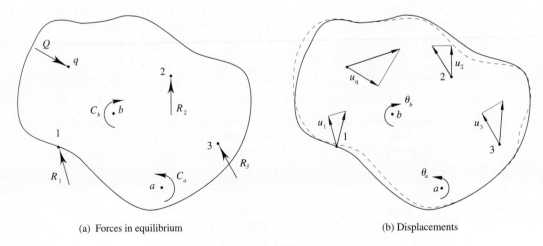

(a) Forces in equilibrium   (b) Displacements

FIGURE A.1 Notation.

Section A.1 Supplemental Notes on the Principle of Virtual Work

and θ_c is the rotation at the point of application of couple C in the sense of C. By the assumptions regarding the internal forces and the deformations, δW_σ can be evaluated, i.e., δW_σ can be treated as a known quantity. If the displacements u_r and the rotations θ_c are known, then each of the terms Ru_r and $C\theta_c$ can be evaluated. Under the foregoing conditions, with the additional assignment of $Q = 1$, the work equation becomes

$$u_q = \delta W_\sigma - \sum_r ru_r - \sum_c c\theta_c \tag{a}$$

in which r and c are the values of R and C for $Q = 1$. Because all quantities on the right hand side of Equation (B.4) are known, the equation can be used to calculate u_q. Equation (a) is the formulation of an extended version of the unit-load method or the virtual-work method of calculating displacements. For a plane-framed structure in which the members undergo axial and bending deformation

$$\delta W_\sigma = \sum \int_0^L (n\varepsilon + m\kappa)\, dx$$

where n and m represent any distribution of internal axial forces and bending moments such that the structure is in equilibrium under the external forces $Q = 1, r_1, r_2, \ldots, r_r, c_1, c_2, \ldots, c_c$. If the deformations are due to applied loads and the material obeys Hooke's law

$$\delta W_\sigma = \sum \int_0^L \left(\frac{nN}{EA} + \frac{mM}{EI}\right) dx$$

and the expression for the displacement takes the working form

$$u_q = \sum \int_0^L \left(\frac{nN}{EA} + \frac{mM}{EI}\right) dx - \sum_r ru_r - \sum_c c\theta_r$$

Frame Example

The members of the frame shown in Figure A.2 are assumed to be axially rigid. A sidesway analysis of the frame was performed using moment distribution and superposition. The sidesway displacement u was found to be such that $EIu = -84.0$ k-ft³ and the calculated values for the end-moments, positive in the counterclockwise sense, are

$M_{12} = -13.33$ k-ft, $M_{21} = -21.1$ k-ft, $M_{23} = 21.1$ k-ft, $M_{32} = -4.66$ k-ft

$M_{43} = -1.88$ k-ft, $M_{34} = 4.66$ k-ft

From the geometric analysis (which is an essential preliminary part of the sidesway analysis), it was determined that the vertical components of displacement at joints 2 and 3 are related to u as shown in Figure A.2. Calculate the vertical component of displacement at q, the mid-point of member 23.

Because the horizontal and vertical components of displacement at joints 2 and 3 are known, any set of values for $r_1, r_2, \ldots, r_{10}$ (shown in Figure A.2b) for which the frame is in equilibrium under the unit load defines internal forces n and m such that the principle of virtual work will yield the required displacement. Clearly, $r_1, r_2, \ldots, r_{10}$ should be chosen to keep the effort required to evaluate δW_σ small. A convenient choice which satisfies the equilibrium requirement is $r_i = 0$ for $i = 1, 2, \ldots, 8$, $r_9 = r_{10} = 0.5$. For this choice $n = 0$ for all members

426 Appendix A

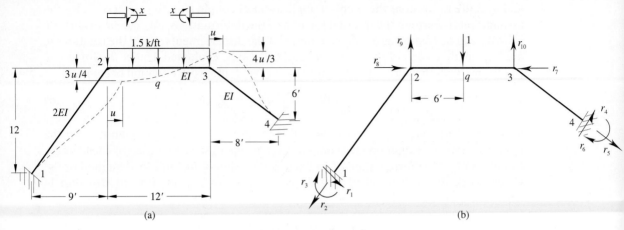

FIGURE A.2 Frame example.

and $m = 0$ for members 12 and 43. With coordinates and sign conventions as shown in Figure A.2, the bending moments for member 23 are

for segment $2q$: $M = -21.1 + 10.37x - 0.75x^2$, $m = 0.5x$ $0 \leq x \leq 6$
for segment $3q$: $M = -4.66 + 7.63x - 0.75x^2$, $m = 0.5x$ $0 \leq x \leq 6$

By the principle of virtual work, the vertical displacement at q reduces to the evaluation of

$$u_q = \sum_m \int_0^L \frac{mM}{EI} dx - \left(-\frac{1}{2}\right)\left(\frac{3u}{4}\right) - \left(\frac{1}{2}\right)\left(\frac{4u}{3}\right)$$

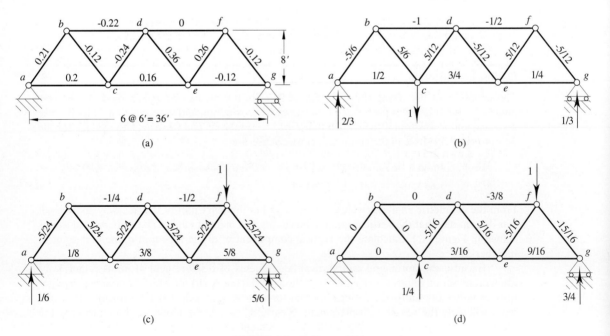

FIGURE A.3 Truss example.

from which $EIu_q = 197.7$ k-ft³. The result is readily verified by application of the moment-area method.

TABLE A.1

Bar	ΔL	$n_{(b)}$	$n_{(b)}\Delta L$	$n_{(c)}$	$n_{(c)}\Delta L$	$n_{(d)}$	$n_{(d)}\Delta L$
	in	—	in	—	in	—	in
(1)	(2)	(3)	(4)	(5)	(6)	(5)	(6)
ab	0.21	−5/6	−0.175	−5/24	−0.04375	0	0
ac	0.2	1/2	0.1	1/8	0.025	0	0
bd	−0.22	−1	0.22	−1/4	0.055	0	0
bc	−0.12	5/6	−0.1	5/24	−0.025	0	0
cd	−0.24	5/12	−0.1	−5/24	0.05	−5/16	0.075
ce	0.16	3/4	0.12	3/8	0.06	3/16	0.03
df	0	−1/2	0	−1/2	0	−3/8	0
de	0.36	−5/12	−0.15	5/24	0.075	5/16	0.1125
ef	0.36	5/12	0.15	−5/24	−0.075	−5/16	−0.1125
eg	−0.12	1/4	−0.03	5/8	−0.075	9/16	−0.0675
fg	−0.12	−5/12	0.05	−25/24	0.125	−15/16	0.1125
			0.085		0.17125		0.15

Truss Example

The differences, measured in inches, between the theoretical lengths of the members of a truss and the lengths as fabricated are as shown in Figure A.3a. Calculate the vertical component of displacement of joint c and of joint f due to the fabrication errors. By the principle of virtual work, $u_c = \delta W_\sigma(n_{(b)}) = \Sigma n_{(b)}\Delta L$ in which $n_{(b)}$ is the tension in a member due to a unit load applied at joint c as shown in Figure A.3b. Similarly, $u_f = \delta W_\sigma(n_{(c)}) = \Sigma n_{(c)}\Delta L$, where $n_{(c)}$ is the tension in a member due to a unit load applied at joint f as shown in Figure A.3c. As indicated by the sums shown in columns 4 and 6 of Table A.1, the displacements are $u_c = 0.085$ inches and $u_f = 0.17125$ inches. By the principle of virtual work, the vertical component of displacement of joint f is also given by $u_f = \delta W_\sigma(n_{(d)}) - r_c u_c$ in which $n_{(d)}$ is the tension in a member due to the loading shown in Figure A.3d, and r_c is the external force at c, measured positive in the direction of u_c. From column 8 of Table A.1, $\delta W_\sigma(n_{(d)}) = 0.15$ inches, and from Figure A.3d, $r_c = -1/4$. Therefore,

$$u_f = 0.15 - (-0.25)(0.085) = 0.17125 \text{ in.}$$

which agrees with the previously calculated value.

Appendix B
Sidesway Analysis

B.1 GENERAL FORMULATION AND SUPERPOSITION

The general formulation of a sidesway analysis, in terms of unknown joint rotations and displacements, has three characteristic features that detract from its utility in practical applications in which calculations are carried out manually. The first aspect is that of formulating appropriate independent equations of equilibrium that correspond to the independent joint displacement variables. The simple fact is that the formulation is not an easy task. The second aspect relates to the number of simultaneous equations to be solved. The analysis of a structure with q degrees of freedom with respect to joint rotation and r degrees of freedom with respect to joint translation requires one solution of one set of $q + r$ simultaneous equations for the $q + r$ unknown rotations and displacements. Because the number of operations required to solve a set of simultaneous linear algebraic equations is proportional to the cube of the number of equations, the work required in a manual solution becomes excessive, even with relatively small values of q and r. The third characteristic is the poorer conditioning of a set of n equations that involve displacements as well as rotations as unknowns relative to the conditioning of a set of n equations that contain only unknown rotations.

The intent of a sidesway analysis based on superposition is to avoid the objectionable features of the general formulation. No attempt is made herein to consider or set out details for conditions under which the superposition procedure is preferable to the general formulation. Rather, the goals of the discussion to follow are to distinguish between the two approaches to the sidesway problem and to show how they are related. Elements of matrix algebra are employed to achieve compactness. Differences between the matrix formulations presented here and those given in the section of Chapter 10, "**Equations of Equilibrium by Virtual Work/Matrix Methods**," are entirely superficial. Notational changes were introduced primarily to sidestep any discussion of partitioning matrices. Because there is no loss in generality, all structural members are treated as being axially rigid.

B.2 GENERAL FORMULATION

The elements $z_1, z_2, \ldots, z_s$ of the displacement vector z are split into two groups. The first group forms a joint rotation vector identified by $x^T = (z_1, z_2, \ldots, z_q) = (x_1, x_2, \ldots, x_q)$ in which each element is an independent joint rotation. The second group forms a joint translation vector identified by $y^T = (z_{q+1}, z_{q+2}, \ldots, z_{q+r}) = (y_1, y_2, \ldots, y_r)$ where $y_1, y_2, \ldots, y_r$ are the independent physical components of joint translation that define the sidesway displacements. The member end-force vectors are joined to form a single structure member end-force vector R, the elements of which are the end-moments and end-shears for all members of the structure. With the aid of the slope-deflection equations and the geometric relations that connect member end-displacements to joint rotations and translations, R can be expressed in the form

$$R = Dx + Ey + R_F \tag{B.1}$$

where D is a matrix of $4m$ rows and q columns, E is a matrix of $4m$ rows and r columns, R_F is a column vector of $4m$ elements, and m is the number of members that make up the structure. The elements in D and E are properties of the structure. The elements are determined by the mechanical properties of the members and how the members are joined. The elements in R_F are the fixed end-forces due to the loading between the ends of the members.

The moment equations of equilibrium for the joints corresponding to the rotations in the joint rotation vector x are

$$GR - a_{(p)} = 0 \tag{B.2}$$

in which G is a matrix of q rows and $4m$ columns, and $a_{(p)}$ is a column vector of q rows. The elements in G are constants that characterize the geometry of the structure. The elements in $a_{(p)}$ are the prescribed external couples applied to the joints.

In application of the superposition method, the prescribed forces at the joints are divided into two groups. The first group consists of the r forces $F_1, F_2, \ldots, F_j, \ldots, F_r$ where F_j is the prescribed force at the point where displacement y_j is measured. F_j is measured positive in the direction of y_j. The second group contains all remaining prescribed forces. By application of the principle of virtual work, it can be shown that the force equations of equilibrium corresponding to the independent displacements $y_1, y_2, \ldots, y_r$ can be written in the form

$$HR - b_{(p)} - b_{(y)} = 0 \tag{B.3}$$

where $b_{(p)}$ is a column vector of r rows whose elements are determined by all prescribed joint forces except for the r forces $F_1, F_2, \ldots, F_r$; and $b_{(y)}$ is a column vector whose elements are the r forces $F_1, F_2, \ldots, F_r$, i.e., $b_{(y)}^T = (F_1, F_2, \ldots, F_r)$.

The equations of the general formulation of the sidesway problem in terms of unknown rotations and displacements are derived by using Equation (B.1) to eliminate R from Equations (B.2) and (B.3). With the aid of the principle of virtual work, it can be shown that the equations can always be cast in the form

$$\begin{aligned} Ax + By &= a = a_{(F)} + a_{(p)} &\text{(a)} \\ B^T x + Cy &= b = b_{(F)} + b_{(p)} + b_{(y)} &\text{(b)} \end{aligned} \tag{B.4}$$

in which A and C are symmetric matrices, $a_{(p)}, b_{(p)},$ and $b_{(y)}$ are vectors as have been defined previously, and $a_{(F)}$ and $b_{(F)}$ are vectors whose elements are defined completely by the fixed end-forces due to loads between the ends of the members. The solution of Equations (B.4) can be expressed in various ways and the results can be given physical interpretations. One arrangement leads to the basic idea of the superposition method of solution. Let $b_{(0)}$ define the particular forces in $b_{(y)}$, which, when acting with the forces that define $a_{(F)}, b_{(F)}$ and $b_{(p)}$, produce no joint displacements, i.e., the elements in $b_{(0)}$ are the forces that are required to act simultaneously with the loads that define $a_{(F)}, b_{(F)}$ and $b_{(p)}$ to maintain the structure in the equilibrium position defined by $y = 0$. Let x_0 be the vector of joint rotations for this case of loading. Then, from Equations (B.4)

$$Ax_0 = a \qquad (a)$$
$$B^T x_0 = b_{(F)} + b_{(p)} + b_{(0)} \qquad (b) \qquad (B.5)$$

By using the foregoing results, the solution of Equation (B.4a) can be expressed as

$$x = x_0 - A^{-1}By \qquad (B.6)$$

and Equation (B.4b) is reduced to

$$(C - B^T A^{-1} B)y + b_{(0)} = b_{(y)}$$

which can be written as

$$K_y y + b_{(0)} = b_{(y)} \qquad (B.7)$$

where $K_y = C - B^T A^{-1} B$. The rotation vector x as given by Equation (B.6) together with the displacement vector y obtained by solving Equation (B.7) define the solution of the sidesway problem via the general formulation. With the displacements known, the member end-forces may be obtained by back substitution into Equation (B.1).

The matrix K_y in Equation (B.7) is the stiffness matrix of the structure with respect to displacement y. Because A and C are symmetric matrices, K_y is also symmetric, as it must be. Clearly, the construction of K_y from the defining relation $K_y = C - B^T A^{-1} B$ relies on the general formulation of the sidesway problem for the construction of the matrices A, B, and C. In addition, evaluation of the elements in K_y requires inversion of matrix A, and the evaluation of the triple product $B^T A^{-1} B$.

B.3 SUPERPOSITION

The essence of the superposition method follows from observations regarding Equation (B.7). It has been noted that the elements in the vector $b_{(0)}$ are nothing more than the forces that must be applied at the points where the displacements $y_1, y_2, \ldots, y_r$ are measured, taken positive in the directions of the displacements, when the displacement is $y = 0$ and all applied loads except for the prescribed forces $F_1, F_2, \ldots, F_r$ (which are the elements in the vector $b_{(y)}$) act on the structure. Because the joint displacements are known ($y = 0$), calculation of the forces in $b_{(0)}$ can be accomplished by means of an uncomplicated analysis in which the slope-deflection method or moment distribution is used first to calculate the end-moments for the given condition. With the end moments known, an easy application of the equations of equilibrium is then used to calculate end-shears. With the end-moments and end-shears known, additional applications of equilibrium principles yield the forces in $b_{(0)}$. Clearly, in this way $b_{(0)}$ is obtained without the application of sophisticated techniques involving matrix algebra.

With the loads that generate the load vector $b_{(0)}$ removed from the structure, Equation (B.7) becomes

$$K_y y = b_{(y)} \qquad (B.8)$$

which can be interpreted as an equation that defines the force vector $b_{(y)}$ that must be supplied to maintain the structure in equilibrium in the deformed state defined by any prescribed sidesway displacement vector y. This observation provides a basis for the calculation of the columns in the stiffness matrix K_y. Let $b_{(i)}$ be the force vector that defines the forces that must be supplied when the elements in the prescribed displacement vector are $y_i = 1$, and $y_j = 0$, for $j \neq i$. Then, the left hand side of Equation (B.8) reduces to column i of the stiffness matrix K_y. Denote column i of matrix K_y by $\{k_y\}_i$. Then

$$\{k_y\}_i = b_{(i)} \tag{B.9}$$

Equation (B.9) is the basis for a direct calculation of column i of K_y by means of an ordinary analysis of the structure that parallels the analysis employed in the calculation of $b_{(0)}$. Because the displacements are prescribed either the slope-deflection equation method or moment distribution can be used to determine the associated end-moments in each member. With the end-moments known, statics can be used to calculate the end-shears and the required forces in the force vector $b_{(i)}$.

Initially, in the superposition method of sidesway analysis Equation (B.7) is interpreted as being a relationship that defines the elements in force vector $b_{(y)}$ as the superposition of the forces that must be supplied under the prescribed loads with no sidesway, and the forces that must be supplied to maintain an arbitrary sidesway condition as defined by the displacement y. Then the actual sidesway displacements are found by choosing y so that Equation (B.7) is satisfied when the forces in $b_{(y)}$ are set equal to their respective assigned values.

Superposition is used to determine the member forces as well as the sidesway displacements. Let R_0 denote the member end-force vector due to the applied loads when $y = 0$, and let R_i denote the member end-force vector that results when the applied loads are removed and the assigned displacements are $y_i = 1$ and $y_j = 0$ for $j \neq i$. Observe that the elements in R_0 and R_i are calculated in the course of the conventional analyses that are employed to generate the force vectors $b_{(0)}$ and $b_{(i)}$. Then, by superposition the end forces are given by

$$R = R_0 + \sum_{i=1}^{r} R_i y_i \tag{B.10}$$

A comparison of Equations (B.1) and (B.10) shows that $R_0 = Dx_0 + R_F$ where x_0 is the vector of joint rotations when there is no sidesway.

Answers

Notation for answers to problems:

Global Coordinates, fixed in space—in this book $\mathbf{i}_X$ is always directed as shown.

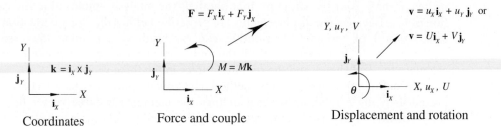

| Coordinates | Force and couple | Displacement and rotation |

Local coordinates—orientation defined by spatial orientation of associated member.

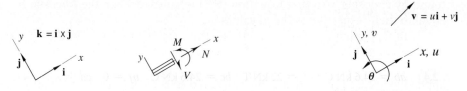

| Coordinates | Bending moment, shear, and axial force | Displacement and rotation |

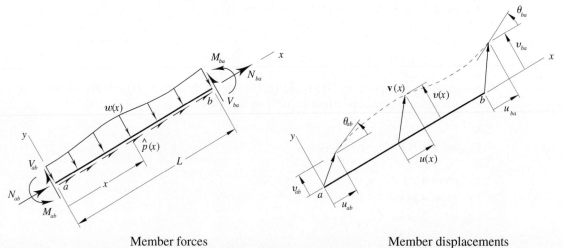

Member forces   Member displacements

The notation $lc(\alpha, \beta)$ is used to denote the origin and orientation of the local coordinates for member $\alpha\beta$. The origin is located at the α-end and the x-axis is directed from the α-end toward the β-end.

Answers

1.1 $A_X = 2.61\,\text{k}$ $A_Y = 2.61\,\text{k}$ $B_X = -2.61\,\text{k}$ $B_Y = 1.29\,\text{k}$

1.2 $B_Y = 15\,\text{k}$ $E_X = -4\,\text{k}$ $E_Y = 0$ $M_E = 10\,\text{k-ft}$

1.3 $A_X = 11.54\,\text{kN}$ $M_A = -67.2\,\text{kN·m}$ $B_X = -11.54\,\text{kN}$ $B_Y = 20\,\text{kN}$

1.4 $A_X = 0$ $A_Y = 23.6\,\text{k}$ $B_Y = -17.56\,\text{kN}$

1.5 $A_X = 0$ $A_Y = 16.13\,\text{k}$ $F_X = 0$ $F_Y = 21.9\,\text{k}$

1.6 $A_X = 9.87\,\text{kN}$ $B_X = -9.87\,\text{kN}$ $B_Y = 19.74\,\text{kN}$

1.7 $A_X = 18.11\,\text{kN}$ $A_Y = 6.58\,\text{kN}$ $C_Y = 25.5\,\text{kN}$ $B_X = 0$ $B_Y = -11.53\,\text{kN}$
$M_B = -2.96\,\text{kN·m}$

1.8 $A_Y = 5.45\,\text{k}$ $D_X = 0$ $D_Y = 6.35\,\text{k}$ $B_X = 0$ $B_Y = -6.35\,\text{k}$ $M_B = -25.4\,\text{k-ft}$

1.9 $A_Y = 0.75\,\text{k}$ $D_X = -0.608\,\text{k}$ $E_X = -1.992\,\text{k}$ $E_Y = 2.25\,\text{k}$ $B_Y = -2.25\,\text{k}$

1.10 $A_X = 3.56\,\text{k}$ $A_Y = 6.16\,\text{k}$ $C_X = 5.44\,\text{k}$ $C_y = 8.24\,\text{k}$ $B_X = 3.56\,\text{k}$
$B_Y = -8.24\,\text{k}$ $M_B = -9.34\,\text{k-ft}$

1.11 $F_X = -6\,\text{kN}$ $F_Y = 13\,\text{kN}$ $E_X = -6\,\text{kN}$ $E_Y = 1\,\text{kN}$ $M_E = -6\,\text{kN·m}$

1.12 $A_X = -2.64\,\text{k}$ $A_Y = 0.568\,\text{k}$ $M_A = 20.7\,\text{k-ft}$ $E_X = -1.757\,\text{k}$ $E_Y = 2.03\,\text{k}$
$D_X = 1.757\,\text{k}$ $D_Y = 0.568\,\text{k}$

2.1 $ab = 6.25\,\text{k C}$ $af = 4.67\,\text{k T}$ $bc = 1.041\,\text{k C}$ $bf = 3.33\,\text{k C}$ $cd = 0.854\,\text{k C}$
$cf = 1.333\,\text{k T}$ $de = 2.13\,\text{k C}$ $df = 2.15\,\text{k T}$ $ef = 1.360\,\text{k C}$

2.2 $ab = 29.1\,\text{kN C}$ $af = 26\,\text{kN T}$ $bc = 15.65\,\text{kN C}$ $bg = 13.42\,\text{kN C}$ $bf = 12\,\text{kN T}$
$cd = 15.65\,\text{kN C}$ $cg = 14\,\text{kN T}$ $de = 33.5\,\text{kN C}$ $dh = 16\,\text{kN T}$ $dg = 17.89\,\text{kN C}$
$eh = 30\,\text{kN T}$ $fg = 26\,\text{kN T}$ $gh = 30\,\text{kN T}$

2.3 $ab = 14.91\,\text{kN C}$ $af = 13.33\,\text{kN T}$ $bc = 5.96\,\text{kN C}$ $bf = 4\,\text{kN T}$
$cd = 17.88\,\text{kN T}$ $ch = 18.86\,\text{kN C}$ $cg = 0$ $cf = 11.31\,\text{kN T}$ $de = 17.89\,\text{kN T}$
$dh = 0$ $eh = 16\,\text{kN C}$ $fg = 5.33\,\text{kN T}$ $gh = 5.33\,\text{kN T}$

2.4 $ab = 24.6\,\text{kN C}$ $af = 22\,\text{kN T}$ $bc = 24.6\,\text{kN C}$ $bf = 0$ $cd = 20.1\,\text{kN C}$
$ch = 11.31\,\text{kN T}$ $cg = 0$ $cf = 16.97\,\text{kN T}$ $de = 20.1\,\text{kN C}$ $dh = 0$
$eh = 18\,\text{kN T}$ $fg = 10\,\text{kN T}$ $gh = 10\,\text{kN T}$

2.5 $ab = 5\,\text{kN C}$ $af = 5.59\,\text{kN T}$ $ae = 5.59\,\text{kN C}$ $bc = 6\,\text{kN C}$ $bg = 1.118\,\text{kN T}$
$bf = 1.118\,\text{kN C}$ $cd = 3\,\text{kN C}$ $ch = 3.35\,\text{kN C}$ $cg = 3.35\,\text{kN T}$ $di = 3.35\,\text{kN C}$
$dh = 3.35\,\text{kN T}$ $ef = 2.5\,\text{kN T}$ $fg = 5.5\,\text{kN T}$ $gh = 4.5\,\text{kN T}$ $hi = 1.5\,\text{kN T}$

2.6 $bc = 57.6\,\text{kN C}$ $bg = 18.78\,\text{kN T}$ $gc = 8.05\,\text{kN T}$

2.7 $dh = 12.92\,\text{k T}$ $ch = 12\,\text{k C}$ $gh = 17.67\,\text{k T}$

2.8 $dg = 26.8\,\text{kN C}$ $fg = 24\,\text{kN C}$ $bg = 48\,\text{kN C}$

2.9 $ce = 4.33\,\text{k C}$ $cf = 2.7\,\text{k T}$ $ef = 0.833\,\text{k T}$

2.10 $fe = 55.9\,\text{k C}$ $gf = 50.9\,\text{k C}$ $ch = 13.42\,\text{k T}$

2.11 $ab = 1.714\,\text{k T}$ $bc = 0.446\,\text{k T}$ $ad = 2.83\,\text{k C}$

2.12 $bc = 3.58\,\text{k C}$ $cf = 9.30\,\text{k T}$

2.13 $ab = 4\,\text{k T}$ $ac = 7.73\,\text{k T}$ $bc = 15\,\text{k C}$

2.14 $ab = 4.24\,\text{kN C}$ $ad = 18\,\text{kN T}$ $bc = 4.24\,\text{kN T}$

2.15 $ab = 27\,\text{k C}$ $ac = 0$ $ad = 58.0\,\text{k C}$

2.16 $ab = 52.2\,\text{kN C}$ $bc = 26.7\,\text{kN C}$

2.17 $ab = 0.0981P\,\text{T}$ $bc = 0.634P\,\text{C}$ $ac = 0.1895P\,\text{C}$

2.18 $ab = 15.54\,\text{k C}$ $ac = 3.13\,\text{k C}$

2.19 $ab = 16.39\,\text{k T}$ $ac = 16.39\,\text{k C}$ $bc = 4.40\,\text{k T}$

2.20 $ab = 5.11\,\text{kN C}$ $ac = 10.82\,\text{kN C}$ $bc = 13\,\text{kN T}$

434 Answers

2.21 $ab = 74.5$ kN T

2.22 $ab = 33.9$ kN C

2.23 $ab = 18.69$ k T $aj = 17.5$ k T $bc = 1.602$ k C $bi = 17.09$ k C $bj = 0$
$cd = 18.69$ k T $cg = 12.62$ k C $ci = 5.41$ k T $de = 18.69$ k T $df = 0$ $dg = 0$
$ef = 17.5$ k C $fg = 17.5$ k C $gh = 14.85$ k C $hi = 2.12$ k C $ij = 17.5$ k T

2.24 Except for fg, all members are in pure axial tension or compression.
$ab = 2.31$ k C $ae = 2.06$ k T $bc = 4.13$ k C $bf = 2.31$ k T $be = 0$
$cd = 4.13$ k C $cf = 0$ $dh = 4.40$ k C $dg = 1.875$ k T $df = 0.21$ k T
$ef = 2.06$ k T $gh = 3.94$ k T
Member $fg: f_X = -3.94$ k $f_Y = 1.125$ k $g_X = 3.94$ k $g_Y = 1.875$ k

2.25 Except for bd, all members are in pure axial tension or compression.
$ab = 5$ k C $ad = 3.54$ k T $ac = 3.54$ k C $be = 4.95$ k C $cd = 2.5$ k T
$de = 3.5$ k T
Member $db: \text{lc}(db)\ d_x = 0.707$ k $d_y = 2.83$ k $b_x = 3.54$ k $b_y = 1.414$ k

3.1 $V = 21$ kN, $M = -15 + 21x$ kN·m $0 < x < 3$
$V = 0$, $M = 48$ kN·m $3 < x < 5$

3.2 $V = 6 - 6x$ kN, $M = 6x - 3x^2$ kN·m $0 < x < 4$
$V = -18$ kN, $M = 6x - 24(x - 2)$ kN·m $4 < x < 6$
$V = -40$ kN, $M = -140 + 40(8 - x)$ kN·m $6 < x < 8$

3.3 $V = 2230 - 450x$ lb, $M = 2230x - 225x^2$ lb-ft $0 < x < 7$
$V = -919$ lb, $M = 919(12 - x)$ lb-ft $7 < x < 12$

3.4 $V = -0.175x^2$ kN, $M = -0.0583x^3$ kN·m $0 < x < 1$
$V = 1.244 - 0.175x^2$ kN, $M = 1.244(x - 1) - 0.0583x^3$ kN·m $1 < x < 4$

3.5 $V = 3.4$ lb, $M = 3.4x$ lb-ft $0 < x < 7$
$V = -10.6$ lb, $M = 98 - 10.6x$ lb-ft $7 < x < 10$

3.6 $V = 54 - 9x$ kN, $M = 54x - 4.5x^2$ kN·m $0 < x < 6$

3.7 $V_{max} = 12$ kN $V_{min} = -35.9$ kN $M_{max} = 29.8$ kN·m $M_{min} = -42$ kN·m

3.8 $V_{max} = 4$ k $V_{min} = -2.78$ k $M_{max} = 0$ $M_{min} = -12$ k-ft

3.9 $V_{max} = 17.08$ k $V_{min} = -24.5$ k $M_{max} = 50.1$ k-ft $M_{min} = 0$

3.10 $V_{max} = 29.2$ kN $V_{min} = -26.3$ kN $M_{max} = 24$ kN·m $M_{min} = -60.4$ kN·m

3.11 $V_{max} = 25.2$ kN $V_{min} = -50.8$ kN $M_{max} = 6.3$ kN·m $M_{min} = -50.4$ kN·m

3.12 $V_{max} = 1.333$ k $V_{min} = -1.45$ k $M_{max} = 0.1361$ k-ft $M_{min} = -5.33$ k-ft

3.13 $V_{max} = 1533$ lb $V_{min} = -267$ lb $M_{max} = 0$ $M_{min} = -4130$ lb-ft

3.14 $V_{max} = 6$ k $V_{min} = -9$ k $M_{max} = 10$ k-ft $M_{min} = -18$ k-ft

3.15 $V_{max} = 9$ k $V_{min} = 0$ $M_{max} = 0$ $M_{min} = -18$ k-ft

3.16 $V_{max} = 3$ k $V_{min} = -2.7$ k $M_{max} = 10.13$ k-ft $M_{min} = -12.38$ k-ft

3.17 $V_{max} = 10.36$ kN $V_{min} = -14.28$ kN $M_{max} = 14.39$ kN·m $M_{min} = -8.78$ kN·m

3.18 Member $AB: \text{lc}(AB)$ $V_{max} = 6.2$ k $V_{min} = -4.6$ k $M_{max} = 16.02$ k-ft $M_{min} = 0$
Member $BC: \text{lc}(BC)$ $V_{max} = 0$ $V_{min} = -4.8$ k $M_{max} = 7.2$ k-ft $M_{min} = 0$

3.19 Member $AB: \text{lc}(AB)$ $V_{max} = 4.93$ k $V_{min} = -2.27$ k $M_{max} = 15.21$ k-ft $M_{min} = 0$
Member $EC: \text{lc}(EC)$ $V_{max} = 0$ $V_{min} = -6$ k $M_{max} = 0$ $M_{min} = -12$ k-ft
Member $DB: \text{lc}(DB)$ $V_{max} = 0$ $V_{min} = 0$ $M_{max} = 0$ $M_{min} = -12$ k-ft

3.20 Member $AB: \text{lc}(AB)$ $V_{max} = 6$ k $V_{min} = -3.6$ k $M_{max} = 22.5$ k-ft $M_{min} = 0$
Member $BC: \text{lc}(BC)$ $V_{max} = -1.61$ k $V_{min} = -1.61$ k $M_{max} = 14.4$ k-ft
$M_{min} = 0$

Answers 435

3.21 Member AB: lc(AB) $V_{max} = 0$ $V_{min} = -12$ k $M_{max} = 0$ $M_{min} = -36$ k-ft
Member BC: lc(BC) $V_{max} = 18$ k $V_{min} = 0$ $M_{max} = 45$ k-ft $M_{min} = -36$ k-ft

3.22 Member DB: lc(DB) $V_{max} = 0$ $V_{min} = 0$ $M_{max} = 0$ $M_{min} = 0$
Member AC: lc(AC) $V_{max} = 0.15$ kN $V_{min} = -6.25$ kN $M_{max} = 0$
$M_{min} = -15$ kN·m
Member CE: lc(CE) $V_{max} = 0$ $V_{min} = -3.75$ kN $M_{max} = 0$ $M_{min} = -15$ kN·m

3.23 Member AC: lc(AC) $V_{max} = 2.25$ k $V_{min} = -2$ k $M_{max} = 7.5$ k-ft
$M_{min} = -6$ k-ft
Member FB: lc(FB) $V_{max} = 0$ $V_{min} = 0$ $M_{max} = 0$ $M_{min} = 0$
Member DE: lc(DE) $V_{max} = 1$ k $V_{min} = 1$ k $M_{max} = 0$ $M_{min} = -3$ k-ft
Member GC: lc(GC) $V_{max} = 0$ $V_{min} = 0$ $M_{max} = 3$ k-ft $M_{min} = 0$

3.24 Member AB: lc(AB) $V_{max} = 2$ k $V_{min} = -2$ k $M_{max} = 4$ k-ft $M_{min} = 0$
Member AD: lc(AD) $V_{max} = -1.66$ k $V_{min} = -1.66$ k $M_{max} = 0$
$M_{min} = -24$ k-ft
Member CE: lc(CE) $V_{max} = 4$ k $V_{min} = 0$ $M_{max} = 16$ k-ft $M_{min} = -8$ k-ft

4.1 $A_Y = (s-5)/10$ $B_Y = (15-s)/10$ $0 < s < 15$
$v_C = (s-5)/10$ $m_C = 0.6(s-5)$ ft $0 < s < 9$
$v_C = -(15-s)/10$ $m_C = 0.4(15-s)$ ft $9 < s < 15$

4.2 $A_Y = 1$ $m_A = -s$ ft $0 < s < 8$
$v_B = 0$ $0 < s < 5$
$v_B = -1$ $5 < s < 8$

4.3 $B_Y = 1$ $m_A = s$ $0 < s < L$

4.4 $A_Y = (s-1.5)/4$ $0 < s < 6.5$
$m_A = 0$ $0 < s < 5.5$, $m_A = 5.5 - s$ m $5.5 < s < 6.5$
$v_C = (s-1.5)/4$ $0 < s < 4$, $v_C = (s-5.5)/4$ $4 < s < 6.5$

4.5 left $v_B = 1 - (5.5-s)/4$ $0 < s < 1.5$, $v_B = -(5.5-s)/4$ $1.5 < s < 6.5$
right $v_B = 1$ $0 < s < 1.5$, $v_B = 0$ $1.5 < s < 6.5$
$m_B = -(1.5-s)$ m $0 < s < 1.5$, $m_B = 0$ $1.5 < s < 6.5$

4.6 $A_Y(0^+) = -1$ $A_Y(24) = -1$ $A_Y(48^-) = 1$
$v_D(0^+) = 1$ $v_D(18^-) = 1$ $v_D(18^+) = 0$ $v_D(48^-) = 1$

4.7 $B_Y(0^+) = 2$ $B_Y(24) = 2$ $B_Y(48) = 0$
$m_D(0^+) = 6$ ft $m_D(18) = 6$ ft $m_D(24) = 0$ $m_D(48) = 0$

4.8 $m_A(0^+) = 15$ ft $m_A(10) = -10$ ft $m_A(20) = 0$
$v_C(0^+) = -1.5$ $v_C(6^-) = 0$ $v_C(6^+) = -1$ $v_C(10) = 0$ $v_C(20) = 0$

4.9 $v_A(0^+) = -1.5$ $v_A(10) = 1$ $v_A(20^-) = 1$
$m_C(0^+) = -6$ ft $m_C(10) = 0$ $m_C(20) = 0$

4.10 $v_A(0) = 0$ $v_A(15) = 0$ $v_A(19^-) = 0.4$ $v_A(19^+) = -0.6$ $v_A(25) = 0$
$m_B(0) = 0$ $m_B(10) = 2.4$ ft $m_B(15) = -3$ ft $m_B(25) = 0$

4.11 $v_B(0) = 0$ $v_B(6^-) = 0.6$ $v_B(6^+) = -0.4$ $v_B(15) = 5$ $v_B(25) = 0$
$m_A(0) = 0$ $m_A(15) = 0$ $m_A(19) = 2.4$ ft $m_A(25) = 0$

4.12 $m_A(0) = 0$ $m_A(8) = 0$ $m_A(12) = -4$ m $m_A(16) = 0$
$m_D(0) = 0$ $m_D(6) = 0$ $m_D(8) = -2$ m $m_D(12) = 0$ $m_D(16) = 0$

4.13 $v_A(0) = 0$ $v_A(8) = 0$ $v_A(12) = 1$ $v_A(16) = 1$
$v_D(0) = 0$ $v_D(6^-) = 0$ $v_D(6^+) = -1$ $v_D(8) = -1$ $v_D(12) = 0$ $v_D(16) = 0$

4.14 $A_Y(0) = 0$ $A_Y(15^-) = 0$ $A_Y(15^+) = 1$ $A_Y(25^-) = 1$
$v_C(0) = 0$ $v_C(10^-) = 0$ $v_C(10^+) = -1$ $v_C(15^-) = -1$ $v_C(15^+) = 0$
$v_C(25) = 0$

4.15 $D_Y(0^+) = 1$ $D_Y(15^-) = 1$ $D_Y(15^+) = 0$ $D_Y(25) = 0$
$m_C(0) = 0$ $m_C(10) = 10$ ft $m_C(15^-) = 10$ ft $m_C(15^+) = 0$ $m_C(25) = 0$

4.16 $m_D(0^+) = 4$ m $m_D(6^-) = -2$ m
$m_B(0^+) = -2$ m $m_B(2) = 0$ $m_B(6) = 0$

4.17 lc(DB) $v_B(0^+) = 0.707$ $v_B(6^-) = 0.707$
$m_B(0^+) = -2$ m $m_B(6^-) = 6$ m

4.18 lc(CD) $v_C(0^+) = -0.5$ $v_C(16^-) = 0.5$ $v_C(16^+) = -0.5$ $v_C(24) = 0$
$m_C(0^+) = -4$ ft $m_C(16) = 4$ ft $m_C(24) = 0$

4.19 lc(AC) $v_C(0^+) = -0.4$ $v_C(24^-) = 0.8$
$m_C(0^+) = -4$ ft $m_C(24^-) = 8$ ft

4.20 $A_X(0) = 0$ $A_X(30^-) = 3$
$v_E(0) = 0$ $v_E(20^-) = 0$ $v_E(20^+) = -1$ $v_E(30^-) = -1$
$m_E(0) = 0$ $m_E(20) = -20$ ft $m_E(30^-) = -30$ ft

4.21 lc(DC) $v_C(0) = 0$ $v_C(30^-) = 3$
$m_C(0) = 0$ $m_C(30^-) = 60$ ft

4.22 $A_X(0) = 0$ $A_X(10) = 3/7$ $A_X(16) = 0$
$m_B(0) = 0$ $m_B(10) = -24/7$ ft $m_B(16) = 0$

4.23 $E_Y(0^+) = 1$ $E_Y(10) = 3/7$ $E_Y(16) = 0$
lc(ED) $m_D(0) = 0$ $m_D(10) = 30/7$ ft $m_D(16) = 0$

4.24 $V_{D\,max} = 1535$ lb $V_{D\,min} = -6200$ lb
$M_{D\,max} = 28700$ lb-ft $M_{D\,min} = -12530$ lb-ft

4.25 $M_{A\,max} = -42$ k-ft $M_{A\,min} = -163$ k-ft
$V_{B\,max} = -3.9$ k $V_{B\,min} = -11$ k

4.26 $V_{A\,max} = 32$ kN $V_{A\,min} = -10.5$ kN
$M_{A\,max} = 42$ kN·m $M_{A\,min} = -43$ kN·m

4.27 $v_{AB}(0) = 0$ $v_{AB}(45) = 0.75$ $v_{AB}(60) = 0$
$m_B(0) = 0$ $m_B(45) = 11.25$ ft $m_B(60) = 0$

4.28 $v_{BC}(0) = 0$ $v_{BC}(30) = 0.5$ $v_{BC}(45) = -0.25$ $v_{BC}(60) = 0$
$m_C(0) = 0$ $m_C(30) = 15$ ft $m_C(60) = 0$

4.29 $v_{AB}(0) = 0$ $v_{AB}(14.4) = 0.8$ $v_{AB}(18) = 0$
$m_B(0) = 0$ $m_B(14.4) = 2.88$ m $m_B(18) = 0$

4.30 $v_{DE}(0) = 0$ $v_{DE}(3.6) = 0.2$ $v_{DE}(7.2) = -0.6$ $v_{DE}(18) = 0$
$m_D(0) = 0$ $m_D(7.2) = 4.32$ m $m_D(18) = 0$

4.31 $v_{BC}(0) = 0$ $v_{BC}(24) = 0.4$ $v_{BC}(48) = -0.2$ $v_{BC}(60) = 0$
$m_B(0) = 0$ $m_B(48) = 9.6$ ft $m_B(60) = 0$

4.32 $v_{DE}(0^+) = -0.25$ $v_{DE}(16) = -0.75$ $v_{DE}(40) = 0$
$m_D(0^+) = -6$ ft $m_D(16) = 6$ ft $m_D(40) = 0$

4.33 $v_{EF}(0^+) = 1$ $v_{EF}(16) = 0$ $v_{EF}(40) = 0$
$m_E(0^+) = -8$ ft $m_E(16) = 0$ $m_E(40) = 0$

4.34 $v_{bc}(0^+) = -0.25$ $v_{bc}(24^-) = 0.5$ $v_{bc}(24^+) = -0.5$ $v_{bc}(40) = 0$
$m_B(0^+) = -2$ ft $m_B(24^-) = 4$ ft $m_B(24^+) = 12$ ft $m_B(40) = 0$

4.35 $v_{CD}(0^+) = -0.25$ $v_{CD}(16) = 0.25$ $v_{CD}(24) = -0.5$ $v_{CD}(40) = 0$
$m_C(0^+) = -4$ ft $m_C(24) = 8$ ft $m_C(40) = 0$

4.36 $v_{BC}(0^+) = -1/6$ $v_{BC}(42^-) = 5/6$ $v_{BC}(42^+) = -1/6$ $v_{BC}(60^-) = 1/3$
$m_B(0) = 0$ $m_B(42^-) = 0$ $m_B(42^+) = 6$ ft $m_B(60^-) = -12$ ft

Answers 437

4.37 $v_{DC}(0^+) = -1/3$ $v_{DC}(24) = 1/3$ $v_{DC}(42^-) = -2/3$ $v_{DC}(42^+) = -1/6$
$v_{DC}(60^-) = 1/3$
$m_D(0^+) = -8$ ft $m_D(24) = 8$ ft $m_D(60^-) = -4$ ft

4.38 $v_{EF}(0) = 0$ $v_{EF}(4^-) = 2$ $v_{EF}(4^+) = 0$ $v_{EF}(12) = 0$
$m_E(0) = 0$ $m_E(2^-) = -4$ m $m_E(4^+) = 0$ $m_E(12) = 0$

4.39 $v_{DE}(0) = 0$ $v_{DE}(4^-) = -0.5$ $v_{DE}(4^+) = 0$ $v_{DE}(6) = -0.75$ $v_{DE}(12) = 0$
$m_D(0) = 0$ $m_D(4^-) = -3$ m $m_D(4^+) = 0$ $m_D(6) = 1.5$ m $m_D(12) = 0$

4.40 $v_{CD}(0) = 0$ $v_{CD}(8) = 0$ $v_{CD}(12) = -1$ $v_{CD}(16) = -2/3$ $v_{CD}(20) = 0$
$m_D(0) = 0$ $m_D(8) = 0$ $m_D(12) = -4$ m $m_D(16) = -4$ m $m_D(20) = 0$

4.41 $v_{DE}(0) = 0$ $v_{DE}(4) = 0.5$ $v_{DE}(8) = 0$ $v_{DE}(12) = 0.5$ $v_{DE}(16) = 0.5$
$v_{DE}(20) = 0$
$m_B(0) = 0$ $m_B(12) = 0$ $m_B(16) = 4/3$ $m_B(20) = 0$

4.42 $v_{CD}(0^+) = -1.5$ $v_{CD}(8^-) = 0$
$m_B(0^+) = 3.75$ ft $m_B(8^-) = 0$

4.43 $v_{BC}(0^+) = -1.5$ $v_{BC}(8^-) = 0$
$m_D(0^+) = -3$ ft $m_D(8) = 0$

4.44 $v_{CD}(0^+) = -1.5$ $v_{CD}(30) = 0$ $v_{CD}(50) = 0$
$m_D(0^+) = 30$ ft $m_D(30) = -10$ ft $m_D(50) = 0$

4.45 $v_{AB}(0^+) = -1.5$ $v_{AB}(30) = 1$ $v_{AB}(50) = 0$
$m_D(0^+) = -12$ ft $m_D(30) = 0$ $m_D(50) = 0$

4.46 $V_{CD\,max} = 125.3$ kN $V_{CD\,min} = 39.3$ kN
$M_{C\,max} = 1148$ kN·m $M_{C\,min} = 509$ kN·m

4.47 $V_{DE\,max} = -15.83$ k $V_{DE\,min} = -50.5$ k
$M_{E\,max} = -38.4$ k-ft $M_{E\,min} = -355$ k-ft

4.48 $V_{BC\,max} = 9.22$ k $V_{BC\,min} = -21.0$ k
$M_{E\,max} = -19.84$ k-ft $M_{E\,min} = -227$ k-ft

4.49 $a(0) = 0$ $a(50^-) = -25/24$ $a(50^+) = 5/8$ $a(100) = -5/12$
$b(0) = 0$ $b(100) = 0$
$c(0) = 0$ $c(20) = -1$ $c(40) = 0$ $c(100) = 0$

4.50 $a(0) = 0$ $a(24) = -8/9$ $a(72) = 0$
$b(0) = 0$ $b(24) = 0.401$ $b(48) = -0.401$ $b(72) = 0$
$c(0) = 0$ $c(24) = 2/3$ $c(48) = 2/3$ $c(72) = 0$

4.51 $a(0^+) = 5/9$ $a(40) = -5/9$ $a(60) = 5/9$ $a(80) = 0$
$b(0^+) = 8/9$ $b(40) = -8/9$ $b(80) = 0$
$c(0^+) = -8/9$ $c(40) = 8/9$ $c(80) = 0$

4.52 $a(0) = 0$ $a(3.2) = 0$ $a(6.4) = 1$ $a(9.6) = 0$ $a(12.8) = 0$
$b(0) = 0$ $b(3.2) = 0.354$ $b(6.4) = -0.707$ $b(12.8) = 0$
$c(0) = 0$ $c(6.4) = 1$ $c(12.8) = 0$

4.53 $a(0) = 0$ $a(480) = -8/9$ $a(144) = 0$
$b(0) = 0$ $b(48) = -5/18$ $b(72) = 5/12$ $b(144) = 0$
$c(0) = 0$ $c(48) = 0$ $c(72) = -1/2$ $c(96) = 0$ $c(144) = 0$

4.54 $V_{BC\,max} = 89.7$ kN $V_{BC\,min} = -14$ kN
$M_{C\,max} = 720$ kN·m $M_{C\,min} = 0$

4.55 $V_{BC\,max} = 88$ kN $V_{BC\,min} = -17$ kN
$M_{C\,max} = 682$ kN·m $M_{C\,min} = 0$

438 Answers

4.56 $V_{BC\,max} = 18.42$ k $V_{BC\,min} = -5.33$ k
$M_{C\,max} = 221$ k-ft $M_{C\,min} = -114$ k-ft

4.57 $V_{BC\,max} = 17.83$ k $V_{BC\,min} = -4.75$ k
$M_{C\,max} = 214$ k-ft $M_{C\,min} = -128$ k-ft

4.58 $V_{BC\,max} = 13.75$ k $V_{BC\,min} = -6.25$ k
$M_{C\,max} = 290$ k-ft $M_{C\,min} = -100$ k-ft

4.59 $V_{BC\,max} = 11.25$ k $V_{BC\,min} = -5$ k
$M_{C\,max} = 290$ k-ft $M_{C\,min} = -80$ k-ft

4.60 $a_{max} = 27$ k $a_{min} = -26.7$ k
$b_{max} = 43.2$ k $b_{min} = -61.1$ k

4.61 $a_{max} = 29.3$ k $a_{min} = -19.67$ k
$b_{max} = 36.3$ k $b_{min} = -66.7$ k

5.1 $A_Y = 2.61$ k $B_Y = 1.29$ k

5.2 $B_Y = -15.59$ kN $A_Y = 23.2$ kN

5.3 $B_Y = 15$ k $M_E = 10$ k-ft $E_Y = 0$ $E_X = -4$ k

5.4 $M_A = -67.2$ kN·m $|B| = 23.1$ kN

5.5 $A_Y = 16.13$ k $F_Y = 21.9$ k $F_X = 0$

5.6 $A_Y = 0.75$ k $D_X = -0.608$ k $E_X = -1.992$ k

5.7 $A_X = 9.87$ kN $B_X = -9.87$ kN $B_Y = 19.74$ kN

5.8 $|A| = 7.11$ k $C_X = 5.44$ k $C_Y = 8.24$ k

5.9 $A_X = -2.84$ k $A_Y = 6.87$ k $|D| = 5.70$ k

5.10 $A_Y(0^+) = -1/2$ $A_Y(15^-) = 1$
$v_C(0^+) = -1/2$ $v_C(9^-) = 0.4$ $v_C(9^+) = -0.6$ $v_C(15) = 0$
$m_C(0^+) = -3$ ft $m_C(9) = 2.4$ ft $m_C(15) = 0$

5.11 $A_Y(0^+) = -0.375$ $A_Y(6.5) = 1.25$
$m_B(0^+) = -1.5$ m $m_B(1.5) = 0$ $m_B(6.5) = 0$
$v_B(0^+) = -0.375$ $v_B(1.5^-) = 0$ $v_B(1.5^+) = -1$ $v_B(6.5) = 0.25$

5.12 $A_Y(0^+) = -1$ $A_Y(24) = -1$ $A_Y(48) = 1$
$B_Y(0^+) = 2$ $B_Y(24) = 2$ $B_Y(48) = 0$
$v_C(0^+) = 1$ $v_C(18^-) = 1$ $v_C(18^+) = 0$ $v_C(48) = 0$
$m_E(0^+) = 24$ ft $m_E(24) = 0$ $m_E(48) = 0$

5.13 $A_Y(0) = 0$ $A_Y(15^-) = 0$ $A_Y(15^+) = 1$ $A_Y(25^-) = 1$
$C_Y(0) = 0$ $C_Y(15^-) = 1.5$ $C_Y(15^+) = -1$ $C_Y(25) = 0$
$m_C(0) = 0$ $m_C(10) = 0$ $m_C(15^-) = -5$ ft $m_C(15^+) = 10$ ft $m_C(25) = 0$
$v_C(0) = 0$ $v_C(10^-) = 0$ $v_C(10^+) = -1$ $v_C(15^-) = -1$ $v_C(15^+) = 0$
$v_C(25) = 0$

5.14 $v_A(0^+) = -1$ $v_A(8^-) = -1$
$m_B(0) = 0$ $m_B(5) = 0$ $m_B(8^-) = -3$ ft

5.15 $m_A(0^+) = 15$ ft $m_A(10) = -10$ ft $m_A(20) = 0$
$v_A(0^+) = 1.5$ $v_A(10) = 1$ $v_A(20^-) = 1$
$m_B(0^+) = -6$ ft $m_B(6) = 0$ $m_B(20) = 0$

5.16 $A_Y(0) = 0$ $A_Y(6) = 0$ $A_Y(10^-) = 1$
$C_Y(0) = 0$ $C_Y(6) = 1.5$ $C_Y(10) = 0$
$v_D(0) = 0$ $v_D(2^-) = 1/2$ $v_D(2^+) = -1/2$ $v_D(6) = 1/2$ $v_D(10) = 0$
$m_D(0) = 0$ $m_D(2) = 1$ $m_D(6) = -1$ $m_D(10) = 0$

Answers 439

5.17 $m_A(0) = 0 \quad m_A(L^-) = L$
$v_B(0^+) = -1 \quad v_B(L^-) = -1$

5.18 $A_Y(0) = 0 \quad A_Y(8) = 0 \quad A_Y(12) = 1 \quad A_Y(16^-) = 1$
$D_Y(0) = 0 \quad D_Y(8) = 4/3 \quad D_Y(12) = 0 \quad D_Y(16) = 0$
$m_D(0) = 0 \quad m_D(6) = 0 \quad m_D(8) = -2\,\text{m} \quad m_D(12) = 0 \quad m_D(16) = 0$

5.19 $v_{BC}(0) = 0 \quad v_{BC}(30) = 1/2 \quad v_{BC}(45) = -1/4 \quad v_{BC}(60) = 0$
$m_B(0) = 0 \quad m_B(45) = 45/4\,\text{ft} \quad m_B(60) = 0$

5.20 $v_{BC}(0) = 0 \quad v_{BC}(10.8) = 0.6 \quad v_{BC}(14.4) = -0.2 \quad v_{BC}(18) = 0$
$m_B(0) = 0 \quad m_B(14.4) = 2.88\,\text{m} \quad m_B(18) = 0$

5.21 $v_{CD}(0) = 0 \quad v_{CD}(7.2) = 2/5 \quad v_{CD}(10.8) = -2/5 \quad v_{CD}(18) = 0$
$m_C(0) = 0 \quad m_C(10.8) = 4.32\,\text{m} \quad m_C(18) = 0$

5.22 $v_{CD}(0^+) = -0.25 \quad v_{CD}(16) = 0.25 \quad v_{CD}(24) = -0.5 \quad v_{CD}(40) = 0$
$m_E(0^+) = -8\,\text{ft} \quad m_E(16) = 0 \quad m_E(40) = 0$

5.23 $v_{CD}(0^+) = -1/4 \quad v_{CD}(24^-) = 1/2 \quad v_{CD}(24^+) = -1/2 \quad v_{CD}(40) = 0$
$m_B(0^+) = -2\,\text{ft} \quad m_B(24^-) = 4\,\text{ft} \quad m_B(24^+) = 12\,\text{ft} \quad m_B(40) = 0$

5.24 $v_{CD}(0^+) = -0.25 \quad v_{CD}(16) = 0.25 \quad v_{CD}(24) = -1/2 \quad v_{CD}(40) = 0$
$m_C(0^+) = -4\,\text{ft} \quad m_C(24) = 8\,\text{ft} \quad m_C(40) = 0$

5.25 $v_{BC}(0^+) = -1/3 \quad v_{BC}(42^-) = 5/6 \quad v_{BC}(42^+) = -1/6 \quad v_{BC}(60^-) = 1/3$
$m_D(0^+) = -8\,\text{ft} \quad m_D(24) = 8\,\text{ft} \quad m_D(60^-) = -4\,\text{ft}$

5.26 $v_{CD}(0^+) = -1/3 \quad v_{CD}(24) = 1/3 \quad v_{CD}(42^-) = -2/3 \quad v_{CD}(42^+) = -1/6$
$v_{CD}(60^-) = 1/3$
$m_C(0^+) = -4\,\text{ft} \quad m_C(42^-) = 10\,\text{ft} \quad m_C(42^+) = 4\,\text{ft} \quad m_C(60^-) = -8\,\text{ft}$

5.27 $v_{EF}(0) = 0 \quad v_{EF}(4^-) = 2 \quad v_{EF}(4^+) = 0 \quad v_{EF}(12) = 0$
$m_E(0) = 0 \quad m_E(4^-) = -4\,\text{m} \quad m_E(4^+) = 0 \quad m_E(12) = 0$

5.28 $v_{DE}(0) = 0 \quad v_{DE}(4^-) = -1/2 \quad v_{DE}(4^+) = 0 \quad v_{DE}(6) = -3/4 \quad v_{DE}(12) = 0$
$m_D(0) = 0 \quad m_D(4^-) = -3\,\text{m} \quad m_D(0^+) = 0 \quad m_D(6) = 1.5\,\text{m} \quad m_D(12) = 0$

5.29 $v_{CD}(0) = 0 \quad v_{CD}(8) = 0 \quad v_{CD}(12) = -1 \quad v_{CD}(16) = 2/3 \quad v_{CD}(20) = 0$
$m_D(0) = 0 \quad m_D(8) = 0 \quad m_D(12) = -4\,\text{m} \quad m_D(16) = -4\,\text{m} \quad m_D(20) = 0$

5.30 $v_{DE}(0) = 0 \quad v_{DE}(4) = 1/2 \quad v_{DE}(8) = 0 \quad v_{DE}(12) = 1/2 \quad v_{DE}(16) = 1/2$
$v_{DE}(20) = 0$
$m_B(0) = 0 \quad m_B(12) = 0 \quad m_B(16) = 4/3\,\text{m} \quad m_B(20) = 0$

5.31 $v_{CD}(0^+) = -1.5 \quad v_{CD}(30) = 0 \quad v_{CD}(50) = 0$
$m_D(0^+) = -12\,\text{ft} \quad m_D(30) = 0 \quad m_D(50) = 0$

5.32 $v_{AB}(0^+) = -1.5 \quad v_{AB}(30) = 1 \quad v_{AB}(50) = 0$
$m_A(0^+) = 30\,\text{ft} \quad m_A(30) = -10\,\text{ft} \quad m_A(50) = 0$

5.33 $a(0) = 0 \quad a(50^-) = -25/24 \quad a(50^+) = 5/8 \quad a(100^-) = -5/12$
$b(0) = 0 \quad b(100) = 0$
$c(0) = 0 \quad c(20) = -1 \quad c(40) = 0 \quad c(100) = 0$

5.34 $a(0) = 0 \quad a(24) = -8/9 \quad a(72) = 0$
$b(0) = 0 \quad b(24) = 0.401 \quad b(48) = -0.401 \quad b(72) = 0$
$c(0) = 0 \quad c(24) = 2/3 \quad c(48) = 2/3 \quad c(72) = 0$

5.35 $a(0^+) = 5/9 \quad a(40) = -5/9 \quad a(60) = 5/9 \quad a(80) = 0$
$b(0^+) = 8/9 \quad b(40) = -8/9 \quad b(80) = 0$
$c(0^+) = -8/9 \quad c(40) = 8/9 \quad c(80) = 0$

5.36 $a(0) = 0 \quad a(3.2) = 0 \quad a(6.4) = 1 \quad a(9.6) = 0 \quad a(12.8) = 0$
$b(0) = 0 \quad b(3.2) = 0.354 \quad b(6.4) = -0.707 \quad b(12.8) = 0$
$c(0) = 0 \quad c(6.4) = 1 \quad c(12.8) = 0$

5.37 $a(0) = 0$ $a(48) = -8/9$ $a(144) = 0$
$b(0) = 0$ $b(48) = -5/18$ $b(72) = 5/12$ $b(144) = 0$
$c(0) = 0$ $c(48) = 0$ $c(72) = -1/2$ $c(96) = 0$ $c(144) = 0$

6.1 $(0, L)$
$v(0) = 0$ $v'(0) = 0$
$EIv_B = M_e L^2/8$ $EI\theta_B = M_e L/2$ $EIv_C = M_e L^2/2$ $EI\theta_C = M_e L$

6.2 $(0, L)$
$v(L) = 0$ $v'(L) = 0$
$EIv_A = -2M_e L^2/9$ $EI\theta_A = M_e L/3$ $EIv_B = 0$ $EI\theta_B = 0$

6.3 $(0, 12)$
$v(12) = 0$ $v'(12) = 0$
$EIv_A = -9072$ k-ft^3 $EI\theta_A = 1080$ k-ft^2 $EIv_B = -3020$ k-ft^3 $EI\theta_B = 864$ k-ft^2

6.4 $(0, 7.2)$
$v(0) = 0$ $v'(0) = 0$
$EIv_C = -1209$ kN·m^3 $EI\theta_B = -196$ kN·m^2

6.5 $(0, L)$
$v(0) = 0$ $v'(L) = 0$
$EIv_B = -PL^3/6$ $EIv_C = -11PL^3/48$ $EI\theta_A = -3PL^2/8$

6.6 $(0, 6)$
$v(0) = 0$ $v(6) = 0$
$EI\theta_A = 12$ kN·m^2 $EI\theta_B = -4$ kN·m^2 $EIv_B = 80/3$ kN·m^3

6.7 $(0, 18)$
$v(0) = 0$ $v(18) = 0$
$EIv_B = -384$ k-ft^3 $EI\theta_B = 96$ k-ft^2

6.8 $(0, 14)$
$v_B = 0$ $v_D = 0$
$EI\theta_B = 1000/9$ kN·m^2 $EIv_A = -14000/9$ kN·m^3 $EIv_C = -800/9$ kN·m^3

6.9 $(0, 21)$
$v_B = 0$ $v_D = 0$
$EIv_A = -10.8$ k-ft^3 $EIv_C = -172.8$ k-ft^3 $EIv_E = -124.2$ k-ft^3

6.10 $(0, 7^-)$ $(7^+, 12)$
$v(0) = 0$ $v(7^-) = v(7^+)$ $v(12) = 0$ $\theta(12) = 0$
$EIv_B = -87.1$ k-ft^3 $EIv_C = -107.1$ k-ft^3 $EI\theta_A = -34.2$ k-ft^2

6.11 $(0, 4^-)$ $(4^+, 8)$
$v(0) = 0$ $\theta(0) = 0$ $v(4^-) = v(4^+)$ $v(6) = 0$
$EIv_B = 70.4$ kN·m^3 $EIv_D = -134.4$ kN·m^3 $EI\theta_C = -51.2$ kN·m^2

6.12 $(0, 12^-)$ $(12^+, 24^-)$ $(24^+, 42)$
$v(0) = 0$ $\theta(0) = 0$ $v(12^-) = v(12^+)$ $v(24^-) = v(24^+)$ $v(30) = 0$ $v(42) = 0$
$EIv_B = -5180$ k-ft^3 $EIv_C = -3240$ k-ft^3 $EIv_D = -648$ k-ft^3

6.13 $EIv_{B(AB)} = 3890$ k-ft^3 $EI\theta_B = 648$ k-ft^2 $EIv_{B(BD)} = -11340$ k-ft^3
$EIv_C = -6480$ k-ft^3

6.14 $EIv_{D(ACD)} = 640$ kN·m^3 $EI\theta_D = 232$ kN·m^2 $EIv_{D(DF)} = -1586$ kN·m^3
$EIv_E = -728$ kN·m^3

6.15 $EIv_B = -3460$ k-ft^3 $EIv_E = -864$ k-ft^3

6.16 $EIv_B = -10940$ k-ft^3 $EIv_D = -17500$ k-ft^3

6.17 $v_B = -0.393$ in. $v_D = -0.583$ in.

6.18 $U_B = 0.00604$ in. $V_B = 0.01207$ in.

6.19 $EIU_B = -11.25 \text{ kN·m}^3 \quad EIV_B = 15 \text{ kN·m}^3 \quad EIU_D = 101.25 \text{ kN·m}^3$
$EIV_B = -135 \text{ kN·m}^3$
6.20 $EIU_D = -577 \text{ k-ft}^3 \quad EIV_B = 1385 \text{ k-ft}^3$
6.21 $EIU_B = 547 \text{ kN·m}^3 \quad EIV_B = 547 \text{ kN·m}^3$
6.22 $EI\theta_B = 288 \text{ k-ft}^2 \quad EIU_C = -3460 \text{ k-ft}^3 \quad EIV_C = 1152 \text{ k-ft}^3$
6.23 $EI\theta_B = 258 \text{ kN·m}^2 \quad EIU_A = -566 \text{ kN·m}^3 \quad EIV_A = -1371 \text{ kN·m}^3$
6.24 $EI\theta_B = -248 \text{ k-ft}^2 \quad EIU_C = 3560 \text{ k-ft}^3 \quad EIV_C = -4260 \text{ k-ft}^3$
6.25 $EI\theta_B = -160 \text{ kN·m}^2 \quad EIU_B = 1067 \text{ kN·m}^3 \quad EIV_C = -175 \text{ kN·m}^3$
6.26 $EIU_B = 667 \text{ k-ft}^3 \quad EIV_A = -747 \text{ k-ft}^3$
6.27 $EIU_A = 859 \text{ kN·m}^3 \quad EIV_C = 0$
6.28 $EIV_B = 314 \text{ k-ft}^3 \quad EIU_D = -1735 \text{ k-ft}^3$
6.29 $EIV_A = 256 \text{ k-ft}^3 \quad EIV_D = -664 \text{ k-ft}^3$
6.30 $EIV_A = -72 \text{ kN·m}^3 \quad EIV_C = -12 \text{ kN·m}^3$
6.31 $EIU_B = 877 \text{ k-ft}^3 \quad EIU_C = 1515 \text{ k-ft}^3 \quad EIV_D = -1216 \text{ k-ft}^3$
6.32 $EIU_B = 877 \text{ k-ft}^3 \quad EIU_C = 1515 \text{ k-ft}^3 \quad EIV_D = -1216 \text{ k-ft}^3$
6.33 $EIU_B = 877 \text{ k-ft}^3 \quad EIU_C = 1515 \text{ k-ft}^3 \quad EIV_D = -1216 \text{ k-ft}^3$
6.34 $EIV_B = -528 \text{ k-ft}^3 \quad EIV_C = -928 \text{ k-ft}^3 \quad EIU_D = 133.3 \text{ k-ft}^3$
6.35 $EIV_B = -528 \text{ k-ft}^3 \quad EIV_C = -928 \text{ k-ft}^3 \quad EIU_D = 133.3 \text{ k-ft}^3$
6.36 $EIV_B = -528 \text{ k-ft}^3 \quad EIV_C = -928 \text{ k-ft}^3 \quad EIU_D = 133.3 \text{ k-ft}^3$
6.37 $EIU_B = 699 \text{ kN·m}^3 \quad EIU_F = 1407 \text{ kN·m}^3 \quad EIV_D = -203 \text{ kN·m}^3$
6.38 $EIU_B = 699 \text{ kN·m}^3 \quad EIU_F = 1407 \text{ kN·m}^3 \quad EIV_D = -203 \text{ kN·m}^3$
6.39 $EIU_B = 699 \text{ kN·m}^3 \quad EIU_F = 1407 \text{ kN·m}^3 \quad EIV_D = -203 \text{ kN·m}^3$
6.40 $EIV_B = -745 \text{ k-ft}^3 \quad EIV_C = -1260 \text{ k-ft}^3 \quad EIU_D = 180 \text{ k-ft}^3$
6.41 $EIV_D = -434 \text{ kN·m}^3 \quad EIU_B = 1260 \text{ kN·m}^3 \quad EIU_F = 2690 \text{ kN·m}^3$
6.42 $EIV_C = -2860 \text{ k-ft}^3 \quad EIU_D = -1807 \text{ k-ft}^3 \quad EIU_E = 7830 \text{ k-ft}^3$
7.1 $EIV_B = M_e L^2/8 \quad EIV_C = M_e L^2/2 \quad EI\theta_B = M_e L/2 \quad EI\theta_C = M_e L$
7.2 $EIV_A = -2M_e L^2/9 \quad EIV_B = 0 \quad EI\theta_A = 2M_e L/3 \quad EI\theta_B = 0$
7.3 $EIV_A = -9070 \text{ k-ft}^3 \quad EIV_B = -3020 \text{ k-ft}^3 \quad EI\theta_A = 1080 \text{ k-ft}^2 \quad EI\theta_B = 864 \text{ k-ft}^2$
7.4 $EIV_B = -1209 \text{ kN·m}^3 \quad EI\theta_B = -196 \text{ kN·m}^2$
7.5 $EIV_B = -PL^3/6 \quad EIV_C = -11PL^3/48 \quad EI\theta_A = -3PL^2/8$
7.6 $EI\theta_A = 12 \text{ kN·m}^2 \quad EI\theta_B = -4 \text{ kN·m}^2 \quad EIV_B = 26.7 \text{ kN·m}^3$
7.7 $EIV_B = -384 \text{ k-ft}^3 \quad EI\theta_C = 96 \text{ k-ft}^2$
7.8 $EI\theta_B = 111.1 \text{ kN·m}^2 \quad EIV_A = -1556 \text{ kN·m}^3 \quad EIV_B = -88.9 \text{ kN·m}^3$
7.9 $EIV_A = -10.8 \text{ k-ft}^3 \quad EIV_C = -172.8 \text{ k-ft}^3 \quad EIV_E = 124.2 \text{ k-ft}^3$
7.10 $EIV_B = -87.1 \text{ k-ft}^3 \quad EIV_C = -107.1 \text{ k-ft}^3 \quad EI\theta_A = -34.2 \text{ k-ft}^2$
7.11 $EIV_B = 70.4 \text{ kN·m}^3 \quad EIV_D = -134.4 \text{ kN·m}^3 \quad EI\theta_C = -51.2 \text{ kN·m}^2$
7.12 $EIV_B = -5180 \text{ k-ft}^3 \quad EIV_D = -648 \text{ k-ft}^3 \quad EIV_C = -3240 \text{ k-ft}^3$
7.13 $EI\theta_B = 648 \text{ k-ft}^2 \quad EIV_{B(AB)} = 3890 \text{ k-ft}^3 \quad EIV_{B(BD)} = -11340 \text{ k-ft}^3$
7.14 $EI\theta_D = 232 \text{ kN·m}^2 \quad EIV_{D(ACD)} = 640 \text{ kN·m}^3 \quad EIV_{D(DF)} = -1586 \text{ kN·m}^3$
$EIV_E = -728 \text{ kN·m}^3$
7.15 $EIV_B = -3460 \text{ k-ft}^3 \quad EIV_E = -864 \text{ k-ft}^3$
7.16 $EIV_D = -17500 \text{ k-ft}^3 \quad EIV_B = -10940 \text{ k-ft}^3$
7.17 $EIV_D = -0.583 \text{ in.} \quad EIV_B = -0.393 \text{ in.}$

7.18 $EIU_B = 181.1$ k-ft^3 $EIV_B = 362$ k-ft^3

7.19 $EIU_B = -11.25$ kN·m^3 $EIV_B = 15$ kN·m^3 $EIU_D = 101.3$ kN·m^3
$EIV_B = -135$ kN·m^3

7.20 $EIU_D = -577$ k-ft^3 $EIV_B = 1384$ k-ft^3

7.21 $EIU_B = 547$ kN·m^3 $EIV_B = 547$ kN·m^3

7.22 $EI\theta_B = 288$ k-ft^2 $EIU_C = -3460$ k-ft^3 $EIV_C = 1152$ k-ft^3

7.23 $EI\theta_B = 258$ kN·m^2 $EIU_A = -566$ kN·m^3 $EIV_A = -1371$ kN·m^3

7.24 $EI\theta_B = -248$ k-ft^2 $EIU_C = 3560$ k-ft^3 $EIV_C = -4260$ k-ft^3

7.25 $EI\theta_B = -160$ kN·m^2 $EIU_B = 1067$ kN·m^3 $EIV_C = -175$ kN·m^3

7.26 $EIU_A = 667$ k-ft^3 $EIV_A = -747$ k-ft^3

7.27 $EIU_A = 859$ kN·m^3 $EIV_C = 0$

7.28 $EIV_B = 314$ k-ft^3 $EIU_D = -1735$ k-ft^3

7.29 $EIV_A = 256$ k-ft^3 $EIV_D = -664$ k-ft^3

7.30 $EIV_A = -72$ kN·m^3 $EIV_C = -12$ kN·m^3

7.31 $EIU_B = 877$ k-ft^3 $EIU_C = 1515$ k-ft^3 $EIV_D = -1216$ k-ft^3

7.32 $EIV_B = -528$ k-ft^3 $EIV_C = -928$ k-ft^3 $EIU_D = 133.3$ k-ft^3

7.33 $EIU_B = 699$ kN·m^3 $EIU_F = 1407$ kN·m^3 $EIV_D = -203$ kN·m^3

7.34 $EIV_B = -745$ k-ft^3 $EIV_E = -1260$ k-ft^3 $EIU_D = 180$ k-ft^3

7.35 $EIV_D = -434$ kN·m^3 $EIU_B = 1263$ kN·m^3 $EIU_F = 2690$ kN·m^3

7.36 $EIV_C = -2860$ k-ft^3 $EIV_D = 1808$ k-ft^3 $EIU_E = 7830$ k-ft^3

7.37 $V_g = 0.378$ in. $V_h = 0.756$ in.

7.38 $V_g = -3.09$ mm $U_c = -0.690$ mm

7.39 $U_f = -0.0238$ in. $V_f = -0.0268$ in.

7.40 $V_b = -6.39$ mm $V_h = -7.64$ mm

7.41 $V_f = -1.336$ mm $V_g = -1.835$ mm

7.42 $V_b = -0.0281$ in. $U_f = 0.01854$ in.

7.43 $V_b = -4.29$ mm $V_f = -1.089$ mm

7.44 $EIv = -w(6L^2x^2 - 4Lx^3 + x^4)/24$ $EIv = -P(3Lx^2 - x^3)/6$
$W^{wP} = wPL^4/(8EI)$

7.45 $EIW^{pq} = -10400/9$ kN^2·m^3

7.46 $EIW^{pq} = -11520$ k^2-ft^3

7.47 $EIW^{pq} = -41472$ k^2-ft^3

7.48 $W^{pq} = -4.027$ kN·m

7.49 $EI\mu_{12} = -64/3$ m^3

7.50 $EI\mu_{12} = -1$ ft

7.51 $EI\mu_{12} = -18$ ft^2

7.52 $\mu_{12} = -1.117 \times 10^{-3}$ in-k^{-1}

9.1 $B_Y = 31.5$ kN $EIV_C = 105$ kN·m^3

9.2 $M_D = 32.4$ kN·m $EIV_A = -700$ kN·m^3

9.3 $A_Y = -2.25$ k $EIV_C = -621$ k-ft^3

9.4 $B_Y = 16.5$ k $EI\theta_C = -13.5$ k-ft^2

9.5 $A_Y = 6.75$ kN $EI\theta_A = -4.5$ kN·m^2

Answers 443

9.6 $lc(AC)$ $M_C = -31.5$ kN·m $EI\theta_A = 74.3$ kN·m²
9.7 $A_Y = -0.75$ kN $EIV_B = 4.64$ kN·m³
9.8 $C_Y = 19.5$ kN $EIV_D = -101.3$ kN·m³
9.9 $C_Y = 13.5$ k $EIV_B = 292$ k-ft³
9.10 $M_D = 32.4$ k-ft $EIV_B = -1555$ k-ft³
9.11 $B_Y = 13$ kN $EI\theta_B = -24$ kN·m²
9.12 $D_Y = 10.5$ kN $EIV_C = -126$ kN·m³
9.13 $B_Y = 18.43$ kN $D_Y = 9.43$ kN
9.14 $M_A = -15.43$ k-ft $D_Y = 9.43$ k
9.15 $C_Y = 2.4$ kN $M_E = -14.85$ kN·m
9.16 $A_Y = 2.4$ kN $E_Y = 9.68$ kN
9.17 $A_Y = 23$ kN $D_Y = -1$ kN
9.18 $B_Y = 37$ kN $D_Y = -1$ kN
9.19 $M_A = 3.75$ kN·m $C_Y = 21.8$ kN
9.20 $A_Y = 10.88$ kN $C_Y = 21.8$ kN
9.21 $M_A = 42.7$ kN·m $M_C = -42.5$ kN·m
9.22 $M_A = 42.7$ k-ft $C_Y = 5.40$ k
9.23 $M_A = 18.62$ k-ft $C_Y = 13.03$ k
9.24 $M_D = 22.4$ k-ft $C_Y = 13.03$ k
9.25 $D_Y = 0.75$ k $EIV_C = -273$ k-ft³
9.26 $M_A = 58.5$ k-ft $EIU_B = 2673$ k-ft³
9.27 $A_X = 3.38$ kN $EIV_B = -116.4$ kN·m³
9.28 $E_Y = 21.4$ kN $EIU_D = 45.6$ kN·m³
9.29 $C_Y = -11.05$ kN $EIU_A = 501$ kN·m³
9.30 $M_A = -63.8$ kN·m $EI\theta_C = 29.5$ kN·m²
9.31 $A_X = 1.826$ k $EIU_D = -2780$ k-ft³
9.32 $E_Y = -1.087$ k $EIU_B = -2955$ k-ft³
9.33 $A_X = -1.294$ k $EIU_C = 216$ k-ft³
9.34 $D_Y = 1.256$ k $EIV_C = 288$ k-ft³
9.35 $M_A = 10.91$ kN·m $EIV_C = -31.1$ kN·m³
9.36 $D_Y = 9.59$ kN $EIU_B = -31.1$ kN·m³
9.37 $A_Y = 13.3$ kN $D_X = -14.09$ kN
9.38 $A_X = 11.74$ kN $E_X = 2.35$ kN
9.39 $M_A = 73.9$ kN·m $E_Y = 3.12$ kN
9.40 $D_Y = 23.3$ kN $E_Y = 3.12$ kN
9.41 $M_E = -172.9$ k-ft $A_X = 4.23$ k
9.42 $M_A = 7.70$ k-ft $E_X = -4.23$ k
9.43 $D_X = -5.07$ k $D_Y = 0.826$ k
9.44 $M_A = 42.2$ k-ft $D_Y = 0.825$ k
9.45 $D_X = 0.686$ k $E_Y = -0.494$ k
9.46 $E_X = -8.07$ k $E_Y = -0.494$ k
9.47 $D_Y = 33.5$ k $A_X = -6.70$ k $M_A = 12.36$ k-ft

444 Answers

9.48 $C_X = -1.838$ k $E_X = -2.30$ k $M_E = 16.91$ k-ft
9.49 $D_X = -3.30$ kN $D_Y = 7.82$ kN $M_D = -8.88$ kN·m
9.50 $d_Y = 4.23$ k $V_e = -0.248$ in.
9.51 $a_X = -17.30$ k $U_c = -0.0335$ in.
9.52 $c_Y = 12.45$ kN $V_d = -2.09$ mm
9.53 $a_X = 35.2$ kN $U_b = -0.1747$ mm
9.54 $h_Y = 31.1$ k $V_g = -0.1383$ in.
9.55 $f_Y = 9.94$ k $V_i = -0.1898$ in.
9.56 $a_X = -14.8$ k $V_e = -0.0929$ in.
9.57 $d_Y = 3.2$ k $V_e = -0.0550$ in.
9.58 $h_Y = 14.07$ kN $V_f = -1.292$ mm
9.59 $h_X = -22.9$ kN $V_b = -0.552$ mm
9.60 $ab = 9.98$ k $bc = 9.98$ k $cd = -6.02$ k $da = -11.02$ k $ea = 15.58$ k
$eb = -14.12$ k $ec = 8.51$ k $ed = -7.04$ k
9.61 $ab = 9.98$ k $bc = 9.98$ k $cd = -6.02$ k $da = -11.02$ k $ea = 15.58$ k
$eb = -14.12$ k $ec = 8.51$ k $ed = -7.04$ k
9.62 $ab = -21.9$ kN $bc = -8.57$ kN $cd = 6.44$ kN $da = -0.232$ kN $ea = -8.58$ kN
$eb = 0.242$ kN $ec = -7.74$ kN $ed = -2.62$ kN
9.63 $ab = -21.9$ kN $bc = -8.57$ kN $cd = 6.44$ kN $da = -0.232$ kN $ea = -8.58$ kN
$eb = 0.242$ kN $ec = -7.74$ kN $ed = -2.62$ kN
9.64 $ab = -14.95$ kN $bc = -0.578$ kN $cd = -0.501$ kN $df = 4.52$ kN $fa = 4.52$ kN
$ea = 2.18$ kN $eb = -8.08$ kN $ec = 0.602$ kN $ed = -4.306$ kN $ef = -4.65$ kN
9.65 $ab = -14.95$ kN $bc = -0.578$ kN $cd = -0.501$ kN $df = 4.52$ kN $fa = 4.52$ kN
$ea = 2.18$ kN $eb = -8.08$ kN $ec = 0.602$ kN $ed = -4.306$ $ef = -4.65$ kN
9.66 $ab = 1.6$ k $bc = -9.24$ k $cd = 6.53$ k $de = 1.191$ k $ea = -5.34$ k $fa = 7.55$ k
$fb = -11.5$ k $fd = 7.55$ k $fe = -11.55$ k $bd = -5.34$ k
9.67 $a_X = -7.37$ k
9.68 $ab = 7.11$ kN
9.69 $C_Y = -9.38$ k
9.70 $A_Y = 12.5$ k
9.71 $A_Y = -18.44$ kN
9.72 $C_Y = -30.1$ kN
9.73 $A_Y = -80$ kN
9.74 $A_Y = -160$ kN
9.75 $A_Y = -11.11$ k $B_Y = 24.1$ k
9.76 $A_Y = 16.05$ k $B_Y = -32.7$ k
9.77 $C_X = 2.5$ kN
9.78 $A_Y = -18.75$ kN
9.79 $A_Y = -0.526$ k
9.80 $D_Y = -0.631$ k
9.81 $M_B = -27$ k-ft $EIV_C = -621$ k-ft^3
9.82 $M_B = -31.5$ kN·m $EI\theta_C = 74.3$ kN·m^2
9.83 $M_A = 18.81$ k-ft $M_B = -37.6$ k-ft $M_C = -29.7$ k-ft $EI_0V_E = -631$ k-ft^3
9.84 $M_A = 15.42$ k-ft $M_B = -30.9$ k-ft $EIV_C = -586$ k-ft^3

9.85 $M_C = -9.64 \text{ kN·m}$ $M_D = -15.43 \text{ kN·m}$ $EI_0 V_B = -1.326 \text{ kN·m}^3$
9.86 $M_B = -13.75 \text{ kN·m}$ $M_D = -14.85 \text{ kN·m}$ $EI_0 V_C = -62.8 \text{ kN·m}^3$
9.87 $M_A = -23.8 \text{ k-ft}$ $M_B = -24.4 \text{ k-ft}$ $M_D = -36.3 \text{ k-ft}$ $EI_0 V_C = -421 \text{ k-ft}^3$
9.88 $M_A = 6 \text{ k-ft}$ $M_B = -12 \text{ k-ft}$ $M_D = -18 \text{ k-ft}$ $EI_0 \theta_B = -36 \text{ k-ft}^2$
9.89 $M_B = -110.7 \text{ kN·m}$
9.90 $M_B = -80 \text{ k-ft}$ $M_C = 40 \text{ k-ft}$
9.91 $M_B = -63.6 \text{ k-ft}$ $M_C = 54.5 \text{ k-ft}$
9.92 $B_Y = 31.5 \text{ k}$
9.93 $M_C = 32.4 \text{ kN·m}$
9.94 $B_Y = 16.5 \text{ k}$
9.95 $A_Y = -2.25 \text{ k}$
9.96 $A_Y = 6.75 \text{ kN}$
9.97 $B_Y = 38.8 \text{ kN}$
9.98 lc(AC) $M_C = -18 \text{ kN·m}$
9.99 lc(AC) $M_B = -32.4 \text{ k-ft}$
9.100 lc(AC) $M_B = -12 \text{ kN·m}$
9.101 lc(AC) $M_A = 15.42 \text{ k-ft}$ $M_B = -30.9 \text{ k-ft}$
9.102 lc(AC) $M_A = -3.75 \text{ kN·m}$ $M_B = -46.5 \text{ kN·m}$
9.103 $D_Y = 0.75 \text{ k}$
9.104 lc(BC) $M_B = -44.1 \text{ kN·m}$
9.105 $D_X = -7.18 \text{ kN}$
9.106 lc(AC) $M_B = 39.9 \text{ kN·m}$ lc(EC) $M_D = 7.04 \text{ kN·m}$
9.107 $M_A = 7.7 \text{ k-ft}$ $A_X = 4.23 \text{ k}$
9.108 $B_Y = 30.7 \text{ kN}$
9.109 $A_Y = -18.44 \text{ kN}$
9.110 $A_Y = -160 \text{ kN}$
9.111 $B_Y = 320 \text{ kN}$
9.112 $C_Y = -1.235 \text{ k}$
9.113 $A_Y = 1.235 \text{ k}$
9.114 $C_X = 1.235 \text{ k}$
9.115 $d_Y = 3.20 \text{ k}$
9.116 $h_Y = 14.07 \text{ kN}$
9.117 $bc = 9.98 \text{ k}$
9.118 $bc = -8.57 \text{ kN}$
9.119 $a_X = -6.94 \text{ k}$ $ab = 1.6 \text{ k}$
10.1 $k = 1$ $d = 2$
  $M_{12} = -2.4 \text{ k-ft}$ $M_{21} = -4.8 \text{ k-ft}$ $M_{23} = 4.8 \text{ k-ft}$ $M_{32} = -5.6 \text{ k-ft}$
  $EIV_A = -102.4 \text{ k-ft}^3$
10.2 $k = 2$ $d = 2$
  $M_{12} = 2.5 \text{ kN·m}$ $M_{21} = -31 \text{ kN·m}$ $M_{23} = 31 \text{ kN·m}$ $M_{32} = 0$
  $EIV_A = -173 \text{ kN·m}^3$
10.3 $k = 2$ $d = 4$
  $M_{12} = -12 \text{ kN·m}$ $M_{21} = -24 \text{ kN·m}$ $M_{23} = 24 \text{ kN·m}$ $M_{32} = -24 \text{ kN·m}$
  $M_{34} = 24 \text{ kN·m}$ $M_{43} = 12 \text{ kN·m}$
  $EI_0 V_A = 48 \text{ kN·m}^3$

446 Answers

10.4 $k = 3$ $d = 1$
$M_{12} = 0$ $M_{21} = -10.53$ k-ft $M_{23} = 10.53$ k-ft $M_{32} = 0$

10.5 $k = 2$ $d = 2$
$M_{12} = 44.5$ k-ft $M_{21} = -8.17$ k-ft $M_{23} = 8.17$ k-ft $M_{32} = -20$ k-ft

10.6 $k = 2$ $d = 4$
$M_{12} = -10.1$ kN·m $M_{21} = -20.2$ kN·m $M_{23} = 20.2$ kN·m $M_{32} = -23.8$ kN·m
$M_{34} = -0.2$ kN·m $M_{43} = -0.1$ kN·m

10.7 $k = 2$ $d = 2$
$M_{12} = -6.35$ k-ft $M_{21} = -12.7$ k-ft $M_{23} = 12.7$ k-ft $M_{32} = 0$

10.8 $k = 2$ $d = 2$
$M_{12} = 0$ $M_{21} = -19.29$ kN·m $M_{23} = 19.29$ kN·m $M_{32} = -3.86$ kN·m

10.9 $k = 2$ $d = 2$
$M_{12} = 23.9$ kN·m $M_{21} = -6.26$ kN·m $M_{23} = 6.26$ kN·m $M_{32} = -5.09$ kN·m
$M_{34} = 16.04$ kN·m $M_{43} = -32.5$ kN·m $M_{35} = -10.96$ kN·m $M_{53} = -5.48$ kN·m

10.10 $k = 1$ $d = 3$
$F_0 = -3.15$ kN

10.11 $k = 3$ $d = 2$
$F_0 = 6.99$ k

10.12 $K = 3$ $d = 5$
$F_{02} = -3.66$ k $F_{03} = -4.06$ k

10.13 $v_{12} = 0$ $v_{21} = -u_2$ $u_{12} = u_{21} = 0$
$v_{23} = -u_2$ $v_{32} = 0$ $u_{23} = u_{32} = 0$

10.14 $v_{12} = 0$ $v_{21} = 0$ $u_{12} = u_{21} = u_2$
$v_{23} = u_2$ $v_{32} = 0$ $u_{23} = u_{32} = 0$

10.15 $v_{12} = 0$ $v_{21} = -u_3$ $u_{12} = u_{21} = 0$
$v_{23} = 0$ $v_{32} = 0$ $u_{23} = u_{32} = u_3$
$v_{34} = u_3$ $v_{43} = 0$ $u_{34} = u_{43} = 0$

10.16 $v_{12} = 0$ $v_{21} = -u_3$ $u_{12} = u_{21} = 0$
$v_{23} = -0.6u_3$ $v_{32} = -0.6u_3$ $u_{23} = u_{32} = -0.8u_3$

10.17 $v_{12} = 0$ $v_{21} = 0.75u_1$ $u_{12} = u_{21} = u_1$
$v_{23} = 1.25u_1$ $v_{32} = 0$ $u_{23} = u_{32} = 0$

10.18 $v_{12} = 0$ $v_{21} = -u_3$ $u_{12} = u_{21} = 0$
$v_{23} = 0$ $v_{32} = u_3$ $u_{23} = u_{32} = u_3$
$v_{34} = 1.414u_3$ $v_{43} = 0$ $u_{34} = u_{43} = 0$

10.19 $v_{12} = 0$ $v_{21} = 1.077u_3$ $u_{12} = u_{21} = 0$
$v_{23} = -0.4u_3$ $v_{32} = 0$ $u_{23} = u_{32} = u_3$
$v_{34} = -u_3$ $v_{43} = 0$ $u_{34} = u_{43} = 0$

10.20 $v_{12} = -0.707u_1$ $v_{21} = 0$ $u_{12} = u_{21} = 0.707u_1$
$v_{23} = 0.707u_1$ $v_{32} = 0$ $u_{23} = u_{32} = 0$

10.21 $v_{12} = 0$ $v_{21} = 1.083u_2$ $u_{12} = u_{21} = 0$
$v_{23} = -1.002u_1$ $v_{32} = 0.412u_2$ $u_{23} = u_{32} = -0.412u_2$
$v_{34} = 0.583u_3$ $v_{43} = 0$ $u_{34} = u_{43} = 0$

10.22 $M_{21} = -10.53$ k-ft $M_{23} = 10.53$ k-ft

10.23 $M_{12} = 2.5$ kN·m $M_{21} = -31$ kN·m $M_{23} = 31$ kN·m

10.24 $M_{12} = 44.5$ k-ft $M_{21} = -8.17$ k-ft $M_{23} = 8.17$ k-ft $M_{32} = -20$ k-ft

10.25 $M_{21} = -19.29$ kN·m $M_{23} = 19.29$ kN·m $M_{32} = -3.86$ kN·m

Answers 447

10.26 $M_{12} = -6.35$ k-ft $M_{21} = -12.7$ k-ft $M_{23} = 12.7$ k-ft

10.27 $M_{12} = 24.6$ kN·m $M_{21} = -4.71$ kN·m $M_{23} = 4.71$ kN·m $M_{32} = -10.5$ kN·m
$M_{34} = 27.6$ kN·m $M_{43} = 0$ $M_{35} = -17.14$ kN·m $M_{53} = -8.57$ kN·m

10.28 $EI_0 u_2 = 72.6$ k-ft^3
$M_{12} = 2.10$ k-ft $M_{21} = 1.169$ k-ft $M_{23} = -1.169$ k-ft $M_{32} = -11.93$ k-ft

10.29 $EI_0 u_3 = 699$ kN·m^3
$M_{12} = -28.8$ kN·m $M_{21} = -57.6$ kN·m $M_{23} = 57.6$ kN·m $M_{32} = 35.5$ kN·m
$M_{34} = -35.5$ kN·m $M_{43} = -56.6$ kN·m

10.30 $EI_0 u_2 = 1645$ k-ft^3
$M_{12} = 86.9$ k-ft $M_{21} = 31.0$ k-ft $M_{23} = -31.0$ k-ft $M_{32} = -20$ k-ft

10.31 $EI_0 u_2 = 90.7$ kN·m^3
$M_{12} = 15.84$ kN·m $M_{21} = 5.76$ kN·m $M_{23} = -5.76$ kN·m $M_{32} = -2.88$ kN·m

10.32 $EI u_3 = 366$ k-ft^3
$M_{12} = 0$ $M_{21} = 1.231$ k-ft $M_{23} = -1.231$ k-ft $M_{32} = -4.92$ k-ft
$M_{34} = 4.92$ k-ft $M_{43} = 19.08$ k-ft

10.33 $EI_0 u_3 = 13.89$ kN·m^3
$M_{12} = 9.30$ kN·m $M_{21} = -5.10$ kN·m $M_{23} = 5.10$ kN·m $M_{32} = -8.39$ kN·m
$M_{34} = 8.39$ kN·m $M_{43} = 6.80$ kN·m

10.34 $EI_0 u_3 = 592$ k-ft^3
$M_{12} = 0$ $M_{21} = 11.81$ k-ft $M_{23} = -11.81$ k-ft $M_{32} = -3.38$ k-ft
$M_{34} = 3.38$ k-ft $M_{43} = 0$

10.35 $EI u_3 = 9.08$ kN·m^3
$M_{12} = 0$ $M_{21} = 1.384$ kN·m $M_{23} = -1.384$ kN·m $M_{32} = -2.83$ kN·m
$M_{34} = 2.83$ kN·m $M_{43} = 5.77$ kN·m

10.36 $EI_0 u_1 = -186.7$ kN·m^3
$M_{12} = 0$ $M_{21} = 7.4$ kN·m $M_{23} = -7.4$ kN·m $M_{32} = -9.2$ kN·m

10.37 $EI_0 u_2 = 1125$ k-ft^3
$M_{12} = 8.33$ k-ft $M_{21} = -8.33$ k-ft $M_{23} = 8.33$ k-ft $M_{32} = 0$

10.38 $EI u_2 = 6250$ k-ft^3 $EI u_3 = 12430$ k-ft^3
$M_{12} = 74.6$ k-ft $M_{21} = 33.4$ k-ft $M_{25} = -61.8$ k-ft $M_{52} = 0$
$M_{23} = 28.4$ k-ft $M_{32} = 25.6$ k-ft $M_{34} = -25.6$ k-ft $M_{43} = 0$

10.39 $M_{BA} = -112.5$ k-ft

10.40 $M_{BA} = -158.6$ kN·m

10.41 $M_{BA} = -80$ k-ft $M_{BC} = 80$ k-ft $M_{CB} = 106.6$ k-ft

10.42 $M_{BA} = 55.6$ k-ft

10.43 $u_3 = -0.1126$ in.
$M_{12} = -19.03$ k-ft $M_{21} = 3.66$ k-ft $M_{23} = -3.66$ k-ft $M_{32} = -3.29$ k-ft
$M_{34} = 3.29$ k-ft $M_{43} = 0$

10.44 $M_{12} = -2.4$ k-ft $M_{21} = -4.8$ k-ft $M_{23} = 4.8$ k-ft $M_{32} = -5.6$ k-ft
$EIV_A = -102.4$ k-ft^3

10.45 $M_{12} = 2.5$ kN·m $M_{21} = -31$ kN·m $M_{23} = 31$ kN·m $M_{32} = 0$
$EI\theta_2 = -28.5$ kN·m^2

10.46 $M_{12} = -12$ kN·m $M_{21} = -24$ kN·m $M_{23} = 24$ kN·m $M_{32} = -24$ kN·m
$M_{34} = 24$ kN·m $M_{43} = 12$ kN·m
$EI_0 V_A = 48$ kN·m^3

10.47 $M_{12} = 0$ $M_{21} = -10.53$ k-ft $M_{23} = 10.53$ k-ft $M_{32} = 0$
$EI_0\theta_2 = -16.2$ k-ft^2

10.48 $M_{12} = 44.5$ k-ft $M_{21} = -8.17$ k-ft $M_{23} = 8.17$ k-ft $M_{32} = -20$ k-ft
$EI_0 V_4 = -492$ k-ft^3

10.49 $M_{12} = -10.1$ kN·m $M_{21} = -20.2$ kN·m $M_{23} = 20.2$ kN·m $M_{32} = -23.8$ kN·m
$M_{34} = -0.22$ kN·m $M_{43} = -0.11$ kN·m

10.50 $M_{12} = -6.35$ k-ft $M_{21} = -12.7$ k-ft $M_{23} = 12.7$ k-ft $M_{32} = 0$

10.51 $M_{12} = 0$ $M_{21} = -19.29$ kN·m $M_{23} = 19.29$ kN·m $M_{32} = -3.86$ kN·m

10.52 $M_{12} = 24.7$ kN·m $M_{21} = -4.71$ kN·m $M_{23} = 4.71$ kN·m $M_{32} = -10.52$ kN·m
$M_{34} = 27.7$ kN·m $M_{43} = 0$ $M_{35} = -17.15$ kN·m $M_{32} = -8.58$ kN·m

10.53 $M_{15} = 0$ $M_{51} = -8.49$ k-ft $M_{25} = -11.79$ k-ft $M_{52} = -23.6$ k-ft
$M_{56} = 32.1$ k-ft $M_{65} = -44.8$ k-ft $M_{63} = 17.94$ k-ft $M_{36} = 8.97$ k-ft
$M_{67} = 26.8$ k-ft $M_{76} = 13.22$ k-ft $M_{74} = -0.096$ k-ft $M_{47} = 0$
$M_{78} = -0.192$ k-ft $M_{87} = -0.064$ k-ft $M_{79} = -12.93$ k-ft $M_{97} = -25.63$ k-ft
$M_{9,10} = 25.6$ k-ft $M_{10,9} = -51.2$ k-ft

10.54 $EIu_3 = 366$ k-ft^3
$M_{12} = 0$ $M_{21} = 1.22$ k-ft $M_{23} = -1.22$ k-ft $M_{32} = -4.92$ k-ft
$M_{34} = 4.92$ k-ft $M_{43} = 19.07$ k-ft

10.55 $EI_0 u_3 = 592$ k-ft^3
$M_{12} = 0$ $M_{21} = 11.81$ k-ft $M_{23} = -11.81$ k-ft $M_{32} = -3.38$ k-ft
$M_{34} = 3.38$ k-ft $M_{43} = 0$

10.56 $EIu_1 = -733$ kN·m^3 $EIu_3 = -119.8$ kN·m^3
$M_{12} = 0$ $M_{21} = 31.7$ kN·m $M_{23} = -31.7$ kN·m $M_{32} = -28.0$ kN·m
$M_{34} = 28.0$ kN·m $M_{43} = -43.8$ kN·m

10.57 $EI_0 u_2 = 1125$ k-ft^3
$M_{12} = 8.33$ k-ft $M_{21} = -8.34$ k-ft $M_{23} = 8.33$ k-ft $M_{32} = 0$

10.58 $EIu_2 = 6230$ k-ft^3 $EIu_3 = 12410$ k-ft^3
$M_{12} = 74.3$ k-ft $M_{21} = 33.2$ k-ft $M_{25} = -61.1$ k-ft $M_{52} = 0$
$M_{23} = 27.9$ k-ft $M_{32} = 25.7$ k-ft $M_{34} = -25.7$ k-ft $M_{43} = 0$

10.59 $M_{21} = 6.76$ kN·m $M_{23} = -6.76$ kN·m $M_{32} = -47.3$ kN·m

10.60 $M_{21} = -8.57$ kN·m $M_{23} = 8.57$ kN·m $M_{32} = -0.851$ kN·m $M_{34} = 0.851$ kN·m

10.61 $M_{12} = 44.3$ kN·m $M_{21} = 14.8$ kN·m $M_{23} = -14.8$ kN·m $M_{32} = -28.7$ kN·m
$M_{34} = 28.7$ kN·m $M_{43} = 0$ $M_{35} = -0.05$ kN·m $M_{53} = 7.47$ kN·m

10.62 Member 1 lc(1, 2) $R_1^T = (-2.4$ k-ft, -4.8 k-ft, -0.6 k, 0.6 k$)$
Member 2 lc(2, 3) $R_2^T = (4.8$ k-ft, -5.6 k-ft, 1.956 k, 1.044 k$)$

10.63 Member 1 lc(1, 2) $R_1^T = (2.5$ kN·m, -31 kN·m, 7.25 kN, 16.75 kN$)$
Member 2 lc(2, 3) $R_2^T = (31$ kN·m, $0, 21.44$ kN, 14.55 kN$)$

10.64 Member 1 lc(1, 2) $R_1^T = (-12$ kN·m, -24 kN·m, -6 kN, 6 kN$)$
Member 2 lc(2, 3) $R_2^T = (24$ kN·m, -24 kN·m, 18 kN, 18 kN$)$
Member 3 lc(3, 4) $R_3^T = (24$ kN·m, 12 kN·m, 4 kN, -4 kN$)$

10.65 Member 1 lc(1, 2) $R_1^T = (44.5$ k-ft, -8.18 k-ft, 12.82 k, 8.78 k$)$
Member 2 lc(2, 3) $R_2^T = (8.17$ k-ft, -20 k-ft, -0.986 k, 0.986 k$)$

10.66 Member 1 lc(1, 2) $R_1^T = (-10.1$ kN·m, -20.2 kN·m, -5.05 kN, 5.05 kN$)$
Member 2 lc(2, 3) $R_2^T = (20.2$ kN·m, -23.8 kN·m, 17.7 kN, 6.3 kN$)$
Member 3 lc(3, 4) $R_3^T = (-0.2$ kN·m, -0.1 kN·m, -0.0333 kN, 0.0333 kN$)$

10.67 Member 1 lc(1, 2) $R_1^T = (20.3$ kN·m, -13.5 kN·m, -13.13 kN, 10.88 kN$)$
Member 2 lc(2, 3) $R_2^T = (13.5$ kN·m, -6.75 kN·m, 10.13 kN, 7.88 kN$)$

10.68 Member 1   lc(1, 2) $R_1^T = (-6.35 \text{ k-ft}, -12.7 \text{ k-ft}, -1.058 \text{ k}, 1.058 \text{ k})$
Member 2   lc(2, 3) $R_2^T = (12.7 \text{ k-ft}, 0, 6.18 \text{ k}, 1.82 \text{ k})$

10.69 Member 1   lc(1, 2) $R_1^T = (8.18 \text{ kN·m}, -1.636 \text{ kN·m}, 3.60 \text{ kN}, 2.06 \text{ kN})$
Member 2   lc(2, 3) $R_2^T = (1.636 \text{ kN·m}, 0, 0.1928 \text{ kN}, -0.1928 \text{ kN})$

10.70 Member 1   lc(1, 2) $R_1^T = (1.174 \text{ kN·m}, 2.35 \text{ kN·m}, 0.587 \text{ kN}, -0.587 \text{ kN})$
Member 2   lc(2, 3) $R_2^T = (-2.35 \text{ kN·m}, -8.22 \text{ kN·m}, -1.409 \text{ kN}, 1.409 \text{ kN})$
Member 3   lc(3, 4) $R_3^T = (17.61 \text{ kN·m}, -31.7 \text{ kN·m}, 24.7 \text{ kN}, 29.3 \text{ kN})$
Member 4   lc(3, 5) $R_4^T = (-9.39 \text{ kN·m}, -4.70 \text{ kN·m}, -2.34 \text{ kN}, 2.34 \text{ kN})$

10.71 Member 1   lc(1, 2) $R_1^T = (15.84 \text{ kN·m}, 5.76 \text{ kN·m}, 7.2 \text{ kN}, 0)$
Member 2   lc(2, 3) $R_2^T = (-5.76 \text{ kN·m}, -2.88 \text{ kN·m}, -0.96 \text{ kN}, 0.96 \text{ kN})$

10.72 Member 1   lc(1, 2) $R_1^T = (0.706 \text{ k-ft}, 1.412 \text{ k-ft}, 0.235 \text{ k}, -0.235 \text{ k})$
Member 2   lc(2, 3) $R_2^T = (-1.412 \text{ k-ft}, -4.94 \text{ k-ft}, -0.706 \text{ k}, 0.706 \text{ k})$
Member 3   lc(4, 3) $R_3^T = (19.06 \text{ k-ft}, 4.94 \text{ k-ft}, 4 \text{ k}, 0)$

10.73 Member 1   lc(1, 2) $R_1^T = (9.31 \text{ kN·m}, -5.10 \text{ kN·m}, 5.20 \text{ kN}, 3.80 \text{ kN})$
Member 2   lc(2, 3) $R_2^T = (5.1 \text{ kN·m}, -8.39 \text{ kN·m}, 3.45 \text{ kN}, 8.55 \text{ kN})$
Member 3   lc(4, 3) $R_3^T = (6.8 \text{ kN·m}, 8.39 \text{ kN·m}, 3.8 \text{ kN}, -3.8 \text{ kN})$

10.74 Member 1   lc(1, 2) $R_1^T = (-30.4 \text{ k-ft}, -16.94 \text{ k-ft}, -5.25 \text{ k}, 5.25 \text{ k})$
Member 2   lc(2, 3) $R_2^T = (16.94 \text{ k-ft}, 25.6 \text{ k-ft}, 0.366 \text{ k}, -6.37 \text{ k})$
Member 3   lc(4, 3) $R_3^T = (-23.8 \text{ k-ft}, -25.6 \text{ k-ft}, -2.75 \text{ k}, 2.75 \text{ k})$

10.75 Member 1   lc(1, 2) $R_1^T = (8.33 \text{ k-ft}, -8.33 \text{ k-ft}, 0, 0)$
Member 2   lc(2, 3) $R_2^T = (8.33 \text{ k-ft}, 0, 8.55 \text{ k}, 3.44 \text{ k})$

10.76 Member 1   lc(1, 2) $R_1^T = (6.87 \text{ kN·m}, 1.789 \text{ kN·m}, 3.73 \text{ kN}, 0.269 \text{ kN})$
Member 2   lc(2, 3) $R_2^T = (-1.789 \text{ kN·m}, 0, -0.358 \text{ kN}, 0.358 \text{ kN})$

10.77 Member 1   lc(1, 2) $R_1^T = (2.64 \text{ k-ft}, 12.57 \text{ k-ft}, 0.585 \text{ k}, -0.585 \text{ k})$
Member 2   lc(3, 2) $R_2^T = (-4.34 \text{ k-ft}, -12.57 \text{ k-ft}, 1.555 \text{ k}, 3.25 \text{ k})$
Member 3   lc(4, 3) $R_3^T = (10.19 \text{ k-ft}, 4.34 \text{ k-ft}, 1.453 \text{ k}, -1.453 \text{ k})$

10.78 Member 1   lc(3, 4) $R_1^T = (-24.2 \text{ k-ft}, -12.1 \text{ k-ft}, -2.02 \text{ k}, 2.02 \text{ k})$
Member 2   lc(2, 3) $R_2^T = (29.8 \text{ k-ft}, 24.2 \text{ k-ft}, 6 \text{ k}, 0)$
Member 3   lc(2, 5) $R_3^T = (-67.0 \text{ k-ft}, -33.5 \text{ k-ft}, -5.59 \text{ k}, 5.59 \text{ k})$
Member 4   lc(1, 2) $R_4^T = (70.8 \text{ k-ft}, 37.2 \text{ k-ft}, 6 \text{ k}, -6 \text{ k})$

11.1 $R_A = 1 - \dfrac{3}{2}\left(\dfrac{x}{L}\right) + \dfrac{1}{2}\left(\dfrac{x}{L}\right)^3 \quad \dfrac{M_B}{L} = \dfrac{1}{2}\left(\dfrac{x}{L}\right) - \dfrac{1}{2}\left(\dfrac{x}{L}\right)^3, \quad s = L - x$

11.2 $B_Y = 1 + \dfrac{3}{7}\left(\dfrac{x}{L}\right) - \dfrac{15}{7}\left(\dfrac{x}{L}\right)^2 + \dfrac{5}{7}\left(\dfrac{x}{L}\right)^3, \quad x = L - s \quad 0 < s < L$

$B_Y = \dfrac{18}{7}\left(\dfrac{x_1}{L}\right)^2 - \dfrac{11}{7}\left(\dfrac{x_1}{L}\right)^3, \quad x_1 = 2L - s \quad L < s < 2L$

11.3 $C_Y = \dfrac{3}{7}\left(\dfrac{x}{L}\right) + \dfrac{6}{7}\left(\dfrac{x}{L}\right)^2 - \dfrac{2}{7}\left(\dfrac{x}{L}\right)^3, \quad x = L - s \quad 0 < s < L$

$C_Y = -\dfrac{3}{7}\left(\dfrac{x_1}{L}\right)^2 + \dfrac{3}{7}\left(\dfrac{x_1}{L}\right)^3, \quad x_1 = 2L - s \quad L < s < 2L$

11.4 $B_Y = -\dfrac{4}{5}\left(\dfrac{x}{L}\right) + \dfrac{6}{5}\left(\dfrac{x}{L}\right)^2 - \dfrac{2}{5}\left(\dfrac{x}{L}\right)^3, \quad x = L - s \quad 0 < s < L$

$B_Y = 1 - \dfrac{1}{5}\left(\dfrac{x_1}{L}\right) - \dfrac{9}{5}\left(\dfrac{x_1}{L}\right)^2 + \left(\dfrac{x_1}{L}\right)^3, \quad x_1 = 2L - s \quad L < s < 2L$

$B_Y = \dfrac{8}{5}\left(\dfrac{x_2}{L}\right) - \dfrac{3}{5}\left(\dfrac{x_2}{L}\right)^3, \quad x_2 = 3L - s \quad 2L < s < 3L$

11.5 $M_C/L = -\dfrac{8}{15}\left(\dfrac{x}{L}\right) + \dfrac{4}{5}\left(\dfrac{x}{L}\right)^2 - \dfrac{4}{15}\left(\dfrac{x}{L}\right)^3$, $x = L - s$ $0 < s < L$

$M_C/L = -\dfrac{2}{15}\left(\dfrac{x_1}{L}\right) - \dfrac{1}{5}\left(\dfrac{x_1}{L}\right)^2 + \dfrac{1}{3}\left(\dfrac{x_1}{L}\right)^3$, $x_1 = 2L - s$ $L < s < 2L$

$M_C/L = \dfrac{1}{15}\left(\dfrac{x_2}{L}\right) - \dfrac{1}{15}\left(\dfrac{x_2}{L}\right)^3$, $x_2 = 3L - s$ $2L < s < 3L$

11.6 $C_Y = 1 - \dfrac{3}{13}\left(\dfrac{x}{L}\right) - \dfrac{33}{13}\left(\dfrac{x}{L}\right)^2 + \dfrac{23}{13}\left(\dfrac{x}{L}\right)^3$, $x = L - s$ $0 < s < L$

$C_Y = \dfrac{12}{13}\left(\dfrac{x_1}{L}\right) + \dfrac{18}{13}\left(\dfrac{x_1}{L}\right)^2 - \dfrac{17}{13}\left(\dfrac{x_1}{L}\right)^3$, $x_1 = 2L - s$ $L < s < 2L$

$C_Y = -\dfrac{6}{13}\left(\dfrac{x_2}{L}\right) + \dfrac{6}{13}\left(\dfrac{x_2}{L}\right)^3$, $x_2 = 3L - s$ $2L < s < 3L$

11.7 $M_D/L = -\dfrac{7}{26}\left(\dfrac{x}{L}\right) - \dfrac{6}{13}\left(\dfrac{x}{L}\right)^2 + \dfrac{19}{26}\left(\dfrac{x}{L}\right)^3$, $x = L - s$ $0 < s < L$

$M_D/L = \dfrac{1}{13}\left(\dfrac{x_1}{L}\right) + \dfrac{3}{26}\left(\dfrac{x_1}{L}\right)^2 - \dfrac{5}{26}\left(\dfrac{x_1}{L}\right)^3$, $x_1 = 2L - s$ $L < s < 2L$

$M_D/L = -\dfrac{1}{26}\left(\dfrac{x_2}{L}\right) + \dfrac{1}{26}\left(\dfrac{x_2}{L}\right)^3$, $x_2 = 3L - s$ $2L < s < 3L$

11.8 $M_B/L = -\dfrac{4}{11}\left(\dfrac{x}{L}\right) + \dfrac{6}{11}\left(\dfrac{x}{L}\right)^2 - \dfrac{2}{11}\left(\dfrac{x}{L}\right)^3$, $x = L - s$ $0 < s < L$

$M_B/L = -\dfrac{7}{11}\left(\dfrac{x_1}{L}\right)^2 + \dfrac{7}{11}\left(\dfrac{x}{L}\right)^3$, $x_1 = 2L - s$ $L < s < 2L$

11.9 $V_B = 1 - \dfrac{9}{11}\left(\dfrac{x}{L}\right) - \dfrac{12}{11}\left(\dfrac{x}{L}\right)^2 + \dfrac{4}{11}\left(\dfrac{x}{L}\right)^3$, $x = L - s$ $0 < s < L$

$V_B = \dfrac{3}{11}\left(\dfrac{x_1}{L}\right)^2 - \dfrac{3}{11}\left(\dfrac{x_1}{L}\right)^3$, $x_1 = 2L - s$ $L < s < 2L$

References

Hohn, F. E. 1958. *Elementary Matrix: Algebra*. New York: Macmillan.

Popov, E. P. 1978. *Mechanics of Materials*. 2d Ed. Upper Saddle River, NJ: Prentice-Hall.

Shames, I. H. 1997. *Engineering Mechanics: Statics*. 4th Ed. Upper Saddle River, NJ: Prentice-Hall.

Timoshenko, S. P. and D. H. Young. 1945. *Theory of Structures*. New York: McGraw-Hill Book Company.

Timoshenko, S. P. 1953. *History of Strength of Materials*. New York: McGraw-Hill Book Company.

Index

Admissible displacement, 95
Approximations, successive, 346
Assumptions, fundamental, 2

Beams
    compound, 244
    continuous, 244, 261
Bending moment diagrams, 38
    beams, 39
    floor girders, 67
    frames, 47
Bending moment, influence line for, 59, 60, 62, 68
    by virtual work, 110, 111
Betti's theorem, 200

Carryover factor, 342
Castigliano's theorems, 203
    and compatibility, 204
    formulated for Hooke's law, 206, 209
    of least work, 207
    on deflections, 204
Center, instantaneous, 99
Coefficients
    flexibility, 235
    stiffness, 237
Compatibility equations, 1, 233
    and Castigliano's theorems, 204
Complementary strain energy, 198
    and force deformation relations, 198
    Hooke's law, 199
Complex truss, 16, 24
Compound
    beam, 244
    truss, 16
Conservative applied forces, 195
Constrained mechanical system, 93
Constraint-ideal—See contact surfaces
Contact surfaces, ideal, 96
Contact forces, 96
Continuous beams, 244, 261
Convergence of moment distribution, 340
Coordinates
    generalized, 93
    global, 143
    local, 47, 143
Crossbeam, 67
Cross, H, 340
Cut-back
    structure, 5
    system, 241

Dead load, 66
Degrees of freedom, 94
    mathematical significance, 94
Degree of statical indeterminacy, 5
Deflection of pin-jointed trusses, 128, 188
Deflections, influence lines for, 417
Design codes, 1
Determinantal criterion,
    for full or partial restraint, 4
    for statical determinacy, 4
    for statical indeterminacy, 5

Displacements
    admissible, 95
    prescribed, 2, 203, 257, 277, 324, 354
    rectangular components, 99
    rigid, 97
    sidesway, 326
    small, 98
    virtual, 95
Displacements and virtual displacements in
    small displacement theory, 175, 178
Displacement methods, 234, 308
    moment distribution, 234, 339–359
    slope-deflection equations, 234, 310–333
    stationary potential energy, 234, 378
Displacement-deformation analysis, 126
    differential equations, 128
    double integration, 130
    energy methods, 128
    fundamental problems, 126
    geometric methods, 128
    of trusses, 128
Displacement vector (column matrix), 362
Displacement calculations
    by Castigliano's theorem, 209
    by moment-area method and superposition, 135–155
    by virtual work, 173–191
Displacement-deformation equations, 2, 233
Distribution factor, 342

Effective applied load vector, 367
End
    displacement vector, 362
    displacements, 311
    force vector, 362
    moments, 311
    rotations, 311
    shears, 311
Energy
    strain, and force-deformation relations, 194
    complementary strain, and force-deformation relations, 199
    strain, 194
    complementary strain, 198
    potential of applied loads, 195
Energy methods, basic concepts
    configuration, position, or placement, 93
    constrained mechanical system, 93
    degrees of freedom, 94
    generalized coordinates, 93
    particle, 93
    position variables, position coordinates, 93
Equations of equilibrium, 2, 3, 231
    by stationary potential energy, 378
    by virtual work, 359–366
    differential equations, 45
    in small deflection theory, 2, 205, 207
    integral equations, 44, 45

Fixed end
    force vector, 366
    moments, 313
    shears, 313

Index

Fixed loads of variable placement, extreme effects, 72–79
Flexibility coefficients, 235
Floor beam—See crossbeam
Free-body diagrams, 7
Force analysis, 2
   fundamental problem of, 2
   general procedure of, 2
   piecemeal methods, 5
Force-deformation equations, 2, 194, 199, 233
Forces—See also Loads
   of reaction, 2
   of interaction, 1, 2
   prescribed (applied), 1
   stresses or internal, 37
Force methods, 234
   consistent deformation, 234, 239, 241–261
   least work, 234, 273–279
   three-moment equation, 234, 261–271
Frame displacements, 143
   combination approach, 156
   method of virtual work, 185
   superposition of rigid displacements, 143–155
   moment-area theorems, 143
Framed structure, 1
Freedom, degrees of, 94
Fundamental assumptions, 1

Girders, with floor beams, 67
Global coordinates, 143

Henneberg, 25
Hyperelastic body, 194

Influence function/lines, 58
   applications, 63–67, 72–76
   beams and frames, 59–63
   definitions, 58
   floor girders, 67
   for statically indeterminate structures, 415
   trusses, 69
Independent position variables (coordinates), 94
Instantaneous center, 99
Internal-external load relationships, 43–46
   discontinuity relations, 45
   integral equations of equilibrium, 44, 45
   differential equations of equilibrium, 45
Internal forces
   in trusses, 18
   in beams and frames, 37
Internal force resultants, 37–38
   bending moment, 37
   representations, 37
   shear, 37
   sign convention, 37
   tension or axial force, 37
Internally indeterminate structures, 241
Iteration
   and moment distribution, 339
   Gauss-Seidel, 346

Joint displacements, 1
   of trusses, 128
Joint (node), 1
Joints, method of, 19

Kinematic indeterminacy, 234
   and independent equation of equilibrium, 235

Least work, method of, 273
Live load, 66
Load functions
   axial, 38
   transverse, 38
Loaded surface, 67
Load train, 72
Loads of variable placement, 57
   extreme effect due to, 72
Loads—See also Forces
   arbitrary applied, 3
   dead, 66
   live, 66
   prescribed (applied), 1
   truss, 15
Local coordinates, 47

Matrices
member stiffness, 366
structure stiffness, 367
Matrix formulations
   potential energy of prescribed loads, 376
   principle of stationary potential energy, 378
   strain energy of bending, 373
   virtual work, 362–363
Maximum and minimum effects due to combined dead and live load, 66
Maxwell's reciprocal theorem, 201
Members, redundant—see internally indeterminate structures
Method
   of consistent deformation, 234, 239, 241–261
   of joints, 19
   of least work, 234, 273–279
   of sections, 19
Minimum potential energy, principle of, 197
   and Hooke's law, 197
   and stability, 198
Model, n degree-of-freedom, 94
Moment distribution, 234, 339–359
   and sidesway analysis, 356
   and prescribed displacements, 354
   carryover factor, 342
   convergence, 339
   distribution factor, 342
   joint stiffness, 341
   member stiffness, 341
Moment-area theorems, 128, 130
   applied to frames, 143–146
   applied to beams, 135–142
   first theorem, 133
   second theorem, 134
   superposition of terms, 134
Müller-Breslau, 415

Panel point, 67
Panel shear, 68
   by virtual work, 111
Particle, 93
   position, 94
   displacement, 94
Pinned joint, 15, 69

Plane-framed structures, 6
Prescribed displacements, 2, 203, 257, 277, 324, 354
Primary structure, 5, 232
Principle of superposition, 2
Principle of virtual work, 100
    conditions on displacements and deformations, 95
    conditions on forces, 174
    displacements by, 173
    influence lines by, 108

Radius of gyration, 248
Reactions, influence lines by virtual work, 109
Reaction force vector, 370
Reciprocal theorem, 201
Redundant reaction components, 5
Redundant mechanical devices, 5
Reference axis, 37
Relationships between internal and external loads, 43–46
    discontinuity relations, 45
    integral equations of equilibrium, 44, 45
    differential equations of equilibrium, 45
Rigid displacements, 97

Sections, method of, 19
Shear, influence line for, 59, 68, 417
    by virtual work, 110, 111
Shear diagrams, 38
    for beams, 39
    for frames, 47
Sidesway, 326–330, 332
    and superposition, 333-339
    and moment distribution, 356
Slope-deflection equations, 234
    application to beams and frames, 310–333
    modified equations, 330
Small displacements, 98
    rectangular components, 99
Stability and determinacy, 3–5
Statically indeterminate structures, 231
    combined methods, 271
    internally indeterminate, 241
    displacement methods, 234
    force methods, 234
    general procedure of analysis, 233
    solution strategies, 233
    and member rigidity, 233
    reaction force vector, 370
Stationary potential energy, principle of, 196
Statical determinacy, determinantal criterion, 4
Stiffness coefficients, 237
Stiffness matrix
    member, 366
    structure, 367
Strain energy, 194
    and internal force-deformation relations, 194
    Hooke's law, 195
Stringers, 67, 70
Structural analysis, general problem of, 1

Structures
    fully restrained, 3
    incompletely/partially restrained, 3
    kinematically unstable, 3
    geometrically unstable, 3
    statically determinate, 3
    statically indeterminate, 3, 231
    useful, 4
    cut-back, 5
    primary, 5
    planar models, 7
    plane framed, 6
    kinematically indeterminate, 234
Supports, settlement of, 258
    See also prescribed displacements

Theorem
    Betti's, 200
    Castigliano's, on deflections, 204
    Castigliano's, of least work, 207
    reciprocal (Maxwell), 201
Truss, plane, 15
    internal force resultants, 18
    statically determinate, 17
    statically indeterminate, 251, 253, 255, 278
    influence lines for, 70, 112
    complex, 15, 16
    loads on, 15
    simple, 15
    compound, 15, 16
    connections, 15
Two-force member, 17

Vector representations, 18, 38, 102
    and work formulations, 102
Vectors-column matrices
    axial load, 362
    end force, 362
    end displacement, 362
    joint displacement, 362
    displacement, 363
    joint applied load, 362
    axial displacement, 362
    effective applied load, 367
    fixed end-force, 366
    reaction force, 370
Virtual displacement, 95
Virtual work, 95
    of forces on a rigid displacement, 100
    in terms of position variables, 96
Virtual work, displacements by, 173–191
    basic applications, 178–181
    evaluation of δW_σ, 175
    displacements due to applied loads, 181
    Hooke's law, 181

Wichert truss, 9
Williot-Mohr, 129
Work
    virtual, 95
    least, 207, 209